Strategies
for Planned Change

Strategies for Planned Change

GERALD ZALTMAN
University of Pittsburgh

ROBERT DUNCAN
Northwestern University

A WILEY-INTERSCIENCE PUBLICATION

JOHN WILEY & SONS, New York • Chichester • Brisbane • Toronto • Singapore

Library of Congress Cataloging in Publication Data:

Zaltman, Gerald.
  Strategies for planned change.

  "A Wiley-Interscience publication."
  Bibliography: p.
  Includes indexes.
  1. Social change. 2. Planning. 3. Organiza-
tional change. 4. Diffusion of innovations.
I. Duncan, Robert, joint author. II. Title.

HM101.Z287   301.24   76-39946
ISBN 0-471-98131-1

Printed in the United States of America

10 9

This book is dedicated to
Everett M. Rogers (G.Z.)
and
Chris Argyris (R.D.)

PREFACE

This book has several objectives. The first objective is to present a number of ideas about the broad topics represented by chapter titles. A second objective is to bring together some of the literature about innovation diffusion and organizational change. Each reader may find his or her particular specialty area underrepresented. Here we caution not to expect a comprehensive literature review but rather a treatment of selected ideas in the two areas. The two main areas of concern have traditionally been treated separately and need to be drawn together. Organizational dynamics and the dynamics of field work interact. A change in the structure of an organization for example can influence how well that organization may service its clients. Also, a change among the clientele of an organization may require the adoption of new organizational structures. Thus change agents concerned with altering organizations must have familiarity with change processes among individuals as well, and change agents concerned with altering the behavior of individuals must know something about change processes in organizations. A third objective is to identify the experiences in social change we and other change agents have had that have implications for creating change or illustrate an important point and that would not ordinarily come to the readers' attention. Much of what is new in this book is derived from these experiences. A fourth objective is to be prescriptive—to identify the various considerations a change agent must be sensitive to and to state a number of principles which may serve as guidelines. Thus the conclusions to each chapter consist of basic principles for social change. Many of the principles are rather obvious. It is equally obvious that they are often forgotten or ignored by practitioners. The principles are far from exhaustive, and undoubtedly, readers will find missing some principles that they have found essential. Moreover, some of the statements may be debated in terms of their importance or on other dimensions. We will be satisfied, however, if our presentation of principles does nothing more than stimulate the reader to think about additional or better principles that might not otherwise have been crystalized. A fifth objective is to highlight topics very much related to planned social change but which, in our

vii

judgment, are often given too little attention in that context. We refer particularly to the issues of resistance to change, problem definition, research use, and values.

GERALD ZALTMAN
ROBERT DUNCAN

Pittsburgh, Pennsylvania
Evanston, Illinois
July 1976

CONTENTS

Strategies
for Planned Change

PART ONE

Basic Concepts in Change

Part I is intended to introduce several major concepts and caveats with regard to planned change. Chapter 1 introduces some conceptual and terminological issues. Participants in the change process are introduced, and common pitfalls change program managers and agents may encounter are discussed. Stimuli for social change are also treated. Chapter 2 of Part I focuses exclusively on the task of defining social problems. This topic is one that we feel is of great importance and one that is very inadequately treated in most of the social change literature and in training programs for change agents. We feel that defining problems is one of the very first critical decision activities in planned social change and hence must be addressed early in this book.

CHAPTER ONE

Introduction to Social Change

At no other point in human history has there been as much concern with
social change than at the present. Social change is occurring rapidly and
with profound effects in many sectors of social life throughout the world. In
most instances where change is not taking place or is taking place only
slowly there is often great concern about how to stimulate change. Because

the consequences of social change, as well as the consequences of no social change, can have great impact, there is great interest in managing change to maximize its benefits and minimize its unfortunate effects. Of course, benefits and detriments are often relative: a particular change that is beneficial to one party may be detrimental to another party, and thus it sometimes becomes necessary to be better at change management than a competitor. Population control is debated, advocated, and resisted at village levels and in high government councils; auto safety is promoted through driver education courses, advertising, and legislation, yet the impact is disappointing and even antagonizes the automobile driving public. Court-ordered school desegregation is found in some areas to reinforce what it is intended to abolish. At the same time, many diseases have been conquered or are about to be eradicated; important strides have occurred in the delivery of health care; advances in space technology and agriculture have been substantial, as has been our growth in medical technology; sex discrimination is lessening; and we have improved our skills in helping children and adults with learning disabilities to be more effective students and employees.

Significant undertakings in planned social change that have met with various degrees of success or failure are too numerous to count. Those mentioned above are an exceedingly small sample of the varied and dynamic undertakings in social change around the world. It is not surprising that planned change and its successes and failures is a dominant feature— perhaps the dominant feature—of global society today. The stakes are very high. The spectre of nuclear war and mass starvation is very real. The losses in human dignity, satisfactions, and resources, through drug addiction, pollution, crime, and through the inadequacies of the institutions to which we commit our elderly and disturbed are incalculable. It is easy, given the magnitude of the challenges we face, to turn away from the problems posing those challenges. Indeed, there are few problems that do not have people associated with them who want the problem left alone. The motivation of such people is not always inappropriate or unjustified. As one sage put it, "The will to order can make tyrants out of those who merely aspire to clean up a mess." Efforts to do good can also do bad. Nevertheless, our own values, which we readily admit are eminently disputable, advocate intervening in the affairs of others. (This issue is treated in a later chapter of this book.)

It does not quite seem appropriate to use the word "exciting" in association with the vast and serious problems and challenges we face in today's society. Yet this word does reflect the mood we feel ourselves and sense among our colleagues in the social change area. Dual advances in our understanding of human behavior and in scientific technology have created unique opportunities to redress social problems more effectively than ever before. This mood among social change specialists is reflected in the titles of their books. Consider just a few of the most recent of these: *Creating Social Change* (Zaltman, Kotler and Kaufman, Eds., 1972); *The Practice of Social Intervention: Roles, Goals and Strategies* (Loewenberg and Dolgoff, 1972); *Communes: Creating and Managing the Collective Life* (Kanter, Ed., 1973); *Training For Change Agents* (Havelock and Havelock, 1973); *Planning For Health: Development and Applications of Social Change Theory* (Blum, 1974); *Planning and Organizing for Social Change* (Rothman, 1974); *The Laboratory Method of Changing and Learning* (Benne, Bradford, Gibb, and Lippitt, Eds., 1975); *Promoting Innovation and Change in Organizations and Communities* (Rothman, Erlich, and Teresa, 1976); and *Dynamic Educational Change* (Zaltman, Florio, and Sikorski, 1977).

It is not, of course, the problems that are exciting, but rather it is the responses to the problems that pose excitement. This obviously implies (or assumes) that social problems are remedial in ways that are generally acceptable to those affected by them. This general position, that planned social change can relieve stress from social problems, is a basic foundation of this book. This position would be naive if it assumed that all stress can be eliminated or that all problems can be dispatched effectively. Indeed, we see significant barriers in the form of resistance to change (Chapter 3), difficulties in diagnosing problems (Chapter 2) (including what our colleague Professor Ian Mitroff refers to as the Error of the Third Kind or the probability of solving the wrong problem), serious ethical and value issues (Chapter 13), and difficulties in generating useful knowledge and applying it effectively (Chapter 12). Still, accumulated experiences in planned social change have generated a large repertoire of strategies and information about their use (Chapters 4–8); and our knowledge base about effective change agentry (Chapter 9), inducing change in individuals (Chapter 10), and inducing change in organizations (Chapter 11) is substantial enough to offer the prescriptions for social change at the end of the various chapters.

Before we can really study and understand change processes, it is necessary to be clear about what is meant by social change and to have effective indicators of whether and how change has occurred. This chapter focuses primarily on conceptual issues that should be kept in mind as the reader proceeds through this book. The very first topic concerns the concept of change, followed by a discussion of the participants in the change process. Certain questionable assumptions widely found in the social change literature are then discussed briefly. We also examine, under the umbrella of the "performance gap" concept, some of the reasons why social change occurs. This chapter concludes with a discussion of the goals of social change.

We feel that the various conceptual and methodological issues discussed in this book and particularly this chapter are important to both the practitioner and the change theorist. These issues are important to the practitioner in that they facilitate the planning and implementation of change. One cause of failure for many change programs is that the intention of the change is never clearly specified. The objective is simply to "change" with little discussion about what this change means. For example, some management development programs have the objective of "changing manager behavior so that they better understand their subordinates." Such a lack of specification of what behavior is to be changed and along what dimensions understanding is to increase creates other problems, such as difficulty in identifying and selecting criteria that determine whether change has occurred. It causes further difficulties in evaluating whether the change objective has been accomplished satisfactorily. Thus if the nature of the change is not clearly defined, it becomes difficult to plan, implement, and evaluate. These processes are discussed in greater detail later in the book.

CONCEPTUALIZING CHANGE

Defining Social Change

Perhaps the most difficult conceptual issue in studying social change is to adequately define social change. This definitional problem becomes apparent when we attempt to differentiate between change and nonchange. Many things, perhaps all things, are always in some state of fluctuation; thus in some absolute sense "things are always changing." Certainly when talking about human behavior one can make the case that most behavior fluctuates. For example, a client may have positive feelings about a consultant, but this feeling of positive regard may be enhanced or lessened or changed in character during subsequent interactions, as familiarity increases, as other parties enter the picture, or as advice received is implemented and evaluated.

Given that behavior is always fluctuating, what differentiates change from the status quo? What constitutes a significant modification or alteration in behavior that we can call change? This latter question involves the notion of critical threshold and the question "when does *more* become *different*?" It is known, for example, that, as stimulus ambiguity is changed in a monotonic fashion, the degree of felt information need and information-seeking behavior changes nonmonotonically. In most cases, nonmontonicity results from qualitative alterations occurring as changes in strength take place. For instance, "as the perceived threat posed by an instance of social change increases in salience to an individual, different psychological mechanisms (attributes) of resistance are called into play" (Zaltman, Pinson, and Angelmar, 1973). Thus, as familiarity with an advocated change increases, not only may resistance (acceptance) increase, but different forms of resistance (acceptance) may be displayed or experienced (Zaltman, Duncan, and Holbek, 1973).

Rogers and Shoemaker (1971) have addressed the question of critical changes in causal variables. They propose that, as the level of knowledge and adoption in a social system increases, there is a cumulatively increasing pressure on the nonadopter to adopt. They refer to this as the "diffusion effect" and tie it directly to thresholds: ". . . as the rate of awareness-knowledge of the innovation increased up to 20–30%, there was almost no adoption. Then, once this threshold was passed, each additional percentage of awareness-knowledge in the system was associated with several percentage increases in the rate of adoption" (p. 163). The social phenomena after this threshold differ from those before it was reached in the sense that there is more social interaction taking place and distinctly different types of people are becoming involved.

With regard to the differentiation of change from the status quo, it is useful to present some definitions of social change that others have developed. Exhibit 1 is a summary of a few definitions. Formal definitions of social change as a *concept* are relatively few, although there is a wide array of theories focusing on the *process* of social change, leaving the definition implicit in the theory or the description of social change processes. Some selected definitions of social change are presented in Exhibit 1 (some definitions are our interpretations of implicit definitions of some authors). Etzioni appears to define social change as a reformulation of social structure involving initial disequilibrium, forces for establishing social equilibrium, and the occurrence of a new equilibrium (Etzioni and Etzioni, 1973, p. 76). Social change can also be viewed as the adjustment of one or more components of a social system to a change in another component of that system.

Lippitt (1973, p. 37) defines change as "any planned or unplanned alteration in the status quo in an organism, situation, or process." He further distinguishes between planned change and organizational change in which the former is any "intended, designed, or purposive attempt by an individual, group, organization, or larger social system to influence directly the status quo of itself, another organism, or a situation" (p. 37). He defines organizational change as "any planned or unplanned alteration of the status quo which effects the structure, technology, and human resources of the total organization" (p. 37). Following Bhola (cited in Lippitt, p. 39), Lippitt distinguishes between two types of social change. The first is *transmitted* social change—evolutionary change that occurs without deliberate guidance.

Exhibit 1   Sample Definitions of Social Change

Gerlach and Hines	Developmental social change is change within an ongoing social system, adding to it or improving it rather than replacing some of its key elements (p. 2).
	Revolutionary social change is change that replaces existing goals with an entirely different set of goals, steering society in a very different direction (p. 14).
Hamblin, Jacobsen, Miller	Quantitative processes that occur through time.
Abcarian	Structural tensions that result in widespread patterns of deviant norms and behavior.
Rogers	Alteration in the structure and function of a social system.
Etzioni	Reformulation of a social structure involving disequilibrium, forces for establishing equilibrium, and the occurrence of a new equilibrium.
Lippitt	Any planned or unplanned alteration in the status quo in an organism, situation, or process.
Smith	Differentiation, reintegration, and adaptation.
Triandis	A new set of social relationships and social behavior that is most likely to lead to rewards.
Lenski	Innovation through discovery or invention or diffusion or alteration.
Dobny, Boskoff, Pendleton	Alterations in the patterns of interactions or social behavior among individuals and groups within a society.
Niehoff	The implementation of a plan as mediated by actions of change agents and reactions of the community of (potential) adopters.
Schien	The induction of new patterns of action, belief, and attitudes among substantial segments of a population.

Transformed social change "occurs when individuals, groups, or organizations change themselves or others through conscious action or decisions." (p. 38).

Smith (1973), in his critique of the functionalist theory of social change, presents his readers with a view of social change as the differentiation, reintegration, and adaptation of a social system. Triandis (1972, p. 129) defines social change from a more social psychological perspective: social change is "a new set of social relationships and social behaviors that is most likely to lead to rewards." An important element in this definition is the concept of conflict that involves considerations of "*who* controls reinforcements, for *whom, when,* and *where*" (p. 129). Abcarian (1971) suggests a definition of social change as a phenomenon of structural tensions that result in "widespread patterns of deviant norms and behavior" (p. 154). The concept of conflict is also implicit in this view of social change. Rogers (1973) defines social change as the process by which alteration occurs in the structure and function of a social system (p. 7). This definition, we feel, is a good synthesis of the various sociological theories and will form part of the definition used in this book.

In this book, we define change at the individual and system level as an alteration in the way an individual or group of individuals behave as a result of an alteration in their definition of the situation. A person changes his or her behavior when *they* define the situation as being different and now requiring different behavior. This approach is closely related to prevailing definitions of innovation (Rogers and Shoemaker, 1971; Zaltman and Stiff, 1973).

For example, a change in managerial behavior toward subordinates begins when the manager is required to rethink his or her strategy for dealing with subordinates. The manager has to question past assumptions about dealing with subordinates and develop a new process of interaction. This change then requires learning new behavioral patterns in responding to subordinates. The manager is thus redefining how he or she performs his role as a supervisor. Similarly, shifting from a female-oriented birth control method to a male-oriented approach may (further) change husband-wife relationships.

A redefinition of how one performs his role can occur at two levels. The individual may change his or her values or beliefs. This may result in a new definition of how to react in a given situation. This redefinition in individual role behavior may also occur in response to pressures at the systems and social environment level. Other persons in the environment with whom the individual interacts who are dependent on the individual, or are persons the individual is dependent on, comprise the individual's role-set. (Kahn, et al., 1964, p. 14) When other persons are dependent on the individual, they exert pressure on him to conform to their expectations. They exert role pressure on the individual. This role-set can then become an important source of pressure for change.

Role-set members' expectations may change with respect to what they consider to be appropriate supervisor behavior. For example, a manager's superiors, peers, and subordinates may view the manager's supervisory behavior as unacceptable. They may believe that the manager in question could more effectively motivate and supervise subordinates by implementing a more people-oriented management sytle. Role-set members might communicate this desire for change though the kinds of expectations and resulting role pressure thay exert on the individual. In this example, the manager in question is pressured to relearn how he or she interacts with or relates to subordinates.

In summary then, change is defined as the relearning on the part of an individual or group (1) in response to newly perceived requirements of a given situation requiring action and (2) which results in a change in the structure and/or functioning of social systems.

Types of Social Change

Having defined change, it is now important to distinguish between the different types of change that can occur. Most of the change discussed in this book is planned change. However, change can also be unplanned. Planned change as viewed here is a deliberate effort with a stated goal on the part of a change agent to create a modification in the structure and process of a social system such that it requires members of that system to relearn how they perform their roles. Members of a system who are the targets of the change are referred to as the target system. Unplanned change also occurs as a result of the interaction of social system forces. However, it is change "brought about with no apparent deliberateness and no coordinated goal setting on the part of those involved in it" (Bennis, 1966, p. 84). A good example of unplanned change can be found in culture change. Arensberg and Niehoff (1964) have defined cultural borrowing as the passing of ideas and techniques between cultures. This activity is often unplanned in that there is no active change agent operating with a predefined set of goals that are to be accomplished.

All change falls into the planned-unplanned dichotomy. Zaltman, Kotler, and Kaufman (1972, pp. 2-3) have identified two additional dimensions on which to categorize social change. Social change can be categorized on the time dimension and on the level of society that is the target of the change. The time dimension can vary from the short run of a relatively few days or months to a long term of several months or years. Using these two dimensions, Zaltman et al., identify six types of change (Table 1-1).

At the micro or individual level there can be short-term changes in attitudes and behavior (type 1). An example of type 1 change would be the utilization of sensitivity training to change an individual's attitudes about how to interact more effectively with others in interpersonal situations. The process here would be to send a person to a group lab for one to two weeks with the hope that, by changing his or her attitudes regarding inter-

Table 1.1

Time Dimension	Level of Society		
	Micro (Individual)	Intermediate (Group)	Macro (Society)
Short term	Type 1 (1) Attitude change (2) Behavior change	Type 3 (1) Normative change (2) Administrative change	Type 5 (1) Invention-innovation (2) Revolution
Long term	Type 2 Life-cycle change	Type 4 Organizational change	Type 6 Sociocultural evolution

Source: Zaltman, et al., 1972, p. 3.

personal relations, resulting interpersonal behavior would also be changed (Golembiewski and Blumberg, 1973). An example of a more long-term change at the micro level (type 2) would be the training and socialization process of new recruits in an institution. For example, when priests begin their training program, they learn a new set of attitudes and behavior that affects their entire life-cycle. This new behabior is exhibited thoughout their life-style as they devleop an entirely new orientation to society (Hall and Schneider, 1973).

At the group or intermediate level of short-run change, there can be normative or administrative change (type 3). Normative changes can occur when a group alters its norms temporarily to experiment with one innovation. Rogers and Shoemaker (1971) point out that an effective way of getting a group to adopt a new practice is to have them temporarily alter their norms and values toward it. The change agent's strategy here is to get the group to suspend judgment, for a temporary period, toward the practice and to experiment with it. Once they have tried it and found it useful, hopefully they will develop a more permanent positive attitude toward the practice. An example of more long-term change (type 4) at the group level would be when the group has decided to institutionalize the change that has just occurred. In type 4, change focuses on the structure of the group in that it creates a set of role expectations on the part of group members that supports change. When an individual's role requires that he make modifications in his role performance, those people around the individual who are either rewarded by the individual or are dependent on the individual in order to perform their roles, are more likely to support and reinforce the changed behavior.

At the societal or macro level of change, short-term (type 5) change is often characterized by innovations or inventions. For example, the introduction of birth control and family planning practices in a receptive society can quickly alter birth rates and population size. In the long run, this change will result in major changes in the social structure of the society. The long-term ramifications are type 6, or sociocultural change. Thus the short-run success of a population control program can result in changing occupational structures as more women participate in the labor force, or it could facilitate the modernization process of an underdeveloped country.

Social Change and Innovation

At this point, it is necessary to differentiate between change and innovation. An *innovation* is any idea, practice, or material artifact perceived to be new by the relevant unit of adoption. The innovation is the change object. A *change* is the alteration in the structure of a system that requires or could be required by relearning on the part of the actor(s) in response to a given situation. The requirements of the situation often involve a response to an innovation. Often, too, an appropriate response to a new requirement is an inventive process producing an innovation. *However, all innovations imply change, but not all change involves innovations,* since not everything an individual or formal or informal group adopts is perceived as new. For example, an organization might adopt a computer-based management information system. This initial adoption is an innovation because this new process has never been used in the organization. The adoption of the management information system also involves change in that the managers using this new technology have to relearn how to carry out their roles. They have to learn a new set of procedures for interacting with the computer-based system. However, after a trial period, the organization might decide to abandon this computer-based management information system because it creates various problems for the organization. This reversal to the original noncomputer-based system would involve change but not innovation. There is nothing new or innovative about this system to the organization, although it does involve change.

One important issue that is largely untreated in the literature concerns the magnitude of deviation from the status quo required for a person to perceive an idea, object, or practice as an innovation. Similarly, how much alteration in the structure and functioning of a situation is required before a change can be said to have occurred? This is very much related to the issue mentioned earlier: when does *more* become *different*? How *much* of an alteration in a product, social situation, and so forth, is required for that product to be called an innovation or the situation to be described as one of social change? This is discussed at the outset of this chapter. We only add here that the issue of more being different has two dimensions: quantitative and qualitative. Quantitative dimensions involve consideration of physical size or amount of service provided, number of persons performing a particular function or role, and so on. The qualitative dimension involves such considerations as new functions the change object or innovation can perform or improvements in efficiency.

In our discussions in this book we normally use the concept of change to refer to the broader phenomena of change and innovation. Innovation is referred to as a subset of change that is characterized by its being an entirely new situation or phenomenon that the individual, group, or organization is encountering.

The Characteristics of Change

Change may take a variety of forms. Change may simply be an idea or concept that may or may not lead to a different practice and may or may not manifest itself in a physical product or service. Furthermore, a change may be radical or routine, and it may be an ultimate goal or simply a mechanism or instrument to achieve some other specified change. Change has many dimensions independent of its being routine or radical, instrumental or ultimate, and conceptual only or with a physical manifestation. Some of these dimensions are treated in the following discussion, which draws on the work of Zaltman, Duncan, and Holbek (1973), and Lin and Zaltman (1973).

RELATIVE ADVANTAGE. This dimension refers to the unique benefit the change provides that other ideas, practices, or things do not provide at all or as well. This dimension is important at the interest and evaluation stages of the adoption decision process. The relative advantage of a change may take the form of more economy, a lessening of social conflict, psychological security, greater production, and so forth. It is necessary for the change agent to diagnose what improvement the target system needs and wants, incorporate this improvement in the change, and make the target system aware that the needed improvement is inherent in the change.

IMPACT ON SOCIAL RELATIONS. Many changes may have a persuasive impact on social relationships within the target system and those between the target system and persons and groups in the outside environment. An organizational development program may create entirely new relationships and alter communication patterns within a group. New leaders may be established and other roles created by a change (Coughlin, Cooke, and Safer, 1973). The type of contraceptive used can change husband-wife relationships. Similarly, mechanization of farming has produced in the United States, and is producing in less developed nations, changes in migration that in turn have affected family and community relationships.

DIVISIBILITY.   Divisibility refers to the extent to which a change can be implemented on a limited scale. Thus an organization development program might be tried in only a few schools within a district, or a new fertilizer or seed may be used on only a few acres, or a new product may be purchased in a small sample quantity. Generally speaking, divisibility is a desirable dimension to have present in a change. Divisibility is important at the trial stage of adoption. It greatly facilitates the trial use of a change.

REVERSIBILITY.   The reversibility dimension is closely related to divisibility. It refers to the ease with which the status quo ante can be established if a change is introduced but is later rejected. It is an important dimension at the adoption stage. The greater the ease with which a change may be discontinued and the fewer the permanent consequences of having tried or adopted the change, the more likely it is that the change will be accepted. Surgical contraception is generally irreversible, whereas the use of a new book for a class is reversible.

COMPLEXITY.   The greater the degree of difficulty in using and understanding a change, the less the likelihood that it will be adopted voluntarily. This dimension is important at the evaluation and use stages. The change agent should distinguish between complexity-in-use and complexity-in-understanding. Different people may respond differently to the same change. For example, within a firm contemplating the purchase of capital equipment, one may find that the firm engineer may be more concerned with understanding how the equipment works, whereas a production supervisor may be more concerned with ease of operation and maintenance. A nurse may be indifferent to the scientific complexity of diagnostic equipment and more concerned with how easy it is to use. A change agent should be prepared to devote considerable time to explanation about the technical aspects of a change and to instruction in its use.

COMPATIBILITY.   The "goodness of fit" a change has with the situation in which it is to be used is very important. The situation includes psychological, sociological, and cultural factors. The change agent must exercise considerable care in making the change as consistent as possible with such things as group values and beliefs about medicine, the other machinery with which the new equipment must be used, the type of soil available for agricultural and animal feeding purposes, literacy levels, the past history of change in an organization, and so forth.

COMMUNICABILITY. The ease with which information about a change can be disseminated is another critical dimension. While a self-designated healer may describe in great detail how the herbs he is selling will increase sexual powers and fertility, another person not perceived as a healer might be beaten by the same crowd were the other person to discuss the same topic publicly. Some social structures are more amenable to word-of-mouth communication than others (Blau, 1974; Rogers and Argarwala-Rogers, 1976). This is true for nations as well as small groups. Additionally, some topics are more amenable to word-of-mouth communication than others. This may be compounded by different kinds of groups. For example, birth control pills are discussed more readily by both men and women than condoms or surgical contraception; however, women appear to discuss contraception in general more readily than men. Communicability is important at the awareness and interest decision-making stages. Certain industrial structures, as well as organizational structures, are more conducive to communication about innovations than others. (Czepiel, 1973; Zaltman, Duncan, and Holbek, 1973; Rogers and Argawala-Rogers, 1976).

TIME. The speed with which a change is introduced is an important dimension. It is necessary to think in terms of optimal time. The most appropriate rate of change may not correspond to the maximum rate of change possible. Change can be introduced too quickly or too slowly. This particular dimension is treated at length as a criterion for selecting change strategies. One argument used by the grocery industry to stall or defeat a proposed federal trade regulation ruling concerning information disclosure in advertising was that most consumers were not ready or able to interpret the nutrient information that the regulation was to require, and, as a result, it was argued, consumers would react negatively and avoid the information. It was argued further that an extensive consumer education program was needed before the regulation was put into effect.

OTHER DIMENSIONS. There are a large number of additional dimensions. A few of these are presented here. More detailed discussions can be found in Zaltman, Duncan, and Holbek (1973), which discusses these dimensions as applied to organizations, and in Lin and Zaltman (1973), oriented toward individuals. *Risk and uncertainty* are very important and probably increase as divisibility, reversibility, complexity, and communicability decrease. The *commitment* required to adopt and implement change is also important. The greater the commitment, that is, the greater the amount of time, money, and other resources that must be allocated, the less likely change is to take place. *Susceptibility to successive modification* is important where technology is changing rapidly or where different persons or groups use the change or innovation but have somewhat different use patterns.

Having defined and clarified the change process, it now becomes necessary to look at the context of social change in terms of participants, goals, and stimuli to change.

PARTICIPANTS IN THE CHANGE PROCESS

Before discussing the change process in more detail, it is necessary to identify briefly the various participants in the process.

Change Agent Role

The role of the change agent has been defined in a variety of ways. We review these briefly and present a definition of change agents that is used throughout this book. A more detailed analysis of the change agent is presented in Chapter 9. Rogers and Shoemaker have defined the change agent as ". . . a professional who influences innovation decisions in a direction deemed desirable by a change agency" (1971, p. 227). In this definition, the change agent serves as a communication link between the client system or change target and the change agency. Thus the change agent in this view represents an *external* agency trying to communicate the need for change to the target system. Jones takes a somewhat different perspective in indicating that change agents are:

> helping professionals whose roles involve the stimulation, guidance and stabilization of change in organizations. . . . a change agent is employed by the client system to assist . . . (he is) a "helper," "mover," "doer" (1969, p. 19).

In Jones' definition, the change agent is seen as operating at the request and consent of the change target system.

Argyris takes a somewhat different perspective. For Argyris, to intervene is " . . . to enter into an ongoing system of relationship, to come between or among persons, groups, or objects for the purpose of helping them" (1970, p. 15). The interventionist or change agent then

> assists a system to become more effective in problem solving, decision making and decision implementation in such a way that the system can

continue to be increasingly effective in these activities and have a decreasing need for the intervenor" (Argyris, 1970, p. 16).

For Argyris, the key element in interventionist-client relationships is for the client to have autonomy from the interventionist and control over the process. Argyris also emphasizes that change is not a primary task for the interventionist. Rather, the primary tasks are (1) to generate valid information, (2) to help the client make informed and reasonable choices, and (3) to develop internal commitment to these choices (Argyris, 1970, p. 21).

The main distinction among the preceding definitions of the change agent role seems to be on the focus of control and collaboration. Rogers' definition tends to put the change agent outside the client system, representing some third party attempting to communicate something to the client system.* Jones, Bennis, and Argyris all focus on collaboration between client system 'and change agent, with the change agent working at the "pleasure of" the client system. This latter formulation neglects many change situations in which someone is operating either inside or outside the client system without the latter's desire to change. For example, Ralph Nader's representatives certainly operate as a source of change for the many organizations that are the target of their reform efforts. This type of change agent would be closer to Rogers' view.

In this book we take a broad view of the change agent role to include people within as well as outside the client system who are attempting to create some change in that system whether it is sanctioned or not. Specifically, a change agent is any individual or group operating to change the status quo in a system such that the individual or individuals involved must relearn how to perform their role(s).

Change Target and Client Systems

The second major actor or group in the change process is the client or change target. Jones (1969) defines the client system as a ". . . specific social system that requests a change agent to assist in altering its organization with the objective of improved performance" (p. 16). Once again, this definition is limited in scope in that it emphasizes the control and choice of the target system in bringing in the change agent. It ignores the case in which a change agent might be trying, through pressure or even coercion, to get some individual or group to change. Civil rights leaders such as Martin Luther King operated as change agents on many change targets that were unwilling to change and resisted the change.

---

* Not all of Rogers' work reflects this perspective.

Given the problem of reference, we distinguish between the change target system and the change client system. The change target system is the unit in which the change agent(s) is trying to alter the status quo such that the individual, group, or organization must relearn how to perform its activities. The change client system is the individual or group requesting assistance from a change agent in altering the status quo. There may be total identity between the change target and change client systems. Also, the two systems may be totally separate entities, such as parents seeking psychiatric help for their child. In this instance, the parents represent clients, the therapist the change agent, and the child the change target. It is not uncommon in this specific situation for the parent(s) or party other than the child to also become a change target.

In situations in which no one in the change target system is requesting help to change, there is often some coercion and persuasiveness on the part of the change agent, even initially to develop some support in the system for the change. For example, advocates for reform in police departments have often had to operate with no support from any individuals or groups within the police department. The change agents here simply had to put pressure on the department by creating adverse public opinion for the department so that the department would finally be forced to change to keep a positive image in the community.

This type of change is qualitatively different from the situation in which some individuals within the target system are seeking some assistance to change from the change agent. In the police department case, the change agent would likely encounter less resistance and more support and exert more influence on the system if there were some individuals in the department who requested assistance. For example, if the chief of police had called the change agent into the department for help in implementing some change, the change agent would be perceived as perhaps more legitimate. In this instance, the initial client would be the chief. Similarly, acceptance and implementation of new organizational development programs by teachers is materially advanced when key administrators demonstrate strong commitment to the program.

The key point here in distinguishing between the change target system and the change client subsystem is that change agents may be attempting to influence and change some system in which there is no support or wish for assistance in any change. There is no client subsystem in the target system. This type of change is much more difficult for the change agent because of potentially greater resistance. Nonetheless, this is an important type of change activity and is quite frequent as various pressure groups attempt to change different types of organizations. The change agent is simply operating under greater constraints when attempting to alter a change target system in which no client subsystem is seeking assistance to change.

PITFALLS IN SOCIAL CHANGE PROGRAMS

At this point in our discussion, it is important to discuss some of the pitfalls that social change efforts can encounter. We raise these issues here to make the reader aware of these potential problems when considering the analysis of various change strategies that follow in the book. This list of pitfalls is not meant to be exhaustive, but represents what we feel to be some of the key problems in individual, group, organizational, or societal change.

Rationalistic Bias

This problem is characterized by the belief that individuals are guided by reason and are rational according to conventional criteria. Therefore, all the change agent needs to do to create change is to present the information and knowledge regarding the change to the change target system, and the change will be accepted. The tenuous assumption here is simply that, when rational individuals are presented with the rationale for a change, they will accept it. The fact that what a client considers rational for him or herself is not necessarily what the change agent believes rational for the client is ignored. Thus the change agent's job might inappropriately be perceived as less complex than it actually is, since all that is required is to inform the target system of the rationale for the change and the "truth shall let them change." Bennis (1966, p. 104) has pointed out the major oversimplification of this approach by indicating that change requires more than just knowledge. There has to be a program for the implementation of change, as well as some commitment on the part of the change target to accept the change.

Perhaps the best example of the problem of this rationalistic bias has been in the attempt to implement operations research/management science systems into organizations. These are highly rational systems, but they have been met by much resistance and failure because well-developed plans and target system commitment did not exist. The change agents operated under the faulty assumption that a "rational" manager would not resist or reject a system that would give him more accurate information for decision making. Argyris (1971) has pointed out quite clearly how a management information system can threaten a manager because he would no longer have the screen of "insufficient information" to hide behind in avoiding difficult decisions.

Social change in family planning has also been overly reliant on information, education, and communication programs stressing rationality. This approach assumed that, with proper education and ready availability of contraceptives, people would quickly adopt family planning practices. Disappointment with this approach has led to consideration of highly controversial "beyond family planning" techniques. These are discussed elsewhere in this book.

Poorly Defined Change Goals

Change attempts are often characterized by poorly defined goals. In fact, it is often the case that there is ambiguity about the objective beyond that of simply "changing things." For example, an organization may decide to implement a training program for supervisors to make these supervisors more aware of their subordinates' problems. What is the goal here? Is it to change attitudes toward subordinates or to change behavior toward subordinates also? The exact goal is important because it obviously affects how the training program is implemented and evaluated. For example, if the goal is simply to change attitudes, classroom exposure would probably be an effective strategy. However, if the objective is to change actual role behavior on the par. of the supervisors, some kind of role playing or experimental learning would be more appropriate. (See Kolb et al., 1974.)

Another problem with poorly defined change goals is that they are likely to create ambiguity, uncertainty, and anxiety for those who are going to be affected by the change. If there is ambiguity about the outcome of a given change, individuals to be affected might assume that the change is going to have a negative impact on them. For example, the classic Coch and French (1948) study of the implementation of change in a pajama factory indicated that, when people were informed and understood the change goal and had some participation in the change process, they were more receptive to the change itself. When the objective of a change is clear, members of the change target system are likely to have the information and develop the skills to utilize the change or innovation in carrying out their role. The study by Gross et al. (1971) of the attempted change to a new teacher role is a good example of the problems created by poorly defined change goals. The change was to a different role model for teachers to help them deal more effectively with motivational problems of lower-class children. In analyzing some of the factors that prevented the effective implementation of this change, Gross et al. (1971, pp. 139–142) found that teachers were unclear about the kinds of role performance that was required to implement this new teacher role. The resulting ambiguity not only made it difficult for them to try and implement this change, but the ambiguity also created additional frustration and tension for the teachers to cope with.

Poorly Defined Problems

Problem definition is often an intricate and highly partisan task (Chapter 2 explores this issue in greater detail). Too often symptoms are mistaken for causes of problems, resulting in misdirected remedial efforts, ineffective use of resources, and increased levels of frustration. It is not uncommon to find tunnel vision among persons responsible for the diagnosis of problems. Community development leaders place great stress on the lack of infrastructure as the reason that modernization occurs slowly, whereas family planners see the problem primarily as one of overpopulation; agriculturalists define slow regional development mainly in terms of inadequate agricultural production and the associated loss of revenues.

Overemphasis on Individuals

Another major difficulty is in the assumption that individuals behave in a vacuum and are not influenced by those around them in the social system. Change programs geared to changing the attitudes and/or behavior of the individual often forget that the individual's behavior is affected quite dramatically by those around him (Katz and Kahn, 1966, p. 391). If a person is asked to change behavior that is not supported or reinforced by those significant others, behavioral change is unlikely to occur, or at least unlikely to be maintained (Argyris, 1962). Thus change strategies that neglect the social and physical context of the situation in which the change takes place are missing some of the important causes of behavior. It may be important to change the key individuals in the system, but, in order for change to permeate the system, this change must be supported throughout the system. The willingness or openness to change by many persons or groups is too often thwarted simply because ways of implementing change by altering the larger system are not defined for them. For example, the lack of adequate physical distribution systems for contraceptives and nutritional foods greatly hinders their adoption even in the context of strongly motivated consumers.

An example of the overemphasis on individuals is reflected in the following example of change in a large urban police department. The police department in question had gone through periods of upheaval and attempted change in the mid and late 1960s. In the late 1960s, a dynamic individual was made chief. This individual was highly committed to change and reform in the department. Many individuals in the community as well as funding agencies around the country took the view that, "if the chief was committed" to change and innovation in police practice, that department could change quite easily. The major problem here was that, although the chief's attitudes and behavior supported change, the existing norms and values in that organization and the resulting role expectations and pressures supported the status quo. The thrust for change was located mainly in the chief, with no mechanism for translating this into action within the department itself.

There are many instances in the commercial marketing of consumer and industrial goods in which new products become commercial failures or gain acceptance only very slowly because informal interpersonal communication is inadequately understood (Czepiel, 1972). Considerable research is now underway in many firms to learn how to better manipulate informal word-of-mouth communication to facilitate the introduction of new products. In fact, commercial marketing research now represents one of the major sources of new information in the social sciences about such roles as opinion leaders and gatekeepers.

Overemphasis on the individual has sometimes led to a failure to take into account adequately the larger cultural and social milieu within which change is advocated. Successful innovations developed within one context having its own unique kinship systems, religious practices, and so forth, are not foreordained to succeed in very different social contexts. In some countries, for example, condoms are little used. This is partly because they are associated with prostitution. The act of buying a condom is a statement, to the vendor at least, that one is going to engage a prostitute regardless of with whom one will actually use the condom. In relatively small, closed, rural communities of Asia, this can severely restrict condom sales. Consequently, it is not surprising to find differences in condom usage between cities and small towns and between countries in which condoms are not strongly associated with prostitution and those in which it is. Even when two social or cultural situations or contexts are quite similar in important ways with respect to the innovation, the advocated change may fail because of its origins in another country or group. In the case of technology transfer, this is labeled the "not-invented-here" syndrome.

Technocratic Bias

Bennis (1966, p. 104) has identified as technocratic bias the overemphasis on development of a change program without an accompanying plan of collaboration for implementing the program. This problem draws from the overemphasis on rationality in its assumption that all the change agent has to do is present the change program to the client system and the clients will then implement it on their own. This neglects the important need for a strategy for implementing the change and the fact that the change target system may need some assistance in actually using the change. Since change usually results in the alteration of power, status, and role relationships, it may be necessary for the change agent to work with the change target as it moves through this process.

For example, in adopting a new computer-based management information system, it is certainly important for the change agent to thoroughly develop and design the new system and to get it operational. However, it is equally important for the change agents to help with the transition process as system members become familiar with this new technology. After the new system is in operation, it may be useful, if the change agents are still present in the system, for them to work with managers in debugging the system and in getting comfortable with using this new technology. The presence of the change agents could be important in reducing anxiety and misunderstanding (see Coleman and Riley, 1973).

The field of education is replete with examples of change efforts that invested heavily in the research and development of innovations but had little provision for implementation. This is related to the assumption that the physical possession of a product or official acceptance of an idea is synonymous with use. Many innovations sit unused and even unopened on closet shelves in classrooms, and desegregation and compensatory education exist as paper policy formally endorsed by school boards but not as actual practice (Turnbull et al., 1974).

Similarly, there are many examples of nutritional foods developed specifically for consumption by poor and uneducated people in less developed countries that never gained acceptance because of poor marketing practices. While considerable research and development went into product development, very little thought was given to such factors as packaging, merchandising techniques, and distribution strategies. The result was low visibility of the product, unattractive packaging leading to poor impressions about product quality, unavailability in places in which the intended user shopped, and so forth. A related issue is the belief that people will automatically prefer and switch to a technically superior product or service. Again, case histories in technology transfer and other contexts refute this assumption.

STIMULI FOR SOCIAL CHANGE

The stimulus to changing an individual, group, organization, etc., occurs when there is a perceived discrepancy between how the change target is performing and how the change target or someone else believes it ought to be performing. This discrepancy creates a performance gap (Downs, 1967, p. 191). A performance gap is a discrepancy between the criteria of satisfaction in performing some act and the actual performance of that act. The individual, group, or organization simply feels that it ought to be doing better in its performance than it actually is. The performance gap then serves as a stimulus to search for alternative ways of responding.

It should be emphasized that a performance gap can be identified by the change target system or by someone outside the target system. For example, decision makers in an organization may be exposed to a new management information system. Having been exposed to this new technology, these decision makers may decide that they now do not have as much information as they could with a new information system. Here the performance gap is identified by members of the target system. There are also instances in which the performance gap can be identified by individuals or groups outside the change target system. Family planning groups operating as advocates for birth control and other family planning activities often observe some population groups and determine that this particular group could be more effective or do better in its birth control and other family planning activities. In this instance the performance gap is identified by a group external to the change target. This external group is trying to communicate this performance gap to the target system so that system members can become aware of the fact that they could be doing better.

Performance gaps, whether they are internally or externally induced, can occur in various ways.

1. The change target may have high expectations regarding some performance, and yet some time may be required for these expectations to be met or to adjust to the actual level of performance. For example, new management recruits in an organization may have very high expectations about the opportunity in their new jobs to utilize their skills and abilities. Their initial experience in the organization may indicate that these expectations regarding their jobs were unrealistic. However, there is still a performance gap as the new recruits adjust their expectations downward as to what to expect regarding opportunities in their jobs. There is still the feeling that the organization could do more to meet their needs.

2. The criteria of satisfaction in the target system may adjust themselves upward (March and Simon, 1958, p. 183). For example, an organization might be performing quite well but its members may decide that it could probably do better. In this instance, organization members' expectations have been raised, and there is now a discrepancy between the system's

actual performance and what members believe the system could be doing. This performance gap then serves as a stimulus to finding ways to improve performance.

3. There can be changes in the change target system's external environment that can create performance gaps.

   a. There might be a lessening in demand for the change target's output. For example, an electrical engineer working in some space-related industry may find that the demand for his skills have changed. He may have to revise his skills so that he can move into another job. One of the authors has had an experience with a large defense industry organization that also illustrates change in demand for the system's output. In the late 1960s, this organization found that the demand for its weapon systems had decreased with the general decline in defense spending. The change in demand then created a need to search for alternative outputs the organization might develop. This search resulted in the organization transferring its technological expertise into the pollution abatement field.

   b. Technological changes in the larger environment may change organization members' criteria for satisfactory performance (Downs, 1966). For example, the development of computer-based management information systems has raised higher expectations on the part of managers regarding how they should be gathering and processing information. This resulting performance gap then creates a need to change the existing management information system and adopt a new computer-based system.

   c. There may be changes in the change target system's power position in relation to other systems in the environment. The world-wide oil boycotts in 1973 drastically changed the power position of American oil companies. Seeing their dependence on Arab oil-producing countries manipulated as a political pressure to affect American public opinion, the oil companies lost a good deal of control over the source of their raw materials. The change has created stimulus for the oil companies to increase their investments in exploration in non-Arab countries.

   d. Greater satisfaction among reference groups is another cause of performance gaps. Lerner's (1971) hypothesis is that developing nations want what developed nations already have, but they do not possess the means to obtain it. The numerator of the want:get ratio is increasing at a much faster rate than the denominator, leading to considerable frustration and anger among substantial numbers of the world's population.

4. Changes can also take place within the change target system that can create performance gaps that then lead to change and innovation.
   a. In studying change at the group, organizational, or societal level, we may find that the introduction of new personnel into the system can create a stimulus for change. For example, police departments in the last few years have been getting more and more college-educated recruits. These recruits bring with them expectations about what they can hope to accomplish in police work. With their higher level of education, they are more willing to challenge the status quo in the department and to press for reform (Juris and Duncan, 1974). As a result, this new segment of the police organization is going to be a continuing stimulus for raising questions about current operation procedures.
   b. Technological changes may occure within the change target system. The introduction of a new management information system will likely raise managers' expectations with respect to what the new system can do for them. If the new system does not operate effectively, this is likely to create dissatisfaction and a need to change to a more efficient system.

CHANGE GOALS

In the above discussion, we focused on some of the conceptual problems in the change process as well as some of the stimuli to change. This section discusses very briefly some of the instrumental goals or objectives of change programs. It was indicated earlier that one of the difficulties with many change programs, whether they are at the individual, group, organizational, or societal level, is that the goals or objectives of the program, and even the nature of the underlying problem of concern, are unclear. This lack of clarity can create problems in trying to evaluate the change attempt, as well as problems among members in the change target system. Often, resistance to change increases when people are unclear about the objective of the change. When unclear about a proposed change, individuals may often assume that it will have a much more negative effect than might actually be the case.

Change efforts may have three basic instrumental goals or objectives. They may be to (1) change attitudes, (2) change behavior, or (3) change both attitudes and behavior. Changes in attitude and/or behavior are the means or instruments by which higher-level objectives or goals such as client or change agent well-being are established.

Attitude Change and Behavioral Change

The goal here is to change the attitudes of individuals within the change target system. The intent of most attempts to change attitudes is to alter behavior. The assumption is that, if attitudes are changed, behavior will change to support these attitudes. For example, many human relations training programs in organizations are geared to changing supervisors' attitudes toward their subordinates (Slocum and Hand, 1971). If supervisors can become more sensitive to the needs of their subordinates by developing more favorable attitudes toward their personal needs, they will be more responsive to subordinate needs in their supervisory behavior. Similarly, if favorable attitudes toward a new product or service are developed, the purchase and use of the new product may follow. Considerable resources in change programs, ranging from irrigation projects to the synthetic food substitute programs, are expended in obtaining favorable attitudes by changing negative attitudes or formulating new attitudes where none previously existed.

It should be mentioned that, while attitude change is generally sought as a precursor to behavioral change, it may also be an end in itself. One recent nutrition-related change program in a Latin American country allocated modest resources to media (television and selected newspapers) that primarily reached only the well-to-do members of the nation who had little need for information concerning procedures for improving infant nutrition. The avowed purpose of this was to create a favorable attitude among this segment of the population toward the organization sponsoring the nutrition campaign.

Attitude change does not always produce behavioral change. In fact, attitude change efforts intended to create behavioral change often fail. Consequently, many strategies and tactics are adopted to secure behavioral change in the presence of negative or weak positive attitudes toward the advocated change. The use of special incentives, deception, and forced choice are a few ways behavioral change is induced without favorable attitudes.

It is useful to review very briefly some of the reasons why there may be discrepancies between attitudes and behavior. Favorable attitudes may exist but the means for converting those attitudes into behavior may be absent. Higher education is desired by many people who lack the necessary financial resources or the ready physical availability of schools or because teachers are unavailable in their geographical location. Modern attitudes toward medical care exist in many places in which trained medical personnel simply are not readily available. Thus both the change target or client and change agent may lack various resources to convert attitudes into behavior.

Another factor to consider is the time lag between attitude formation and its behavioral expression. Many things can happen to interrupt the attitude-behavior change sequence. In one project concerning organizational development in a school system, the relevant decision makers in that system committed themselves to an experiment. Before the experiment could get underway in that school system, the superintendent who had initiated the project left, creating considerable ambiguity and uncertainty. The teachers and principals then revoked the earlier commitment to the project and insisted on another review of the experiment in light of the new circumstances. This produced considerable delay and caused some parties previously committed to the change to withdraw from participation in the experiment.

Lack of knowledge or awareness of how to implement the logical behavioral consequences of an attitude represent another cause of an attitude-behavior gap. Some change programs fail because they do not specify what a person or group must do to convert attitudes into behavior. Questions concerning who to contact, when, where, and how this should be done may be addressed inadequately. In one mass immunization program, community members were not informed about the date the immunization team would visit, nor their location within the village. Consequently, when the teams arrived at the various villages, only a small percentage of those intended for immunization were ever vaccinated. This failure was made dramatic a few months after the program ended when one of the diseases the vaccine was to guard against swept several of the villages, with devastating results.

One very important consideration concerns the assumption that, with all obstacles removed, people will behave in a manner consistent with their attitudes. This assumption is questionable. First, people maintain several attitudes that could result in conflicting behavior and/or mutually exclusive behavior. Because a person is favorably inclined toward planting any of a variety of crops does not mean that he will plant them all; it may simply be economically unwise. Attitudes that two or more employees are qualified for promotion to a higher organizational rank does not mean that they will all be promoted, particularly when there is only one opening. Also, there is a deeper question concerning the logical necessity that attitude change be followed by a corresponding behavioral change.

Normative Change

The objective of normative change is to alter the actual norms and culture of the change target system. The view here is that the only way to create

Organizational development is A change strategy

enduring change is to alter the underlying values and beliefs of the target system that affect attitudes and behavior. Organizational development is an example of a change strategy that tries to attain this objective. Organizational development is the "... creation of a culture which supports the institutionalization and use of social technologies to facilitate diagnosis and change of interpersonal, group, and intergroup behavior ..." (Hornstein et al., 1971). Thus a new set of values is developed in the system, which now defines dealing with change as an appropriate part of the individual's organizational role. Here change is institutionalized.

This chapter can be summarized by the following principles:

1. In defining the concept of change and innovation, the process is one of relearning on the part of an individual or group that (1) is in response to newly perceived requirements of a given situation requiring action and (2) results in changes in the structure and/or functioning of social systems.

2. In planning for change, the change agent and client system should be aware that, in addition to planned change that is a deliberate effort with a stated goal on the part of a change agent to create a modification in the structure and process of a social system, there is also unplanned change. Unplanned change occurs as a result of the interaction of the forces of the social system. However, it is brought about with no apparent deliberateness and no coordinated goal setting on the part of those involved in it.

3. The change agent should distinguish between those affected by the change and that subpart of those affected who may be seeking assistance. The *change target system* is that unit in which the change agent is trying to alter the status quo such that the individual, group, or organization must relearn how to perform their activities. The *change client system* is that part of the target system comprised of individuals requesting assistance in altering the status quo.

4. An individual or group can be considered to be operating in the change agent role when they are operating to change the status quo in the change target system such that the individuals involved must relearn how to perform their role.

5. Change agents and the client system should avoid the rationalistic bias in designing change programs. The tenuous assumption here is that, when rational individuals are presented with the logic for a change, they will accept it. The fact that what a change target considers rational may not be the same as what a change agent considers rational is ignored.

6. In planning for change and innovation, the change agent and client system should be clear about what the change objectives are; otherwise ambiguity and uncertainty may occur that can cause resistance.

7. Change agents and client systems should avoid poorly defining change problems. Too often symptoms are mistaken for causes of problems, resulting in the misdirection of remedial efforts and ineffective uses of resources.

8. Change agents and client systems should avoid focusing change programs only at the individual level, forgetting that an individual's behavior is quite dramatically affected by those around him and the larger culture.

9. Change agents and the client system should avoid the technocratic bias in designing change programs. The difficulty here is the overemphasis on developing a change program withouti an ccompanying plan of collaboration for implementing the program. This neglects the need for a well-developed strategy for implementing the change.

10. A stimulus to change or innovation is created by performance gaps—a perceived discrepancy between what the change target is doing or how it is performing and how the change target or someone outside the change target believes it ought to be performing.

11. To avoid discrepancies between attitudes and behavior regarding change, it is important that the change target have the means for converting favorable attitudes regarding change into actual behavior.

12. Discrepancies between attitudes and behavior can be reduced by minimizing the time lag between attitude formation and its behavioral expression.

13. Discrepancies between attitudes and behavior can be reduced by minimizing the lack of knowledge or awareness of how to implement the logical behavioral consequences of an attitude.

14. The change agent should identify the most salient need of a target system and establish awareness within the target group that the advocated change better meets those needs (assuming it does). This stresses the relative advantage of the change.

15. The change agent should identify those social relationships most likely to be affected by the change and minimize conflict by adapting the change to be compatible with those characteristics and/or preparing the target group well in advance to receive a potentially disruptive change.

16. The change agent must exercise considerable care in making the change as compatible as possible with existing values, beliefs, capabilities, and situational factors surrounding the change target.

17. It is highly desirable to stimulate word-of-mouth communications to augment other communication channels or to compensate for their absence.

18. Change agents should think in terms of optimum rather than minimum time for introducing change.

19. To the extent possible, a change should be developed to permit limited use, that is, it should be divisible.

20. The introduction of a change should occur in such a way that the discontinuance of it, because of bad experiences, enables the target system to return to an earlier preferred state. This gives the change reversibility.

21. Every effort should be made to simplify the change to make it less complex to understand.

CHAPTER TWO

Defining Social Problems

In Chapter 1, we indicated that one of the serious problems encountered by those interested in accomplishing change is that there may be an incorrect definition of just what the change problem is. This incorrect definition of the problem may then lead change agents astray when developing strategies for dealing with the situation. For example, an organization may call in a change agent to help them deal with the supposed problem of employee morale. The organization may define the problem in terms of employees being lazy and uninvolved in their work. Further investigation may show that they are behaving this way because of poor supervision. Their supervisors do not give them enough information to carry out their jobs effectively. The correct change strategy, given the expanded problem, must focus not only on the employees but also on their supervisors.

There are few activities in applied social change that are more important than the definition of the social problem. Sadly, the mechanical task of defining social problems is one of the most neglected areas of applied social research. The proper definition of the problem is important in social change research for several reasons. The way a problem is defined influences:

1. the design of the research to be conducted
2. the selection of the population and sample to be studied
3. the kind of data to be gathered
4. the statistical modes of analysis
5. the remedial action to be initiated by change agents
6. the future conduct of policy making and regulatory agencies

An incorrect definition of a social problem can render ineffective all subsequent efforts to bring about changes intended to help remedy a problem, resulting in a waste of many scarce resources and a lower level of satisfaction among clients.

PROBLEM DEFINITION

Problem definition is also of great importance from the standpoint of research utilization because the two phenomena are mutually causative. The specific diagnosis or definition of a problem establishes the boundaries within which a person will seek research and knowledge to apply to the problem. Also, it is possible that the definition of a problem may affect our processing of existing information. Thus a relevant item of knowledge may be relevant in different ways, depending on the way the problem is viewed. On the other hand, our existing bank of knowledge certainly places limits on the number of different ways a problem is likely to be defined by any given person or agency. Our customary way of using that knowledge is important in shaping remedial action in response to a problem. Because of the interaction between problem definition and our response to situations in need of change, we devote this chapter to problem analysis.

A social problem is considered here as a situation that is perceived by a group as a source of dissatisfaction for its members and for which preferable alternatives are considered possible. Note there are two important aspects to this definition: first, something is considered a source of dissatisfaction; and second, improvements (preferable alternatives) are believed possible.

Two approaches to problem definition are presented: a metatheory approach and an open systems approach. Techniques for gathering problem-related data are discussed next. Various perspectives on problems are treated next. The discussion of problems concludes with an excursus on the nature of problems, the manifestation of problems, and various possible categories of problems.

Defining Problems: Two Approaches

APPROACH 1: A METATHEORY PERSPECTIVE. The task of defining social problems has been largely treated as intuitive. How, then, can the problem definition task be made more explicit? The main benefit of the metatheory perspective is that it stresses a thinking process and identifies various stages or steps involved in thinking about problems. It appears that the following steps could be followed. The (first) is a statement of the symptom, that is, a certain phenomenon exists. (Second) a statement is required saying that the phenomenon is of the nature Q and is produced by factors X_1, X_2, ... X_n. Q represents the experience of a symptom, and X_1, X_2, etc. are the causal factors whose functioning represents the real problem to be remedied. A (third) step, an analytical activity, is desirable as a means of increasing the likelihood that remedial efforts will be effective. This step is at a level of explanation that states that factors X_1, X_2, ... X_n are interactive or have interacted in manner Y_1, Y_2, ... Y_n to produce in some past or present time a phenomenon of the nature Q. (Finally,) for the most complete understanding of the problem, an analytical level is necessary that states that factors X_1, X_2, ... X_n interact the way they do for reasons W_1, W_2, ... W_n, thus producing a phenomenon of the nature Q.

It is useful to provide an example. At step one the symptom may be poor performance (as judged by a teacher) in the language arts of Black children in school. Step two would state the nature of the problem, Q_1, as: conflict in urban Black children that impedes academic success in the language arts. This would be caused, in part, by dissimilarity between homes and schools of urban Black children with respect to social-cultural contexts and valued speech styles, X_1, and educators not having verbal expectations appropriate for urban Black children, X_2. At the third step it would be observed that the dissimilarity, X_1, interacts in manner Y_1 with educator expectation, X_2, to produce Q. The interaction is operationalized in the form of educators expecting verbal styles consonant with the school-valued social-cultural verbal forms that the Black child cannot provide and hence performs poorly when viewed from the school's value system. At problem definition step four, it might be postulated that the reasons (W_1, W_2, ... W_n) educator expectation coupled with a child's home-oriented cultural verbal style produces poor performance ratings is that the schools are either unaware of or are indifferent to the differences between the school and home-valued verbal styles, W_1, and because the child is left to resolve the difference without adequate assistance from the school and home or community, W_2.

Given the four-step problem definition and analysis above, the interventionist could identify certain information needs and research questions. For example, one need would be for educators to understand better the repertoire of verbal skills urban Black children bring to the classroom. This raises the following research question: what are the important conflict-causing discrepancies between school-valued English and Black home- and community-valued English?

The analysis above might very well differ depending on who is defining the problem of poor performance in the language arts by urban Black children. The same four steps would be involved but the causal factors highlighted, the perceived nature of their interaction, and the supporting rationale for their interaction would likely vary as the viewpoint shifts from (1) the marketing agency, for example, the National Institute for Education, to (2) the target group toward which remedial action is directed, for example, the Black child, his or her parents, and/or teacher, or to (3) third parties such as Black community organizations. An excellent treatise on the partisan diagnosis of social problems can be found in a paper by Guskin and Chesler (1973) who observe: "The scholarly cloak of neutrality worn to protect one's professional status from democratic public consideration and attack is simply not viable. The nature of personal preferences, scientific methodology, consumer patterns and the fabric of the society make value neutrality an impossible activity, let alone a viable conception of one's self." The basic questions they pose are: whose side are you on? and do you know it?

The experience of a drugstore chain in the eastern United States provides another example of this process. The drugstore chain had difficulty in maintaining what it considered an optimal inventory of nonprescription antidepressant drugs. There were seemingly random fluctuations in the demand for this product, resulting in oversupply at some times and undersupply at other times. Thus the symptom experienced, Q, was an inability to maintain "adequate" stocks of a particular high-profit item. The marketing staff of the drugstore chain challenged the hypothesis offered by retail operators that the fluctuations were random. The marketing division hired a medical sociologist as a consultant to work with them on this problem. The consultant postulated the alternative hypothesis that the consumption of antidepressant drugs was associated with particular societal activities.

Sociologists had for a long time noted and studied the covariation between consumption of physician-prescribed and over-the-counter antidepressant drugs on the one hand and societal activities such as holidays and prominent events on the other hand. The relationship observed was that consumption of such drugs declined quite noticeably prior to holidays and other occasions such as elections and major sport events. Thus covariation was found between the symptom and some other independent phenomena. These other phenomena are rationally related to drug consumption by a social theory that need not be analyzed here. The basic notion, however, was tested. First, the drugstore chain records showed that consumption did fall off markedly prior to national holidays and the Christmas-New Year seasons. Moreover, a substantial proportion of the population in the region were Jewish, and, as the theory predicted, there were also sharp declines in antidepressant drug sales prior to the important Jewish holidays falling in September/October.

In this example the immediate causal variables, identified initially by observing covariation, could not be manipulated, but the drugstore chain could—and did—adapt their inventory practice by ordering smaller quantities in advance of major societal events and increasing inventory immediately afterwards. Although this ended the problem exploration activities of the chain, an interesting sequel developed approximately three months later.

A manufacturer of two popular, nationally distributed over-the-counter antidepressant drugs went further in the analysis of this problem. It hired, at the suggestion of the director of research for the drugstore chain, the same medical sociologist to consult with the promotional department. The intention was that a further understanding of the theory behind the association of special events and drug consumption could provide a clue for advertising appeals. The phenomenon, Q, was shown to be affected by a societal event, X_1. Another causal factor, X_2, is also part of the theory. This is the need for association or affiliation, that is, a need to feel that one belongs to or can identify with others in society. The consultant explained that these two factors, X_1 and X_2, interacted in manner Y. The interaction of the manner Y was that prominent events (X_1) provided a salient reference point that individuals felt they shared with other persons, thus satisfying to some extent the need for association (X_2). The reasons (W_1) why this interaction takes place lie in a further elaboration of the general theory that cannot be undertaken here. The promotional department, having the information that social-psychological isolation is associated with—in fact is one of the causes of—the use of antidepressant drugs, decided to utilize this information in their advertising. They also planned to advertise more heavily during periods when no major events were taking place or were about to occur. The manufacturer, in cooperation with an advertising agency, proceeded to conduct behavioral laboratory tests of appeals stressing loneliness and found that such appeals were far more effective (in the laboratory setting) than any of their other advertising when it was pretested. Thus the further analysis by the manufacturer of the theory employed by the drugstore chain was apparently successful. The word "apparent" is used deliberately. The reader might be interested to know that the appeal was never used. The legal staff of the company learned through informal channels outside the firm that the Federal Trade Commission was aware of the promotional plans and was prepared to intervene if the plans were put into effect. The company decided not to spend additional funds on developing the campaign in light of the possible injunction against using the appeals. They did, however, adjust their scheduling of advertising.

Let us introduce a more refined version of the four-step procedure just mentioned.

STEP 1.   Identify indicators of social problems. A social indicator is a measurement of a social phenomena whose movements indicate whether a particular social situation is improving or worsening in terms of some goal.

STEP 2.   Monitor indicators of social problems.

STEP 3.   Determine whether the measurement of the indicator has exceeded a threshold or level signifying a danger.

STEP 4.   Determine what variable or set of variables covaries with the indicator variables. These covarying variables are the potential causes of or explanation for the problem.

SUBSTEP 4-A.   Determine whether there are possible explanatory variables that function in time periods considerably removed from the period when symptoms are first noted.

SUBSTEP 4-B.   Determine whether there are possible explanatory or causal variables related to the symptoms at certain times and not at others.

STEP 5.   Select the most plausible explanatory variables. Plausibility is established by virtue of logic, past experience, past research, and soundness of the underlying theory or model.

STEP 6.   Determine whether variables among the most plausible set are (1) sufficient but not necessary, (2) necessary but not sufficient, (3) neither necessary nor sufficient.

STEP 7.   Determine the structure of relationships among the causal or explanatory variables and between these variables and the problem symptoms. Possible relationships are shown below. Qs represent symptoms; Xs represent explanatory variables.

$$X_1, X_2 \ldots X_n \rightarrow Q$$

Serial relations, single effect

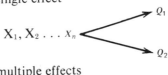

Serial relations, multiple effects

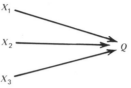

independent relations, single effect

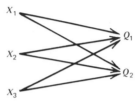

independent relations, multiple effects

STEP 8. Identify "decision variables," that is, those endogenous variables that change agents can manipulate and influence, and exogenous variables, that is, those variables to which change agents can only adjust.

STEP 9. Identify and match appropriate actions with regard to endogenous and exogenous variables.

A good example of this process is provided by the director of a major social welfare agency in the city of Chicago. (The reader will probably find himself disputing the wisdom of much of what this director did, and this is one reason this particular example is used.) Although the director did not describe her problem-solving process in a logical format, she did consider most steps in the process above. One indicator she relied on to assess the current welfare of any particular part of the city or for various ethnic groups was the ratio of welfare cases opened to welfare cases closed. This was step 1. This ratio was reported to her periodically and broken down into various social and demographic categories as indicated above, for example, section of the city, ethnic groups, and type of problem (drug abuse, child abuse, etc.). This was step 2. The operational threshold the director used to indicate a serious (continued) worsening of a problem was a 15% change in any of the ratios over a 12-month period. This was step 3. Factors she associated with the opening and closing of cases were changes in unemployment levels for the particular categories, changes in the absolute numbers of persons within a group or category, availability and effectiveness of caseworkers, and efforts by other social action agencies. The director's selection of these factors constituted step 5. In the director's judgment, none of these or the other factors mentioned in the interview were necessary, but any one was sufficient to cause a problem. This determination corresponds to step 6. Further discussion of these factors revealed in the director's mind the following structure of relationships among some of the factors. The structure presented here is more formal or explicit than its actual verbalization during the interview. The increase in availability of caseworkers enables more cases to be opened and closed. Also, an increase in activity by other agencies can increase the number of cases opened by virtue of increased referrals, while at the same

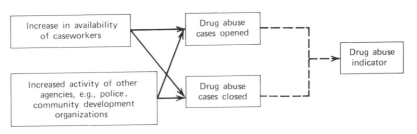

time increase the number of cases closed by helping to find employment, psychiatric care, etc. for clients. The director felt that the availability of caseworkers was within her control in the limited sense that she had (and on occasion exercised) the authority to shift existing caseworkers from one problem area to another, for example, from drug abuse to child abuse, or from the one district to another. On the other hand, the director had little influence on the level or the nature of the activity undertaken by other agencies. The major response to increases in other agencies' activities was to ask them to refer only the most extreme cases or only those which her agency was uniquely equipped to handle.

The recent experience of one of the five largest oil companies provides another concrete example of this nine-step procedure for exploring problems. The particular context of this example involves gas stations in the Detroit metropolitan area. The oil company purchased land, built the gas stations, and then leased the stations to operators. Periodically, oil company representatives visited the stations to check on pump service, cleanliness, and mechanical services. In addition, the company sponsored areawide promotions for their operations.

Executives of the company had selected total area sales and their market share as their most important marketing indicators. This selection of indicators is step 1. Data concerning these indicators were collected quarterly for the Detroit area. This monitoring corresponds to step 2. In the fall of 1971 executives became very concerned when market share dropped substantially for the second consecutive quarter, whereas total area sales remained relatively constant. Observing market share dropping below a predefined acceptable level is step 3. The company then set about step 4 by examining price changes and competitive advertising, which might account for the declines. They noted that a major competing brand had been advertising intensively, stressing station operators' concern for car care. Price was found to be relatively stable. In addition, many complaints had been received concerning cleanliness of rest rooms and lack of courtesy by pump attendants. Past experience, coupled with the results of a small survey of car owners, led executives to believe that the main reason for loss of market share was a deterioration in the public's perception of the quality of services provided by their stations (step 5). This perception may have resulted in part from competitive advertising by the other major oil company, stressing the supposed superior quality of their services, which implied that the company in question had relatively poorer services. This perception was deemed to be a sufficient but not a necessarey condition for the problem to arise (step 6). The situation could be described graphically as follows (step 7):

The competitor's advertising was accepted as an exogenous variable that the company could not manipulate (step 8). However, the company could manipulate the content of its own advertising. To some extent, the company could also pressure stations operators to try to improve their general services in keeping with the new promotional campaign the company was about to launch. This campaign (step 9) was to stress general customer service and expertise in car maintenance offered by the company's stations.

APPROACH 2: AN OPEN SYSTEMS PERSPECTIVE. We present a model for problem definition or diagnosis that provides an open systems perspective. The main benefit of this perspective is that it identifies important substantive components and their interrelationships that are involved in problem diagnosis. The model for diagnosis is presented in Figure 2.1 as follows.

The model presented here is rather inclusive in that it covers five categories of variables. Not all of these variables are important in each change situation. However, they provide a guide for the change agent to follow as he diagnoses a situation. They also provide a checkpoint against ignoring certain important factors that might be affecting the change situation. The variables are defined and discussed below.

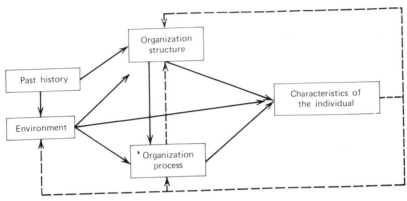

FIGURE 2.1.   Model of Diagnosis for Change

Historical Background (see Levinson 1972)   *Bibliography p. 369*

1. Change targets' past experience with change.
2. Key problems as stated by change target that affect current situation.
3. General circumstances surrounding change situation.

One of the authors had an experience with a police department in a large city that illustrates the importance of the organization's past history as it affects the change process.

In the mid 1960s, the department had an extensive analysis of its structure and functioning as a department undertaken by a large consulting organization. The consulting firm came up with numerous recommendations for improvement. The chief at that time decided to implement the recommendations with little consultation from his staff or the rank and file. The result was not surprising. The proposed changes had little impact on the department, and, in fact, the department became very resistant to change as a result of having these changes "shoved down their throat" by the chief. Soon after this, the chief retired and a new progressive change-oriented chief took over the department.

In 1971 an attempt was made by the new chief to institute some major changes in this department. However, because of the past negative experience with change, departmental personnel were very resistant to change. The prevailing attitude was one of ". . . we tried that before and it's too difficult to change here." This past history thus played a very important role in structuring the second change effort.

Environment of the Change Target

1. Supportiveness of the environment for the change target in general.
2. Supportiveness of the environment for the particular change.
3. The complexity of the environment. This focuses on the number of factors in the environment that must be considered by the change target in its decision making (Duncan, 1972).
4. The dynamics of the environment. This focuses on the extent to which the environment of the change target is changing or remaining the same (Duncan, 1972).
5. General culture—norm, values of the change target's setting.

The environment is important in that it is the setting within which the change target has to operate. For example, if the environment does not provide some support for the change that is being proposed in the change target, this might create some pressures on the change target that could lead to resistance. Therefore, it is important that the change agent makes some assessment during diagnosis as to what the larger environment for the change target is like.

Consider an assessment of the impact of mergers on hospitals that illustrates the importance of the environment (Duncan and Gilfillan, 1975). One system of six hospitals was merged in the late 1960s. This merged system set unrealistically high expectations on the part of the community as to the quantity and cost of services that this hospital system could provide. The planners of this merger were not aware of the high degree of skepticism toward the merger that existed in the environment. As the system developed, planners did not identify strong negative feelings among certain groups. Initially, there was no attempt on the part of the merged system to respond to these negative feelings. The result was hostility and attacks on the credibility and motivation of the merged hospitals. Had there been a better diagnosis before the merger about the supportiveness of the environment, the system might have been better able to respond to the environment before it became hostile. Once the environment "exploded," the merged system became defensive and found itself in a crisis situation, under attack by particular pressure groups.

Organization Structure

1. Patterns of authority.
2. Formulation of rules and procedures.
3. Centralization of decision making.
4. Division of labor.

This set of variables is concerned with the makeup of the organization with which the change agent must deal. Whether the change target is the organization as a whole or groups of specific individuals within the organization, it is important during diagnosis to examine existing rules and procedures to make sure that they are compatible with the change so that individuals who are going to adopt the change do not experience role conflict as they try to use the new change. A study by Gross, et al. (1971) of the attempted adoption of a new teacher role model in a school found that the reason the change was not successful was that teachers were still required to follow the same classroom procedures, which conflicted with using the new teacher role model. The teachers resolved this dilemma by abandoning the change. Proper diagnosis would have identified this incompatibility, which the change agents could have tried to eliminate before attempting to implement the new teacher model.

Organizational Process

1. Communication styles and patterns.
2. Decision making—subordinate involvement, information available for decision making, acceptance of decisions, etc.
3. Coordination of activities with the organization.
4. Intergroup relations—degree of competition, conflict, cooperation among units in systems.
5. Distribution of power and authority within the system.

These process variables focus on the internal functioning of the organization and again set important constraints within which change takes place. For example, it would be important for a change agent, through diagnosis, to have some idea of how information is transferred through an organization contemplating change. If it was difficult to transfer information to various parts of the organization, the change agent might want to be sure that all groups involved in the change have a clear understanding about the change and how it affects them so that rumors and misunderstandings can be reduced. The ability to gather and process information and make decisions are all important during the change process. Therefore, it is important that the change agent be aware of any deficiencies that might exist in these processes at the beginning of the change process.

Individual Characteristics

1. Satisfaction with current situation.
2. Self-concept of individual.

3. Openness to change.

4. Commitment.

These variables at the individual level have important implications for change. The individual's satisfaction with his role can be an important prerequisite to change. Hage and Aiken (1970) point out that job satisfaction is a necessary precondition to change in that it implies some commitment to working through the change situation. If people are very dissatisfied, they might simply withdraw from the situation and make no effort to change. The key here is that satisfaction can affect change, and the individual's self-image can affect his ability to deal with change. A person with a very negative self-image might be more threatened by the uncertainty associated with change (Watson 1967). Also, if the proposed change is not consistent with the individual's self-image, he is more likely to resist change. Thus it is important, when defining a problem in the change situation, that the change agent be sensitive to the characteristics of the individuals involved.

The above model presents a comprehensive list of variables to consider in problem definition in a change situation. Obviously, not every one of these variables is applicable or relevant for every change situation. However, they should serve as a checklist for the change agent to consider when examining each change situation.

Techniques for Gathering Problem-Related Data

In the foregoing section we identified variables that should be considered in approaching a problem situation. We now address the various techniques by which the change agent can gather data to diagnose a problem. We present only a sampling of these techniques. The interested reader is directed to the following for a more comprehensive discussion of diagnosis techniques: Argyris, 1970; Levinson, 1972; Mahler, 1972.

DIRECT OBSERVATION.   Perhaps the most simple form of diagnosis is to observe the behavior of the change target system directly. This methodology has the advantage of not making the change agent dependent on interpretations and explanations given by the change targets about how they "think" they behave. Argyris and Schon (1974) point out that there is often a discrepancy between how a person behaves and how he thinks he behaves. Thus questionnaires and interviews that are dependent on subject interpretations can be inaccurate.

There are various ways in which observation can occur (Filstead, 1970). The change agent can be a participant observer, where he or she is actually a member of the change target system carrying out some role. While carrying out a regular role in the system, change agent also collects information about the change situation. Internal change agents always have this opportunity. For example, a member of the staff of the organizational development group in an organization will have continuous interaction with various people in the organization. This staff member will thus have opportunities to collect information regarding change situations. The change agent can also use direct observation. For example, Argyris, in analyzing the interpersonal skills of executives, observes meetings and also makes tapes that are then analyzed along certain dimensions of interpersonal behavior such as being open to others, owning up to one's own behavior, and experimenting with one's behavior (Argyris 1962, 1974). Argyris then plays segments of the tapes back to the executives to get them to start thinking about how they operate as a group.

INTERVIEWS. Interviews are a somewhat more structured approach to collecting information. One potential pitfall in interviewing is that the change agent may not talk to a representative number of people. For example, in studying an organization, a good rule of thumb would be to talk to at least two people occupying the same organizational role. This would allow the change agent to get two opinions on the questions to see what kind of cross validation there was between respondents. Often, a respondent might be very biased in his perception of the system. Getting two respondents for each position helps to overcome this. Time permitting, it would also be useful to interview people at each level in the organization, since people at different levels might have very different perceptions of the organization.

The following is a list of questions that might be used in a 30–45-minute interview:

1. Could you briefly tell me about yourself? What is your job? What do you do on a typical day?
2. What are the things you like best about your job?
3. What are the major issues facing the organization and you in your job now?
   A. How successful do you feel in dealing with these?
   B. What could be done to better help you deal with these issues?
4. How successful has the organization been in the past in dealing with problems it faced?
   A. What helped the organization deal with these?
   B. What hindered the organization dealing with these?
5. How would you describe the environment of this organization?
   A. Is it supportive of the organization?
   B. Does it change the demands it makes on organization?

6. How would you describe the atmosphere in this organization?

　　A. How do people and groups here relate to one another? Are they cooperative?

QUESTIONNAIRES. Questionnaires as a diagnostic tool have the advantage that they can be given to a large number of people and their results can be presented in a more quantitative format. As in interviews, care should be taken to get a representative sample of respondents. For example, in an organizational setting, one should again select at least two respondents from each organizational role and also sample people from all organizational levels. One interesting thing to look at in the analysis would be to determine what differences there are in the perception of organization problems that different levels or functional units might have. If there were wide discrepancies between different groups in the organization, this would be an additional problem the change agent would have to deal with. He would have to ascertain why different groups saw things differently so that a common approach to the change problem could be launched.

Table 2.1 presents the profile sheet of the results of a survey designed to identify problems in school systems (Cooke et al., 1974). This survey has some 15 categories of information. Table 2.1 presents the profile sheet that allows each school to compare its scores on a specific category or dimension with the district's overall average. From the data presented here, we can see that this particular school was more unfavorable in its perceptions of all of the dimensions than the district as a whole.

Each of the 15 categories can then be broken down into its individual items to present more detailed information. This breakdown is presented for colleague relations in Table 2.2. Here we see that, in this school, faculty perceive quite unfavorably that the faculty is bossy and full of cliques. These might then be some of the first problems to work on in the change program. In fact, the survey feedback change strategy starts out by feeding back this kind of information for problem solving (for a more complete explanation of this strategy, see Chapter Eleven).*

Once the data is collected and a problem defined, the change agent must consider the prognosis for change. The following factors might be considered in weighing the possibility of success in dealing with the change situation.

1. How motivated is the system for dealing with the situation? Is there a sense of dissatisfaction with the present state of affairs? If so, are

---

* There are many other questionnaires that the change agent might use for diagnosis. See Likert, 1967; Robinson et al., 1969; Mahler, 1974; Price, 1972.

Table 2.1 Elementary school survey—Profile.

Group profiled:	Legend: ○ School
Number in group: 13	△ District overall

Percent favorable response

Category	0	10	20	30	40	50	60	70	80	90	100
I. Administrative practices			○	△							
II. Professional work load				○		△					
III. Nonprofessional work load					○		△				
IV. Materials and equipment					○ △						
V. Buildings and facilities					○	△					
VI. Educational effectiveness			○				△				
VII. Evaluation of students			○		△						
VIII. Special services			○		△						
IX. School—community relations			○		△						
X. Supervisory relations			○				△				
XI. Colleague relations			○			△					
XII. Voice in educational program			○ △								
XIII. Performance and development					○	△					
XIV. Students				○	△						
XV. Reactions to survey					○ △						

Table 2.2

		School %?	School % UF	School NA	School Mean	District %?	District % UF	District NA	District Mean
	Overall								
14.	Openness to professional assistance	15.4	46.2	0	2.8	15.8	21.1	1	3.5
48.	Performance of some of staff	15.4	61.5	0	2.4	17.6	30.2	4	3.2
90.	Concern with school welfare	7.7	76.9	0	1.9	18.0	33.7	2	3.0
32.	"Bossy" faculty	0.0	100.0	0	1.5	8.3	36.7	8	3.2
141.	Presence of cliques	9.1	90.9	0	1.6	13.1	25.9	6	3.4
108.	Cooperation	16.7	66.7	0	2.1	11.2	21.4	0	3.5
145.	Freedom to express opinion	15.4	53.8	0	2.5	9.8	32.8	1	3.2
71.	Sharing of ideas and materials	0.0	61.5	0	2.5	9.1	22.8	2	3.5
43.	Support and encouragement	8.3	66.7	0	2.4	13.9	27.1	0	3.4
82.	Interest in helping others	7.7	61.5	0	2.5	15.2	2.8	1	3.5
54.	Old staff helping new	7.7	69.2	0	2.2	10.6	30.2	3	3.3
7.	Time for teacher consultations	0.0	100.0	0	1.7	6.0	36.0	3	2.7

Percent favorable response

0 10 20 30 40 50 60 70 80 90 100

14. Teachers in this school are open to professional assistance from one another.   A

48. The poor work performance of some people in this school makes it difficult for the school to achieve adequate instructional goals.   D

90. Many people here are more concerned with their own personal interests than with the overall welfare of the school.   D

32. Some of the faculty in this school think they run the place.   D

141. There are many cliques or groups in this school that create a nonprofessional atmosphere.   D

108. People in this school cooperate well.   A

145. Most of the time it's safe to say what you think around here concerning school matters.   A

71. Teachers and other professional personnel in this school freely share ideas and materials.   A

43. People in this school provide one another with great support and encouragement.   A

82. Teachers in this school are genuinely interested in helping each other with teaching problems.   A

54. The experienced teachers here do all they can to help new faculty members.   A

members of the system motivated to make an effort to do something constructive to improve the situation?

2. How much commitment is there to maintaining the change process even when short-run failures occur? Motivation is not enough. There also has to be perseverance in working on the change when initial setbacks occur.

3. How competent is the social system? Do members possess the skills and abilities, or the ability to learn, to implement the changes required?

4. How supportive of the change is the larger environment in which the organization or other change target exists? For example, change may focus on making police officers more sensitive to the needs of the minority community in a large city. Does the rest of the larger community support this effort, or does it view it as a capitulation to the minority community? If there is considerable negative reaction from the majority community, it is difficult to make the change.

5. Are the prospects for change limited beforehand by factors unlikely to change (Levinson, 1972, p. 359)? For example, the prognosis for survival among motor home manufacturers has been severely constrained and limited by the fuel shortages that developed in 1973.

Before action is taken, the change agent will want to share the results of the diagnosis with at least some members of the change target system. The change agent has various options. He can provide a written report for the organization, community, etc., or for selected members of the organization. He can also provide verbal feedback of the results to the client. Not uncommonly, both written and verbal feedback occur. Whatever the feedback process, the change agent should pay particular attention to the process he uses. His objective is to facilitate the exchange of information between himself and the change target. Thus he should avoid actions that could impair this exchange of information. The change agent should:

1. Avoid information overload. Do not overwhelm the change target with too much information, since this can cause the target system to view the problem as overly complex and thus become frustrated and anxious about its ability to deal with the situation. An alternative strategy would be to present, at least initially, the minimum essential facts and then respond to the target's request for further information.

2. Attempt to make the data as descriptive and nonevaluative as possible. This has the potential for lessening change target defensiveness. The more the diagnosis can describe the behavior while providing examples, the less likely will the target system feel that it is under attack and thus has to defend itself (Argyris, 1970).

3. Work with the target system in exploring the implications of the prob-

lems defined so that the target system does not become completely dependent on the change agent for analyzing the change situation.

Perspectives on Problems

It is very important to realize that different perceptions of "the problem" may be held by different groups of people. Not uncommonly, there may also be differences among members of any given group. Listed below are four general considerations about which differences among relevant parties may exist.

1. Specific nature of the problem
   a. policy problem
   b. organizational structure problem
   c. people problem
   d. production process problem
   e. product/service problem
   f. channel (distribution/logistics) problem
   g. communication integration problem
2. Manifestations or symptoms of the problem
   a. scope of the problem
      (1) number of individuals and groups affected
      (2) kinds of individuals and groups affected
   b. degree of problem—how dysfunctional is it
   c. factors affecting the magnitude of the problem
3. Location of the problem
   a. geographically
   b. socially
   c. organizationally
   d. social system
4. Past remedial efforts
   a. by whom
   b. problem-solving approach employed
   c. results
      (1) reasons for success
      (2) reasons for resistance
      (3) reasons for failure

The different parties or groups who may differ from one another in their perception and assessment of the above considerations are:

1. The client
2. The change agency (if external to the client system)

3. The change agent
4. Significant others such as government agencies, etc., in the social context of the client system.

As indicated earlier, differences concerning all aspects of the problem may exist within each of the above categories. Decision makers within the client system may differ from gatekeepers and users of an innovation in their perceptions of what the problem is and what remedial change is most desirable. Some examples follow.

One major change agency active in developing countries has experienced considerable debate among its staff about the nature of agricultural problems in certain countries. One staff group has defined the problem to be largely of a channel nature. This perception appears to have two dimensions. First, there is the problem of distributing new agricultural technology or know-how in the form of information and commodities such as fertilizer. Second, they see an inadequate road network impeding the delivery of crops to large areas of consumption. Another group within this agency sees the problem as essentially a people problem, arguing that, even with the ready availability of fertilizer and other commodities, strong ties to traditional farming methods constitute a barrier to change. Undoubtedly both groups are correct, but the competition for funds within the agency causes both groups to exaggerate the importance of their perspective and to minimize the importance or validity of the other perspective.

One community organization in a predominantly Black section of a medium-sized midwestern city found itself constantly arguing without success for much greater police protection at night. The organization had gathered its own statistics showing very widespread drug abuse, rape, and so forth, in their area. The police department, on the other hand, had statistics of its own showing the same problems to be far less than revealed by the agency. The two groups might well be guilty of partisan diagnosis, a topic we turn to shortly. In any event, the two groups disagreed with regard to the scope and degree of the problem.

Change agents are not necessarily always in agreement with their employing agency concerning the nature of problems and do not always act in concert with the agency's perceptions. A management consultant firm recently diagnosed several problems in a mental health hospital as the result of a poor administrative structure. Several recommendations for redesigning this system were made. At the hospital's request, the consulting firm provided an expert to assist the hospital in its administrative reorganization. The consultant, who was not involved in the original diagnosis, began to diagnose the problem as one stemming from ineffective management rather than from structural considerations. Consequently, the consultant began to spend more and more time teaching (in a low-profile way) management skills to senior administrators. (When the gap between the consultant's activities and his firm's recommendations became evident to the hospital trustees, they immediately requested the replacement of the consultant by a person who was involved in the original diagnosis.)

Nature of the Problem

Regardless of perspective, problems may be of one or some combination of six basic types. Table 2.3 contains a listing and definition of each type of problem. Certainly there are interrelationships among problems. An organizational structure problem may be associated with a person problem, for

Table 2.3   Basic Types of Problems

Policy problem	Dissatisfaction rooted in the general operating philosophy of a social action agency. Dissatisfaction may stem from one or both of two aspects of policy practice, the goals set by policy, and/or the means employed to achieve these goals.
Organizational structure problem	Dissatisfaction experienced as a consequence of the particular configuration of individual and group statuses and/or the formal role specification binding on the role occupant.
Person problem	Dissatisfaction associated with the role performance of individuals. The difficulty may lie in the formal role specifications as indicated above in organizational structure problems or, more typically, it may be found in the actual role behavior that is discrepant with the formal role definition.
Production process problem	Dissatisfaction associated with the system used to generate the consumable products of social agencies.
Channel problem	Dissatisfaction resulting from the means used to disseminate consumable products to ultimate users.
Product problem	Dissatisfaction associated with the performance of the idea, service, or product provided by an agency. The dissatisfaction may stem from the product (1) not meeting expectations, or (2) having unanticipated dysfunctional consequences.

example. The problems are defined in the context of a social action agency, but the perception of the problem may be from the perspective of a member of the organization, a member of the target group(s) it is intended to serve, or a third party. Let us turn now to a brief discussion of each problem type.

First, there may be a *policy* problem, that is, a difficulty concerning a general orientation or operating philosophy of some public or private agency. The problem may be experienced by parties within the agency or external to the agency. For example, a family planning clinic in the metropolitan Chicago area recently introduced abortion counseling as part of the general offerings to the public. In this particular case, the new policy was perceived as a danger by groups both within the organization and outside the organization, although the problem the two groups experienced was not the same.

The *organizational structure* of an action agency may be a source of problems inhibiting agency effectiveness. This has been amply demonstrated in many community organizations. Frequently, in newly spawned and rapidly growing community action programs, lines and spans of authority are blurred, resulting in conflicting actions by different members of the same organization.

Related to organizational structure are person problems. Members of action agencies may not be trained or otherwise suited to function effectively in the roles assigned to them. This is particularly likely when informal groups incorporate to form legally constituted social action agencies.

On another dimension, person problems may be found among the target group. The problem is not peculiar to the action agency itself, but may also be found among the clientele it intends to serve. Dysfunctional behavior such as inadequate or unhealthy feeding practices may be the symptoms of problems such as low agricultural production or lack of nutrition-related education.

The *production process* is another source of difficulty in social affairs. Various technologies for educating students pose particular problems. The issues of racial desegregation and bussing in school systems and their (intended) impact on student learning are perceived as problems by some quarters and certainly as controversial from most quarters. Not everyone views desegregation of classrooms as a means of improving academic performance of all groups. The management of family planning programs (Berelson, 1966), systems of infant feeding in hospitals (Littaner et al., 1970), social welfare manpower utilization (Loewenberg, 1970), and the activities of community organizations in mobilizing neighborhood resources (Cunningham, 1965) are all examples of production processes in which social problems are found.

Problems may also be of a *channel* nature. This is the concern of those involved in the delivery of health care services, for example. The surface problems, if not the basic problems, may range from an insufficient number of health care outlets to inadequate servicing or management of those outlets which do exist. The dissemination of innovations and information in education is also conceived by educators as a part of their channel process and is generally acknowledged as an area in need of much remedial activity.

There are also *product or service* problems. The change agency is likely to flounder if its products are not matched with needs as perceived by the target audience or client group. Although needs assessment is of major current concern to education, for example, many agencies in the education sector, both commercial and noncommercial, do not monitor the knowledge, attitudes, and practices of their clientele to determine what client needs currently exist or are emerging for which adequate solutions are not available. In health contexts, problems exist at the conceptual level, particularly in less-developed countries in which poorly educated rural parents do not understand the germ theory of disease and the chemistry of immunology and hence are reluctant to have their children participate in immunization programs (Lin and Burt, 1975). The problem here is that the change agency is unable to communicate effectively a clear understanding of the basic idea behind its product.

Manifestations of Problems

Problems manifest themselves as various symptoms. Poor language skill is not a problem but a symptom; malnutrition is not a problem but a symptom; etc. However, it is very important to recognize the symptoms of problems both because they may be the only indication that a social process is malfunctioning and because the symptoms should be attacked directly for immediate relief while solutions of the basic problem are allowed to gather their effect. In considering the *scope* of the problem, it is necessary to consider the number of individuals and groups affected. Along with this, it is necessary to consider the *kinds* of individuals and groups affected. A given educational innovation, for example, simulation games, may affect both teachers and students but have little impact on administrators, whereas the change team concept may have a profound impact on the relationships between teachers and administators. The innovation examples just mentioned are primarily responses to problems concerning relationships between the groups identified. There is also the consideration of the *seriousness* of the problem, that is, the degree of its dysfunctionality for the different individuals or groups affected. In this regard it is also necessary to isolate or identify those factors affecting the magnitude of the problem. It should be stressed that, when a problem is experienced by different kinds of individuals or groups, the factors influencing the magnitude of the problem for one group may not be the same as those accounting for the magnitude of the problem as experienced by another group. Both teacher and student may experience frustration because of the poor language skills of the latter. For the teacher, high frustration may be due to a lack of understanding of the Black child's home and the communication cues in the community environment. For the child, frustration may result in part from an inadequate understanding of what is expected of him in the school value system. Related to the seriousness of the problem is the consideration of *factors affecting the magnitude* of the problem. This involves identifying both causal variables that can be manipulated in the short run, as well as factors that the change agency must accept as given and unalterable in the immediate future. The change agency must adapt its actions in accordance with these latter variables.

There is, finally, the task of reviewing *past remedial efforts* attacking the same problem. This involves identifying the change agents or agencies, the problem-solving approach employed, and the results of that approach. What were the reasons for success? What were the reasons for resistance and/or failure? (See Watson, 1971; and Zaltman, Duncan, and Holbek, 1973).

Categories of Problems

There appear to be five possible categories for each of the problems discussed. These five categories are recurrent problems, rerecognized problems, current problems, refashioned problems, and unrecognized problems.

A *recurrent problem* is defined as one that is persistent and visible. The high failure rate of new products is an example. The likelihood of a new product failing has been consistently high over time, resulting in considerable resource loss. A *rerecognized problem* is one that has received new attention but has always been present. For example, despite their widespread use, our understanding of why and how incentives influence social behavior is quite poor, and only recently has research attention again been devoted to this topic. A *current problem* is one that has recently been labeled, generally is acknowledged as existing, and is receiving attention. An example of this is the current concern in marketing research with discrepancies between attitudes of consumers toward particular social agencies and their actual behavior with regard to those agencies. A *refashioned problem* is simply one of long standing that has been given a new definition. For example, problems in the relationship between rice farmers and distributors have been redefined in terms of conflict theory. An *unrecognized problem* is one that has not received substantial attention because of low visibility.

The various categories are not mutually exclusive nor necessarily exhaustive. For example, a channel-related problem such as conflict between producers and distributors may be both a current problem and a redefined problem. Moreover, this channel-related problem may also be classified as a communication problem to the extent that the conflict is a result of different understandings of what constitutes an appropriate exchange price. Thus a social problem grid analysis might be useful in analyzing research problems. Such a grid is shown as follows:

				Basic Types of Problems			
Categories	Policy	Organization	Person	Production	Channel	Product	Communication
Recurrent							
Rerecognized							
Current							
Refashioned							
Unrecognized							

This chapter can be summarized by the following general principals:

1. The change agent should realize that, once a particular definition of a problem is accepted, a very major step has been made in determining what the solution will be. Thus considerable care should be given to the problem definition task.

2. The change agent should expect to encounter problem definitions at variance with his or her own, particularly as the symptoms of a problem become more severe and numerous and as the number of different groups involved in the problem area increases.

3. In the great majority of cases—indeed perhaps in all—total objectivity is impossible; hence the change agent should try to clarify his or her own partisanship.

4. The use of causal images or models is very helpful in diagnosing problems. The change agent should try to identify what forces under what conditions produce the forces of immediate concern.

5. The change agent should also seek to understand how one force or factor produces another. This understanding facilitates manipulation of forces to achieve a desired goal.

6. A distinction should be made between forces or factors change agents can manipulate directly and those to which they must adapt.

7. Among the factors that must be considered in diagnosing problems are the past history of a group or person with a problem, the relevant envi-

ronment, organizational structure or processes in the case of a group, and characteristics of individuals. Each of these considerations has a large number of subsidiary factors.

8. Techniques for gathering problem-related data are numerous and diverse. It is often useful to employ more than one technique as a way of allowing different perspectives on problems to arise.

9. The change agent should attempt to specify the nature of the problem as a policy problem, organizational structure problem, and so forth, or as a combination of different kinds of problems, each possibly having different importance weightings.

10. The change agent must carefully document the manifestation or symptoms of the problem and map its scope and severity.

11. The change agent should also examine past remedial efforts concerning the problem at hand and the reasons for their successes and failures.

PART TWO

Strategies for Change

The chapters in Part II are primarily concerned with four strategies for planned change. It is appropriate to preface any discussion of such strategies with a treatment of resistance to change. Quite simply, were it not for resistance, little discussion of strategies would be necessary. Resistance to change is, however, the most commonly encountered response to an advocated change, although, like problem diagnosis, it is typically given much less attention than it deserves.

Each chapter on strategies contains representative case studies, many of which were selected from interviews with many professional change agents. These interviews were conducted as part of the formal activity of preparing this book. In addition to these cases, each chapter discusses various criteria for selecting change strategies. The criteria to consider in selecting change strategies discussed in the following chapters are far from exhaustive but do represent some of the more salient considerations in the task of deciding which strategy or combination of strategies might be most appropriate in a given situation. Still, not every criterion will be applicable or equally so to each strategy.

The various criteria are also interrelated. For example, the greater the magnitude of change (one criterion), the greater the likelihood of resistance to change (another criterion). Similarly, the stronger the felt need for change, the stronger the degree of commitment to an acceptable change. Independently of their implications for the choice of strategies, the various criteria can be used in a checklist fashion for assessing the prospects for successful change. The prospects for successful change will be greater the greater the number of favorable answers to such questions as: Is there a perceived need to change? Are there adequate resources in the client system to initiate change? To sustain change? Of course, different criteria may be weighted differently in importance from situation to situation.

Before beginning our discussion of these cases, several points should be made. First, the four strategies might be considered to lie on a continuum of degree of pressure exerted, as given below.

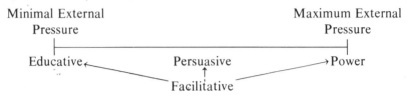

It is often very difficult to determine whether a particular strategy is educative or persuasive or power. In fact, there is no definitive method for categorizing strategies; therefore, the labels are somewhat arbitrary, and the reader should keep this in mind while reading these chapters. Facilitative strategies may be utilized to enhance any of the other three types of strategies.

Second, a change agent is likely to employ multiple change strategies to accomplish the desired change. The change agent may segment the change target into homogenous subgroups on the basis of a number of factors such as stage in the decision-making process, degree of commitment desired, anticipated level of resistance, etc., and use different strategies with each segment. Third, the change agent may employ a sequence of strategies over time. For example, an educative strategy might be used to create awareness, followed by persuasion to achieve adoption of a particular solution. A persuasive strategy might be first tried to achieve change; however, if this strategy fails, a power strategy might be used. The Supreme Court school desegregation decision was a power strategy, later followed by an educational strategy involving research (Coleman et al., 1966) that presented data providing rational, empirically based arguments in support of that decision.

CHAPTER THREE

Resistance to Change

By definition, social change involves an alteration in the status quo. Thus, whenever change is attempted, resistance is likely to appear. This chapter discusses the concept of resistance and focuses specifically on the cultural, social, organizational, and psychological barriers to change.

Resistance is, in fact, a healthy phenomenon, although the personal values of change agents are a partial cause of the fact that the concept of resistance has a pejorative connotation. Resistance is a positive force when, from some relatively objective standpoint, the advocated change is harmful to society. Resistance is justifiable in many cases in which the advocated change is subjectively harmful to a person or group. Threatened persons or groups may want to diffuse resistance to counter the efforts of persons trying to diffuse the change in question. Resistance may also pressure the advocates of change to be more careful in introducing change and perhaps to modify the nature of the change to increase its compatibility with the overall social system.

Another very important point the reader should keep in mind is that resistance may be caused by change agent errors. It is common practice for change agents to attribute "blame" or "fault" for resistance to the client or target group subject to change. This may simply be because change agents are often required to account for failure or strong resistance, whereas the client generally is not asked why he resisted change. One can readily imagine the inventory of change agent errors that clients might generate: the change agent did not communicate competence or trust; the change agent was disdainful of the client's social milieu including norms, values, mores, etc.; the change agent did not fully understand client needs and abilities to adopt and implement change; and the change agent did not communicate clearly the nature of the advocated change and its relevance to the client. This list could be greatly extended. In fact, the violation or insensitivity to nearly any of the principles mentioned at the end of each chapter in this book is a potential source of resistance caused by the change agent. *Go over principles at end of chapter*

RESISTANCE DEFINED

Resistance should be used constructively by change agents: "By resisting, actively or passively, an organization is communicating a message—it is providing data. In a very real sense, an organization is telling us something about 'who it is'—its major resources and limitations, its attitude toward outsiders and change, its important internal norms and values, and the nature of its relationship to other systems in the environment" (Rubin et al., 1974). Thus the display of resistance is an opportunity for insight into the various conditions that should be considered in selecting and shaping intervention strategies.

We define resistance here as any conduct that serves to maintain the status quo in the face of pressure to alter the status quo. Resistant conduct may or may not be controllable within the immediate future. For example, there are varying levels of beliefs with corresponding differential susceptibility to change. The more central a belief, that is, the more it is functionally connected or related to other beliefs and the more strongly it is held, the greater the likelihood of its being a source of resistance when an advocated change is incompatible with it (Rokeach, 1968). Thus a change agent may be able to control resistant conduct resulting from peripheral beliefs and may not be able to control resistant conduct rooted in central beliefs. Also, the pressure for change may or may not be the result of deliberate, planned change.

Resistance is not simply the lack of acceptance or the reverse of acceptance. For example, the process of learning that is sometimes involved in acceptance is not the same as the process of forgetting, which is a phenomenon sometimes involved in resistance. Also, the process of unlearning or "unfreezing" involves dynamics not associated with learning. Although the same factors are relevant for both acceptance and resistance, they may be relevant in different ways. For example, when an innovation is incompatible with a particular norm (the relevant factor), it may be adopted as a symbol of defiance for one person and rejected by another person for fear of social disapproval. Similarly, the uniqueness of an innovation may be a cause of attraction for innovators and a cause of resistance for more conservative people (Zaltman and Pinson, 1974).

In many cases, the same dimension is involved in both hindering and facilitating change. Lippitt, focusing on education, groups various dimensions into four categories. The first category is peer and authority relations involving the dimensions of communication, competition for recognition, openness with respect to ideas, peer support, principal involvement, teacher participation, and provisions for continuing education. The second category is personal attitudes involving such dimensions as openness to innovation, openness to outside assistance, dogmatism, group participativeness, optimism, and risk taking. A third area concerns the characteristics of the practice that includes such dimensions as relevance to needs, commitment, behavioral change, attitude and value congruity, and so forth. The last area is physical and temporal arrangements involving availability of time, clerical assistance, and opportunities for interpersonal interaction. All of these dimensions and examples of how they can function as facilitating and hindering forces are summarized in Table 3.1.

Can function as facilitating + hindering forces!

Table 3.1 Forces Relevant to the Facilitation and Hindrance of Innovation and Diffusion of Teaching Practices

Facilitating Forces	Hindering Forces

1. Peer and Authority Relations

Facilitating Forces	Hindering Forces
A. Sharing sessions or staff bulletins become a matter of school routine.	A. Little communication among teachers.
B. Public recognition given to innovators and adopters; innovation-diffusion seen as a cooperative task.	B. Competition for prestige teachers.
C. Sharing ideas is expected and rewarded; norms support asking for and giving help; regular talent search for new ideas.	C. Norms enforce privatism.
D. Area team liaison supports new ideas.	D. Colleagues reject ideas.
E. Principal or superintendent supports innovation-diffusion activity.	E. Principal not interested in new ideas.
F. Principal helps create a staff atmosphere of sharing and experimentation.	F. School climate does not support experimentation.
G. Staff meetings used as two-way informing and educating sessions.	G. Principal does not know what is going on.
H. Teachers influence the sharing process.	H. Teacher ideas do not matter.
I. In-service training program gives skills needed to innovate and adapt.	I. No continuing education program for staff.

2. Personal Attitudes

Facilitating Forces	Hindering Forces
A. Seeking new ways.	A. Resistance to change.
B. Seeking peer and consultant help.	B. Fear of evaluation and rejection or failure.
C. Always open to adapting and modifying practices.	C. Dogmatism about already knowing about new practices.
D. Public rewards for professional growth.	D. Professional growth not important.
E. See groups as endemic and relevant for academic learning.	E. Negative feelings about group work.
F. Understand connection between mental health and academic learning.	F. Mental health is "extra."
G. Optimism.	G. Pessimism.
H. Test ideas slowly.	H. Afraid to experiment.
I. Suiting and changing practice to fit one's own style and class.	I. Resistance to imitating others.

Table 3.1 (continued)

Facilitating Forces	Hindering Forces

3. Characteristics of the Practice

Facilitating Forces	Hindering Forces
A. Relevant to universal student problems.	A. Does not meet the needs of a class.
B. Can be done a little at a time.	B. Requires a lot of energy.
C. Consultant and peer help available; needed skills are clearly outlined.	C. Requires new skills.
D. Clearly aids student growth.	D. Requires change in teacher values.
E. A behavioral change with no new gimmicks.	E. Requires new facilities.
F. Built-in evaluation to see progress.	F. Will not work.
G. Innovation has tried a new twist.	G. Not new.
H. Student, not subject, oriented.	H. Not for my grade level or subject.
I. No social practice can be duplicated exactly.	I. Effectiveness reduced if practice gains general use.

4. Physical and Temporal Arrangements

Facilitating Forces	Hindering Forces
A. Staff meetings used for professional growth; substitutes hired to free teacher(s) to visit other classrooms; lunchtime used for discussions; students sent home for an afternoon so teachers can all meet together.	A. No time to get together.
B. Extra clerical help provided.	B. Too many clerical duties to have time to share ideas.
C. Staff meetings for everyone to get together occasionally; grade level or departmental meetings.	C. Classrooms are isolated.
D. Meetings held in classrooms.	D. No rooms to meet in.

Source: Lippit et al. in Havelock, 1973.

Barriers to change in any one setting are numerous. For example, speaking of curriculum change, Telfer (1966) scratched only the surface in identifying the following barriers: lack of time, lack of effective means of communication, lack of agreement of what is to be done, lack of money to do the necessary tasks, staff turnover, poor teacher preparation, lack of teacher interest and cooperation, lack of top-level administrative support, and teacher apathy. Notice that these barriers include social system barriers as well as those rooted in individuals. Black (1972), focusing on institutional obstacles in the context of family planning, similarly notes a wide range of barriers including committee management, amateurism, the growing entanglement of the medical profession, over-anticipation ("We can't-possibly-do-that-because-what-if . . . "), territoriality, belief in organization omnipotence, and so forth.

There are several sources of resistance. These are discussed under the headings of cultural, social, organizational, and psychological barriers to change.

CULTURAL BARRIERS TO CHANGE

Resistance can be rooted in ideologies, traditional heritages, social relationships, economic well being, personality needs, health conditions, and so forth. Examples of these are presented in Table 3.2. A somewhat less refined classification scheme is used here.

Table 3.2   Types of Resistances and Corresponding Rationalizations for Non-Use of Contraceptives

Types of Resistance	Corresponding Rationalizations
1. Resistances rooted in ideologies that run counter to population control and beliefs about the negative and positive aspects of specific methods.	1. Examples of rationalizations include the following: (1) elites in developing nations often accuse the highly industrialized nations of an overconcern for controlling their populations; they claim it is a "neocolonialist plot; (2) belief that particular contraceptive methods cause sterility, cancer of the vagina, no enjoy ment of sexual intercourse; (3) nationalistic ideology rationalizes that uncontrolled population contributes to the strength of a nation; etc.
2. Resistances rooted in the traditional heritage of a people.	2. Examples of rationalizations: (1) the community village, family, and clan accept large families; (2) the desire for male children, especially in patrilineal societies in which descent and property are traced through the male line; etc.
3. Resistances rooted in social relationships with particular reference to co-wives, attitudes towards birth control	3. Examples of rationalizations: (1) in polygynous families, co-wives who are favorably oriented to contracep-

Table 3.2 (continued)

Types of Resistance	*Corresponding Rationalizations*
believed to be prevalent among peer and reference groups, and normative values pertaining to "moral-immoral" as well as "natural-unnatural" types of behavior.	tives are regarded by other co-wives as "prostitutes" who deviate from traditional moral standards; (2) to the extent that reference groups give social, economic, and psychological support, one cannot afford to deviate from their scales of values, which may disfavor the use of contraceptives as "unnatural behavior."
4. Resistances rooted in economic well being.	4. Examples of rationalizations: (1) a large number of children is an economic advantage; they are useful in helping a family earn a living; children pay for themselves by working as they grow; (2) a large number of children serves as social security for their parents when they grow old; etc.
5. Resistances that are anchored in the personality needs of the individual with particular reference to sex-role images and sexual relationships.	5. Examples of rationalizations: (1) a demonstration of virility; (2) manifestation of manliness; etc.
6. Resistances that are anchored in health conditions.	6. Examples of rationalizations: (1) from the point of view of underdeveloped communities, high fertility represents a functional adjustment to the high mortality existing in such communities; (2) from the point of view of particular families, it is necessary to raise large numbers of children to guarantee that a few will reach adulthood; etc.

Source: Okediji, 1972, p. 4.

We look first at cultural barriers to change, followed by social, organizational, and then psychological barriers to change. These sources of resistance are quite interrelated, and their separation here is only to facilitate their presentation. The relative importance of cultural, social, and psychological resistance may vary greatly from situation to situation and from innovation to innovation within any given situation. Sandhu and Allen (1974), in a study of the adoption/rejection of modern farm technology found village characteristics to be of greater importance in explaining the adoption/rejection of change in agriculture than individual characteristics. Other researchers looking at other situations have noted the reverse to be true (Williams, 1973).

Cultural Values and Beliefs

One very major barrier to change stems from cultural values and beliefs (Foster, 1962). Weber (1947), for example, ascribed lack of a work ethic, the absence of desires for upward mobility, and so forth, as major barriers to economic development prior to the rise of capitalism in Europe. Rostow (1960), also discussing economic development, notes the absence of entrepreneurship and willingness to accept innovations as sources of resistance to change. Hagen (1962) describes as barriers the lack of socialization in early childhood of key social values such as achievement and autonomy as barriers causing resistance to change. Similarly, McClelland (1961) notes as the major barrier to change the absence of the central social value he calls achievement motivation, which is an inner drive to compete or do well relative to some standard of excellence. Strongly related to the absence of such cultural values and attitudes (such as thrift and achievement motivation) is a trust in traditional ways of doing things. For example, reliance on midwives and folk healers is one of the major obstacles to improved maternal and child health care in rural areas of developing societies. In fact, one of the challenges facing development workers is how to use, in a constructive, creative way, traditional sources of medical and other advice. Several countries are experimenting with the use of traditional midwives to prescribe oral contraceptives. This involves some training concerning contraindications for the use of birth control pills.

Various religious ideologies have impeded change. Confucian teachings, for example, stressed the past as an ideal to be recreated, which, although in itself a form of change from the status quo, did little to permit the introduction of new ideas as ends in themselves or even as means to reestablishing the past. Lauer (1973) suggests that religions with large numbers of religious holidays tend to rechannel social energy from constructive to frivolous activities. Religious institutions also tend to absorb wealth that could otherwise be used in economically more productive ways. Thought control exercised by exponents of political ideologies is also a barrier to change. This is reflected eloquently by A. I. Solzhenitsyn in *The Gulag Archipelago*.

Fatalism is another source of resistance to change that is deeply embedded in culture (Foster, 1962). In many societies, individuals believe that whatever happens is fate, and the individual must adjust to it; he is powerless to change the direction of his life or influence events. This feeling is fatalism, defined as "the degree to which an individual perceives a lack of ability to control his future" (Rogers, 1969, p. 273). It has been suggested by Niehoff and Anderson (1965) that perhaps fatalism is a post hoc rationalization about behavior and not a variable that directly impedes social change. However, considerable research with many varied groups in the United States using the Rotter Internal/External Test suggests that fatalism is indeed important. This test measures the degree to which a person feels he has control over those things that influence his behavior. For example, one study (Zaltman, 1974) found a positive relationship between low degrees of felt control and innovativeness in family planning, that is, the lower the feelings of control, the less innovative the person is. On the other hand, Bonilla (1966) presents evidence that "fatalism did not contribute significantly to the prediction of agricultural innovativeness, aspirations, or achievement motivation."

To the extent that fatalism is important, it is necessary to find ways of circumventing it. One infant nutrition education campaign in Central America found many mothers explaining their children's malnutrition in terms suggesting a strong presence of fatalism, even if only as a post hoc explanation. The campaign utilized one particular advertisement showing a seriously malnourished child in hospital clothing, sitting listlessly on a bed. The caption underneath read in Spanish, "You Can Prevent This." Obviously this advertising would not have a serious impact on fatalistic attitudes, but it did, as evidenced in follow-up studies, successfully plant the idea in mothers' minds that perhaps infant nutrition could be improved with a relatively simple change in their child-feeding practices.

Cultural Ethnocentrism

Cultural ethnocentrism is a barrier to change in two ways. First, the change agent who comes from a different culture may view his or her own culture as superior. Communication of this attitude, often quite indirectly and unintendedly, to the client produces resistance to the change agent and consequently to the advocated change or innovation. In addition, the client may see his own culture as superior to others, at least in certain respects, and hence may passively resist borrowing or adopting artifacts from other cultures. Often underlying this problem is the failure to involve the client system in the change development process. The change agent simply takes the approach, "Here it is, use it." There still may be reluctance by clients to adopt a change they helped formulate if they were not adequately involved in defining the problem to which the change is addressed. Different cultural perspectives held by change agents and clients may cause different perceptions of what the problem is and hence different expectations of what a commonly agreed on change should achieve. In the context of technology transfer involving formal organization, cultural ethnocentrism is referred to as the "not-invented-here" syndrome.

Saving Face

Yet another source of resistance to change surfaces in the form of "face." A contrast of two communities in Paraguay, each approached somewhat differently, provides a good example. In one community, the person in charge of a nutrition program attempted, and largely failed in her efforts, to secure improved feeding practices by mothers during a six-week nutrition education campaign. The community was quite small, and she had considerable contact individually or in small groups with nearly all mothers. As part of the same general nutrition improvement campaign, a colleague with comparable expertise, training, and experience, working in a highly similar community, was very successful. An analysis of the two situations revealed only one major difference. In the former village, the nutritionist in effect argued that current practices, deeply embedden in tradition, were wrong and should be replaced by simple but specific other practices. Her lectures and discussions stressed the wrongness of what mothers were currently doing. This was essentially a confrontation strategy. The other nutritionist, perhaps somewhat deviously, followed a very different approach. The wrongness of current practices was hardly mentioned. Rather, her lectures stressed that, although current practices were possibly adequate—and this is where the deviousness came in, for current practices were not at all adequate—the advocated practices were much better. Here the stress was on the attractiveness of the advocated change, with no censuring of existing practices. Adoption of the advocated behaviors did not imply guilt of wrongdoing in the past.

Another example of "face" as a source of resistance is provided by N. V. Raghu Ram (1972, pp. 89–90).

> While they were being operated, we were talking to other eligible employees. One of them was a smart and well-behaved youth. I engaged him in conversation. I asked him how many children he had got. He did not answer the question straightly. But he said that he had more than he could afford and that he did not want more children. I told him that we were there precisely to offer assistance for his need and advised him to undergo sterilisation to ensure that no more children would be born. He refused point blank. I asked him to explain his self-contradictory stand. He frankly told me that he was not afraid of the operation. He knew that it was perfectly safe and almost painless. However, he was concerned about his social prestige more than his poverty. He explained that he was working at the station while his family stayed at his native village. He was able

to visit his family only once or twice a year when he got leave. Children were born to his wife fairly regularly. Once he got himself sterilized, his village folk would come to know it soon. Should his wife bear a child after his sterilisation, he would not be able to show his face in the village. I was taken aback. I explained to him that he would have to observe certain precautions for about two months after sterilisation. It would be fairly easy in his case as he was staying away from his family. He did not allow me to complete my argument. He broke in: "I know all that. But according to the customs and practices in my village, it is not at all unlikely my wife will bear a child . . . even if I have no relation with her. As long as I am fit to procreate, any child borne by my wife is considered my child. There is no scandal. But if at all I am sterilised . . . oh . . ." Before I could make out his words, he left the place.

Similarly, strategies that involve using *existing power or influence* structures such as midwives to disseminate birth control pills or faith healers to distribute condoms does not cause a loss of pride or dignity among these often highly regarded individuals. To the contrary, their status, which is a source of their pride, is reinforced or enhanced.

Foster (1962), citing Apodoca (1952) as an example, illustrates a very basic tenet most change agents learn rather quickly: the relative values of the change agent are not necessarily shared by the client system.

In a community of Spanish American farmers in the Rio Grande Valley of New Mexico, a Department of Agriculture county extension agent succeeded in 1946 in introducing hybrid corn that produced three times the yield of the traditional seed. After participating in initial test demonstrations, a majority of the growers adopted the innovation. Yet four years later nearly all farmers had reverted to the old corn. Investigation revealed that the farmers' wives had complained about the texture of the dough used to make tortillas, about the color of the finished product, and about the taste. In the system of values of this community, corn quality turned out to be more important than corn quantity; people were willing to sacrifice economic gains for something they esteemed, in this case traditional food characteristics. (Foster, 1962, pp. 74–76)

Recently, a colleague working with a neighborhood improvement association found residents very much opposed to the building of a new, modest-rent apartment building within the area covered by that association. The private and governmental agencies funding the project were very puzzled by the opposition encountered, particularly since the building would be structurally superior to those currently occupied in the neighborhood. An apparently minor reason for the opposition, but the only one that was articulated by residents to the developers, was that various retail stores would be demolished to create room for the building, making shopping more difficult. According to our colleague, however, the real opposition was that the sizes of the proposed apartment units were suited to a life style in which only the nuclear family resided. The cultural tradition for the particular ethnic group populating the community was one where the extended family lived together, requiring larger apartment units. The existing older apartments were much larger than those planned for the new building, and hence the residents could not see themselves moving into the new building, which would entail altering the family structure. Thus the new building presented no particularly real advantages despite its newness and reasonable rents.

Incompatibility of a Cultural Trait with Change

Incompatibility of a cultural trait with a proposed change is one of the most frequent causes of resistance to change. Foster (1962) makes the point when observing that people who practice polytheistic religion have had little difficulty in adopting Christianity in a limited way, at least since it represents just one more deity with special powers and does not replace an existing deity. Practitioners of monotheistic religions, on the other hand, must abandon their God if they are to accept Christianity, since they are conditioned to the concept of only one God. The necessity of displacement in this instance represents a powerful source of resistance.

SOCIAL BARRIERS TO CHANGE

Group Solidarity

Reciprocal or mutual obligation that holds societies together may be incompatible with changes or innovations that stress individualism. For example, making more money may not mean increased consumption but rather an increase in the number of relatives to support.

Related to solidarity is the issue of interdependence (Lippitt, 1958, p. 77). "Readiness for change on one part of a system may be negated by the unwillingness or inability of other interdependent parts to change. A change sequence which would be strong enough to modify a subpart if it existed in isolation may have no effect on the system as a whole, and consequently it may fail to have any real effect on any part of the system." This is also what Watson (1971) refers to as systemic coherence. Coherence or interdependence may be a strong force for resistance if the subsystem that is the object of change satisfies important needs or functions in many other parts of the social system. Also, expectations of the behavior of one group that are held by another group may govern the former group's behavior. Resistance would occur if the reference group influence is a strong one and its expectations of the target group would not be met or would be violated by adopting an advocated change. The more a reference group is threatened by a possible change, the more active it will be in expressing its opposition to the target or client group.

Rejection of Outsiders

Rejection of outsiders is another source of resistance to change that is caused in part by a high degree of in-group identification (Watson, 1973). As Watson concludes: "A major problem in introducing social change is to secure enough local initiative and participation so the enterprise will not be vulnerable as a foreign importation" (p. 129). One way of overcoming this problem is to create a change, the second-order consequence of which is the intended change. Because the advocated change is a second-order consequence, it is less likely to be associated with an external change agent. For example, in Egypt (Mason, 1974), one association created employment opportunities for women with the intention of changing family dynamics in a way conducive to the creation of favorable attitudes toward birth control. This example also illustrates how conflict can be used as a positive force in social change.

> Rural Egyptian women are brought up and taught to become mothers and mothers only. This is their single role in life, with no other occupations or avenues of self-realization open to them. So there is little more for them to do than to have babies and care for them.
> Dr. Bindary considers this situation the greatest single obstacle to winning acceptance of any family planning programs. It is to meet this challenge that his council is now concentrating its efforts.
> "We are trying more and more to bring the woman out of her home into some sort of employment beyond her house and the fields. We are deliberately creating potential conflicts of interests so that she will begin to think seriously about the possibility of limiting the size of her family."
> Already, the council has put this theory to the test in the town of Mahalia in the Nile Delta. Employment for married women is being provided in the town's textile mill for the first time, although the factory is one of the oldest and largest industrial complexes in the country. The result, according to Dr. Bindary, has been startling: the town's birth rate has already been cut in half.
> "This really confirms our conviction that once a woman can escape from her traditional work patterns, the conflict set up in her own mind between the attractions of a richer life and those of a larger family will be resolved

in favor of sensible family planning. And as a bonus she gains a measure of financial independence from her husband which allows her to discuss the whole idea of bringing up children on a basis of what is best for them."

Conformity to Norms

Norms provide stability and behavioral guidelines that define what individuals can expect from one another. They are essential for the conduct of any social system. Consequently, any change that is incompatible with existing norms will tend to be resisted by most members of the social system. The critical question for a change agent to ask is, "Why do people participate in this norm?" Knowing the answer to this question may enable a change agent to modify his change to meet the need satisfied by the norm.

Public opinion is the expression of group feelings about an issue and is based on group norms. In much of the world, the condom is associated with prostitution, and hence the purchase of condoms is interpreted by others, such as vendors, that the buyer is about to engage a prostitute. Strong norms against prostitution, particularly in Latin America, create unfavorable public opinion concerning anything associated with prostitution. Thus it has been difficult to promote condoms among certain groups in developing countries.

Conflict

Conflict has been mentioned above as a means of introducing change. It can also work against change. When conflict and factionalism exist within an organization or society, any change that one faction in the conflict adopts or espouses may automatically be rejected by other groups. The change or innovation suffers guilt by association. There is conflict between two anxieties, which Frye (1969) refers to as conservative anxiety (Let's-be-careful-about-losing-what-we've got) and radical anxiety (Let's-be-careful-and-clear-out-all-this-stuff-and-have-a-fresh-breeze-blow-through). Radical anxiety holders, when they communicate their anxieties effectively, reinforce the anxieties of the conservative and vice versa. When the conservative anxiety holders are more numerous or more powerful, the existing order may become still more deeply entrenched. A good example of this is when an aggressive senior administrator assumes new responsibility. One new Dean in a southern university attempted to revitalize all at once the entire teaching program and the general research atmosphere of his division. His choice of tactics involved a broad assault on the several problems he perceived. His actions were so threatening that early supporters joined in informal opposition to him and effectively neutralized the Dean's actions. As one faculty member put it, "He was throwing the baby out with the bath water."

Group Insight

Lippitt et al. (1958, p. 181) suggest that one of the major barriers to change in small groups is "the members' imperfect awareness of their own interpersonal processes and their lack of a frame of reference in which to judge their performances and their possibilities for improvement." The absence of such feedback may be rooted in resistance to the conduct of evaluation research and/or the reluctance to utilize the findings of evaluation research (Eaton, 1962, pp. 421–442). Concern with this general problem has led to various kinds of organizational development techniques, such as the survey feedback–collective decision making–problem solving technique (Cook et al., 1973) and sensitivity training techniques.

ORGANIZATIONAL BARRIERS TO CHANGE

In this section, we discuss a number of factors at the organizational level that might create resistance to change and innovation.

Threat to Power and Influence

Perhaps one of the most important sources of resistance is that the change or innovation may be seen as a threat to the power or influence of various parts of the organization (Dalton et al., 1968). For example, one of the most difficult problems to overcome when two or more organizations merge is the feeling on the part of the individual organizations that they are going to lose control over decision making (Blumberg et al., 1971; Marrow et al., 1968). In a recent study of hospital mergers, Duncan et al. (1975) found that these fears of reduction in influence were in fact justified. After merger, hospital administrators indicated a significant reduction in influence on anywhere from 8 to 15 of some 20 decision-making areas. Hospital personnel felt that, after the merger, the influence for much decision making was shifting to the centralized corporate staff.

The authors are currently involved in an organizational development program with schools. This intervention is a survey feedback design that involves teachers, principals, and district-level personnel analyzing survey data about their district and then using this information for problem solving. Based on the data, teachers make recommendations to the principals and district-level personnel regarding possible solutions to problems. Some of the strongest initial resistance to this change program has come from principals. This project initially was quite threatening to them, since they see the intervention as reducing their power and authority. A typical response by principals is the following:

> I have a real problem with this program—what you really are doing is giving teachers an input into the decision process. By doing this you are increasing their power—I now have to check things out with them before I make decisions. That really takes my authority away.

The above response quite clearly identifies the feeling that the change poses a threat to the principals' authority and thus, at least from their perspective, it is hard to not view their concern as a legitimate one. Why shouldn't a person be resistant to a change that is likely to threaten his influence? In this project, the interventionists have tried to reduce this resistance by pointing out to the principals that, by getting teachers more involved in the decision-making process, the teachers are likely to be more committed to decisions that are made and to the organization. The result is that it will be easier for the principal to implement decisions with the cooperation of teachers who are going to be affected.

Organizational Structure

In Chapter Eleven of this book, we go into more specific detail about the effects the structure of the organization have on the change and innovation process. In this discussion we only focus on how structure can affect resistance to change per se. For change and innovation to succeed in an organization, it is important that the structure of the organization in terms of authority patterns, channels of communication, division of labor, rules and procedures, etc., be compatible or supportive of the change.

A study by Gross et al. (1971) of the attempt to introduce a new teacher role model in an elementary school illustrates the role of structure as it affects resistance. The new teacher model required that teachers allow students to pursue their interests through the entire day. However, even after the new teacher model was implemented, the school was still organized in departments, which required teachers to direct students' learning in particular areas during particular periods of the day (1971, p. 140). Gross et al. also point out how role conflict and ambiguity can create resistance to change. When the new teacher model was introduced, the teachers indicated they were not clear about the kinds of role performance required to carry out the innovation. As a result, the teachers abandoned the innovation.

Other studies have indicated the importance of making sure that existing rules and procedures in the organizations support change or innovation. Radnor and Neal (1975), in studying the successful adoption of OR/MS in organizations, found that specific formalized procedures for implementation reduced resistance. The development of such procedures as formal project selection, long-range planning, scheduling, and regular progress reports helps the implementation of change by reducing both role conflict and role ambiguity.

The *division of labor* is another structural factor. The division of labor may create interunit competition; for example, groups sharing common tasks may feel that they are in competition with one another for scarce resources in the firm. Different norms and even goals may develop within the different categories, thereby producing conflict and a lessening of cooperation. This in turn makes decisions difficult to reach. Also, if a change is first proposed by one group, competing groups tend to discourage its further development within the organization. As Schein (1970, p. 99) notes: "The fundamental problem of intergroup competition is the conflict of goals and the breakdown of interaction and communication between groups; . . ." For all practical purposes, the different subunits are as distinct as different organizations, and the various factors affecting the flow of innovations between orgainzations or from the external environment to the organization become operative for each subunit. Schein (1970, p. 88) further notes with regard to problems inherent in forming a new interdepartmental committee—which itself is a change—that in considering the adoption of an innovation "Each person is likely to be so concerned abut the group he came from, wishing to uphold its interests as its representative, that it becomes difficult for the members to become identified with the new committee" (p. 88). The dysfunctional consequence of this is that the lack of commitment to the committee engenders a lack of commitment or support to the particular change the specially constituted committee was formed to consider.

Hierarchical and status differentials are additional sources of resistance to change operating during the initial stages of change. Fear of depreciating one's own status within the organization can cause considerable resistance; when such fears are widespread within firms, the organization can be immobilized with regard to change (Burns and Stalker, 1961). This has been noted in several organizational contexts ranging from hospitals to community governments. It has been postulated that "the more hierarchical the structure of an organization the less the possibility of change" (Griffiths, 1964, p. 434).

The reward structure of the organization can also be an important factor in resistance to change. If managers are asked to learn skills to use on their job, they are likely to be resistant if they feel that they are not going to be rewarded for it. Argyris' study of a laboratory education project with a group of executives indicates that the "new behavior" these individuals learned in the T-group experience did not persist once they returned to the organization, because these managers were not rewarded for their new behavior (Argyris, 1962; also Campbell and Dunnette, 1968).

Behavior of Top-Level Administrators

The change process can be initiated at any level in the organization. Various change specialists (Argyris, 1970; Schein, 1970; Beckhard, 1969) have indicated that change should be initiated from the top so that all organizational participants can know that there is support and commitment from the top regarding the change program. Argyris (1970) points out that this top-down change is particularly important when the change is a radical change regarding how people see themselves and behave on their job the logic for top-down change in this case is that, before people are going to try to experiment with new behabior styles, it is important for them to know that there is support from the top. The real issue here comes down to the following: "Why should I really go through the effort of trying to change my behavior if the people at the top don't change theirs?"

Climate for Change in the Organization

The kind of climate for change that exists in an organization has important implications for change and resistance to change. The notion of climate for change focuses on organizational members' perceptions of the change process. What does the change mean to them? (Campbell and Converse, 1972) What are their attitudes toward change? Also important here is, are there differences throughout the organization regarding the climate for change? If some units have different climates for change, problems during the change attempt can be expected.

There are at least three important dimensions of climate for change that can be identified (Duncan, 1972b). The *need for change* focuses on the perception by organizational personnel about the need for change in the organization. The absence of this perceived need can doom many change efforts. For example, Turnbull et al. (1974) suggest that one of the major causes of resistance to educational change is the development of changes or innovations without prior assessment of the potential users' perceived need for the change or even a systematic assessment of whether a perceived need could be established among potential adopters through appropriate communication, demonstration, etc.

The *openness to change* focuses on the perception of organization personnel about the openness or willingness of departmental personnel to change.

The *potential for change* focuses on the perception of organization personnel that the organization has the capabilities for dealing with change—has the organization been successful in past change attempts, is there a commitment to change in the organization, etc.?

Resistance to change is likely to be greater when the need for change is low, and the openness to and potential for change is perceived as being low in an organization.

Another interesting aspect of the climate for change concept that has implications for resistance to change is the relationship between these dimensions. In some earlier work, one of the authors (Duncan, 1972b) found some interesting relationships. The correlation coefficients presented in Table 3.3 show that there is some degree of association among the climate for change dimensions.

The most interesting find here is the fact that the need for change is negatively associated with openness to change ($r = -.26, p < .01$) and potential for change ($r = -.57, p < .01$). This indicates that the greater the need for change as perceived by departmental personnel, the less departmental personnel perceived there to be an openness to change, a potential for change, and participation of non-command staff in change decisions and attempts. (The reverse also holds; that is, the greater the perceived openness to change, potential for change, and participation in change, the lower the perceived need for change.)

These results have some very important implications for change and resistance to change. The results say that the more personnel perceive that there is a need for their department to change to meet the increased demands of society, and so on, the less do they perceive that their department is really able to deal with change. The high need for change may create anxiety with the result that organizational personnel feel that they cannot change. They are potentially less able to deal with change in that they perceive that their department is *less open to change*; there is more skepticism about the success

Table 3.3   Intercorrelations Among Three
Dimensions of Climate for Change in Three
Police Departments

	I	II
1. Need for change		
2. Openness to change	$-.26^a$	
3. Potential for change	$-.57^a$	$.36^a$

[a] $p < .01$.
  From Duncan, 1972b, p. 230.

of change efforts; and so forth. As a result, they may be somewhat less likely to try to change. They may also be potentially less able to deal with change, because they feel that their department has less potential for change; they feel that they have not been successful in the past in change attempts; they feel that there is not much commitment to change within the department; they feel that personnel in the department are resistant to change. Again, the result may be that they are less likely to attempt change or to be able to work through the difficulties that often accompany attempts to implement change.

The above discussion indicates that organizational climate is an important factor that can affect resistance. More importantly, these data indicate that a high need for change may actually create more resistance to change in the sense that people are less open to change. What may be required then is that the proper expectations regarding the change process should be set. If the change agent is trying to create a high need for change, he or she should try to set realistic expectations that help the change target become aware that it has the necessary skills and abilities to make the change work.

Our study of climate for change thus indicates that a high need for change is not enough to overcome resistance to change. In fact, high need for change may actually be inversely related to openness to change. It thus becomes important for the change agent to create a supportive climate in which members of the change target feel confident about their abilities to change.

Technological Barriers for Resistance

A study conducted by Arthur D. Little, Inc. and Industrial Research Inc. (1973, p. 9) has identified some of the specific sources of resistance to technological change and innovation.

One very real source of resistance to change is the absence of the necessary technical human skills to implement the change adequately. The inability to understand the technology involved is another source of resistance that has legendary status in the case of boiling water for sanitary purposes. This is no less true in more complex technological settings. Distrust of outsiders and/or the unwillingness to involve outsiders, such as consultants who could demonstrate or explain the application of the technology, is yet another source of resistance to change.

A proposed new product may be rejected by management for any number of reasons, such as a lack of perceived consumer demand (associated perhaps with a lack of understanding of what needs the product is intended to satisfy), the action of competitors, incompatibility of the new product or line of products with existing products, inability to segment the market readily to reach the target consumer group, distrust of forecasts, inadequate resources for adequate distribution and promotion, and so forth.

Limited mobility of personnel across industries is a restraining factor in the diffusion of new technology. Another barrier is the absence of good communication systems either within an industry or between industries, or alternatively, the inadequate use of an otherwise good communication system. Technological innovations may meet with resistance within a firm as a result of the so-called "not-invented-here" syndrome, whereby changes are not made or adopted if the technology did not originate within the firm. Inadequate mobility and communication within the firm is another cause of resistance to change. Those having insight or experience with a proposed change are unaware of the interest of others within the organization who have need for the insight.

Uncertainty about the actions of regulatory agencies at all levels of government is one source of restraint on innovation in industry, particularly with any change that may have antitrust or consumer and environmental implications. Tax policies are another source of restraint, although they often may be a stimulus to change as well. Patent policies are also important if patents provide inadequate protection, too long a period of protection, or if there are nonexclusive licenses on government-developed technology.

PSYCHOLOGICAL BARRIERS TO CHANGE

Perception

Selective perception and retention may prevent a person from seeing that the status quo is inadequate (Watson, 1971, p. 121). For various reasons, a person may not "see" problems requiring significant change for remedial purposes or not "see" solutions even if a problem is recognized. Moreover, when data suggesting problems and indicating solutions are thrust on someone, the data may become distorted or even forgotten. A study concerning the impact of a mass media nutrition information campaign in Costa Rica found that, among mothers who recalled encountering information about this campaign, those whose children had the highest incidence of nutrition-related problems recalled the messages with the least accuracy, even when controlling for sources of exposure to the campaign and other variables. Moreover, mothers whose child nutrition practices were rated poor were disproportionately more numerous in the category of those who could not recall any encounter with the campaign, indicating selective exposure and/or selective retention among this group.

One important cause of resistance to change occurs when a change agent and a client agree on a problem but do not share common perceptions about its nature and causes and hence have different perceptions of how to remedy the problem. In one country in South East Asia, for example, World Bank officials and country leaders were in concert about the severity of a logistics problem in the distribution of rice. The World Bank officials felt that the problem was of an infrastructure nature, that is, road improvement was necessary. The country officials, on the other hand, felt that the problem was much more one of a lack of mechanized vehicles for transporting the rice. The result was a compromise after nearly a year of negotiation. One fear of the World Bank officials was that the vehicles would be diverted for use in other sectors of the economy. This fear was in fact realized and eventually led to the cessation of the assistance. In another setting, Black ghetto children in very early school years have been found to experience considerable conflict in schools resulting from two different linguistic demands. On one hand, there is the language spoken by family and friends, and on the other hand, the language expected by the teacher. Some Black groups have insisted that Ghetto English be accepted in schools, whereas teachers have insisted on Standard English. Both groups perceive the problem in the same way but have very different perceptions of the appropriate solution.

Lack of clarity about the behaviors required by a change or innovation may be a source of resistance at the trial or adoption stages (Gross et al., 1971). This may be a consequence of inadequate efforts on behalf of the change agent to communicate the nature of the change in a clear manner. Incomplete or inadequate information can initially result in adoption and, because of unexpected negative experiences, serve as a barrier to further change. For example, there is evidence that individuals who obtain a vasectomy without appropriate preoperative counseling and education may, after the operation and learning of its irreversibility, be a source of negative word-of-mouth communication, thus creating resistance to change among other people (Rogers, 1973). In general, the greater the unanticipated punishment or negative consequence attendant on the adoption of an innovation, the greater the resistance to subsequent similar changes (Homans, 1974).

There may also be differential perceptions of what behaviors are expected of a given role. If the change agent engages in activity the client or target system feels is inappropriate, it may show indifference or hostility to all action he or she undertakes. This may be the case when someone perceived to be in an advisory role becomes actively involved in change (Foster, 1962) or when a change agent expands without invitation his activity to include another sector of life in a client system (Niehoff, 1966).

Another source of resistance to change may occur when assistance is provided freely. There has long been noted an association between price and quality in which price has a causal impact on perceived quality. Foster (1962) provides examples in which something given free may be perceived as valueless. In a family planning context (Harvey, 1974), it has been demonstrated that users of condoms and oral contraceptives may prefer to pay for these commodities, and, moreover, such desired payment may be more than just nominal (Harvey, 1974).

Homeostasis

Watson (1971) observes that, although organisms are not naturally complacent, they do seek a comfortable level of arousal and stimulation and try to maintain that state. Many programs of social change involve a level of arousal and stimulation well above what is comfortable to the individual. In such instances, there is a natural tendency to avoid or resist that change effort. Many factors may account for homeostasis. Lippitt et al. (1958, pp. 180–181) suggest that the reluctance to admit having weaknesses, awkwardness and fear of failure associated with doing something new, bad experiences with past change efforts, and concern about the possible loss of present satisfaction, are factors that may account for homeostasis.

Conformity and Commitment

Conformity is a major force working against change (Kiesler and Kiesler, 1970; Homans, 1974). People need to be liked, to be correct, and to participate in the fruits of achieving collective goals. Similarly, commitment is a powerful force working against change (Kiesler, 1971). Financial and social psychological investment in programs and practices help root people in the status quo, and very special efforts, often in the form of incentives, must be used to create alternative investments in the advocated change. For example, one source of resistance in the family planning area is the financial value of children who support parents during their retirement. To overcome this reason for having children, special programs like the following are being instituted.

A non-birth incentive scheme was designed by Dr. Ron Ridker when he was with the U.S. Agency for International Development in India, and this plan (with important modifications) was implemented experimentally on three tea plantations in South India by the United Planters Association of South India (UPASI), starting in July 1971. Each of the tea estates is a relatively closed site, relatively free from possible contamination. The essential idea of the retirement bond incentive is to reward a tea estate worker for limiting her number of births (and abortions) by providing funds for retirement purposes, which are paid in a lump sum at age 45. A monthly deposit of 5 Rupees ($0.67 US) is paid into a bank account for each eligible female worker. Complete forfeiture of these savings and interest is caused by pregnancy after the fourth child; partial forfeiture results from an additional pregnancy after the second or the third child. If no forfeiture occurs, a female employee could earn up to a total of 2,000 Rupees ($267 US) at retirement. It is too early to draw precise conclusions about the effects of the retirement bond incentive, but the Government of India has granted funds to UPASI to extend the incentive scheme to six additional districts in the plantation farming area of South India. (IGCC, 1973)

Professional training and the associated professional socialization process and its attendant pressures for conformity and commitment can also be a source of resistance to change. A study reported by Hendershot and Grimm (1974, pp. 438–441) revealed significant differences between social workers and nurses in their attitudes toward abortion. This is evident in Table 3.4.

The difference in attitudes concerning abortion "is probably not attributable to the different personal characteristics of the two groups" (p. 440). The authors suggest that differential involvement with or commitment to the client and differential settings of interaction with the client may account for the differences in attitudes.

The finding that attitudes toward abortion are different among social workers and nurses even when relevant background characteristics are taken into account suggests that the difference may be attributable to differences between the social structures of the two professions. Two such differences may be of special importance. The first is in the normative orientations of the two groups: whereas social work aims at enabling its clients to "cope" with a wide range of problem situations, nursing aims at preserving the good health of its patients. Thus, social work encompasses a large part of its clients' life experience, while nursing encompasses that smaller part directly concerned with health impairments.

Because abortion is performed, ultimately, by a medical practitioner, it is seen by nurses as a medical procedure, appropriate only when clearly related to a health impairment. While social workers may also regard abortion to be in some sense a medical procedure, their normative orientation toward their clients causes them to see it as appropriate for social, psychological, and economic reasons as well. . . .

A second difference between social work and nursing which may be related to attitudes to abortion is in the conditions under which clients are seen. Social workers are more likely than nurses to see clients under "normal" life conditions; they are, therefore, in a better position to evaluate the net, long term effects of an unwanted pregnancy or an aborted pregnancy.

Table 3.4   Distribution of Respondents by Approval of
Abortion under Each of Six Conditions: Social Workers
and Nurses[a]

Condition	Social Workers (percent)	Nurses (percent)
Don't want	(389)	(146)
Approve	57.8	24.7
Disapprove	25.4	57.5
Don't know	16.7	17.8
Can't afford	(391)	(144)
Approve	63.2	37.5
Disapprove	20.7	47.9
Don't know	16.1	14.6
Not married	(391)	(143)
Approve	63.4	43.4
Disapprove	19.9	42.0
Don't know	16.6	14.7
Deformity	(398)	(145)
Approve	82.4	54.5
Disapprove	5.3	23.4
Don't know	12.3	22.1
Rape	(402)	(150)
Approve	88.8	86.0
Disapprove	3.2	8.0
Don't know	8.0	6.0
Health	(407)	(154)
Approve	96.1	91.6
Disapprove	0.5	2.6
Don't know	3.4	5.8

[a] Number of respondents is given in parentheses.

Source: Hendershott and Grimm, 1974.

Being more often separated by the hospital or clinic setting from the home conditions of patients, nurses are less likely to perceive the net, long term effects of unwanted pregnancies or abortions; furthermore, nurses are more likely than social workers to have firsthand information on illegal abortions, which include many begun by inexpert abortionists and accompanied by medical complications. In sum, the conditions under which clients are seen are such as to increase the probability that social workers will perceive a net benefit to clients in abortion, and decrease the probability that nurses will see a net benefit. (Hendershott and Grimm, 1974)

Rubin et al. (1974) present an interesting discussion of resistance to change in health care systems uniquely associated with the professional training of persons in those settings, particularly physicians. Physicians are taught first "to do no harm," which is a brake on making risky decisions. This, coupled with the greater ability to spot errors relative to successes, creates a conservative attitude. Second, the medical model is one that assumes a unidirectional flow of curative services in which the physician prescribes, often in the face of an emergency or under a crisis environment. This permits relatively little time or effort for preventative medicine. Also, this presumptive curative approach is counter to the sometimes more effective social intervention approach in which the server approaches the problem with the servee in a "let's see how we can solve this together" manner. Thus it is difficult to use this cooperative approach effectively with health care systems dominated by medically trained staff.

Personality Factors

Personality factors are believed to be important as well. For example, Rogers and Shoemaker (1971, pp. 187–188), in their review of the literature, suggest as barriers and factors low empathetic ability, high dogmatism, inability to deal with abstractions, fatalism, and low achievement motivation. Lack of conceptual and inquiring skills may limit the motivation and ability to reexamine, evaluate, and alter behavior (D. W. Johnson, 1969, pp. 143–153). Related to this is the lack of creativity as a barrier to innovation (H. G. Barnett, 1953). Grinstaff (1969, pp. 517–528) suggests the inability to tolerate ambiguity as a cause of resistance to change. Robertson (1971, p. 107), however, concludes with reference to consumers that personality variables should be treated carefully and that their relevance to resistance may be very innovation specific.

Propensity to assume risk is a particularly important personal trait as well as a social system trait. For example, the professional educator is likely to be a follower rather than a leader of change, and hence much change in education comes from outside this field.

> When considering the roles of educational professionals in advancing change, a distinction needs to be made between *initiation,* the function of conceiving and introducing new practices, and *implementation,* a process of enacting changes that others have instigated. By this type of comparison, it is a myth that educational administrators, supervisors, teachers, and professors are leaders for change, despite the fact that they are always involved when changes occur. What school and college personnel do is implement or carry out changes imposed by outside forces. In some cases, educators are sufficiently sensitive to emerging external mandates for

change and so skillful in adapting to them that they, like good politicians, actually take credit for new developments. Riding the groundswells of change, however, involves a kind of brinkmanship that few educators are willing to risk. To move ahead of public opinion is to lose the game, which may mean loss of employment. Chances of failure are so great that most education professionals choose to react to and implement, rather than to lead, change. (Stiles and Robinson, 1973, p. 258)

Carter and Silberman (1965) make a very similar observation: most (educational) changes are originated for the schools rather than by them.

Eichholz and Rogers (1964) have identified many additional sources or causes of resistance: rejection through ignorance or through default (knowledge of an innovation but not interest in it), erroneous logic (rational

Table 3.5 *A Framework for the Identification of Forms of Rejection*

Form of Rejection	Cause of Rejection	State of Subject	Anticipated Rejection Responses
1. Ignorance	Lack of dissemination	Uninformed	"The information is not easily available."
2. Suspended judgment	Data not logically compelling	Doubtful	"I want to wait and see how good it is, before I try."
3. Situational	Data not materially compelling	1. Comparing	"Other things are equally good."
		2. Defensive	"The school regulations will not permit it."
		3. Deprived	"It costs too much to use in time and/or money."
4. Personal	Data not psycho-logically compelling	1. Anxious	"I don't know if I can operate equipment."
		2. Guilty	"I know I should use them, but I don't have time."
		3. Alienated (or estranged)	"These gadgets will never replace a teacher." ("If we use these gadgets they might replace us.")
5. Experimental	Present of past trials	Convinced	"I tried them once and they aren't any good."

Source: Eichholz and Rogers, 1964.

but unfounded reasons), and unsuccessful experience with the change. They provide a useful framework for the identification of various forms of rejection. This is presented in Table 3.5.

This chapter can be summarized by the following general principles:

1. Changes or innovations should be made as compatible as possible with the cultural values of the change target in order to reduce resistance to change.

2. Change or innovation should not be presented in such a way as to threaten the change target's self-esteem or image in the eyes of others or resistance to change will be encountered.

3. Change or innovation should be presented in a way that does not threaten the cohesiveness of the change target system or resistance will be encountered.

4. Changes or innovations should not threaten or jeopardize an individual's position vis a vis reference groups or resistance to the change-innovation will be encountered.

5. Changes or innovations should avoid altering the balance or interdependence between units involved in the change or resistance will be encountered.

6. Change or innovation should be presented in such a way as to avoid creating conflict among individuals or groups involved or resistance will be encountered.

7. Change or innovations should be implemented in a way that does no increase a person's uncertainty or ambiguity as to how the change o innovation will affect the way he or she performs a role or resistance wi be encountered.

8. Change or innovation should be presented in a way to minimize th degree to which it is seen as threatening to the power and influence o individuals or groups in the target system or resistance will bε encountered.

9. Changes and innovations should be implemented in a manner that does not create role conflict for individuals in the target system or resistance will be encountered.

10. Changes and innovations should be implemented so as to avoid increasing intergroup competition in the target system or resistance will be encountered.

11. There should be top-level support in the system for the proposed change or innovation or resistance will be encountered.

12. The system should try to provide rewards—incentives to participants for adopting the change or innovation—that are attractive to them as a way of reducing their resistance.

13. Change or innovation should be implemented in such a way that members of the change target system perceive that there is a good chance of successfully achieving the change or resistance will be encountered.

14. The proper climate for change should exist in the system to reduce resistance to change. Members should have a high need for change, openness to change, commitment to change; they should feel that there is a potential for change in the system, and feel that they have some control or influence over the change process.

15. Those who are going to use the change or innovation should perceive that they have or will be provided with the necessary technical skills to implement the change or resistance will be encountered.

16. Proponents of change and innovation should empathize with opponents of change to understand the causes of their resistance so that some informed action can be taken to reduce or eliminate the causes of resistance.

17. Proponents of change and innovation should strive to reduce the change target's ignorance regarding the change, as a way of reducing resistance.

18. Proponents of change and innovation should be aware of the personality characteristics (such as dogmatism, low tolerance for ambiguity, low risk-taking propensity) of members of the change target system that can create resistance to change.

CHAPTER

FOUR

Facilitative Strategies

Facilitative strategies are those strategies which make easier the implementation of changes by and/or among the target group. The use of facilitative strategies assumes that the target group (1) already recognizes a problem, (2) is in general agreement that remedial action is necessary, and (3) is open to external assistance and willing to engage in self-help. This does not imply that total consensus must exist within the target group as to exactly what the facilitating strategies should be or what the specific end result

90

should be. However, the greater the consensus about both means and ends among the members of the target group, the more effective facilitative strategies will be. The change agent should be aware that, when dis-consensus exists, he is strengthening one group's position relative to another by virtue of facilitating the implementation of one group's goals or social problem solution and not providing assistance to another group with competing approaches to the same problem. To the extent that multiple approaches to a problem within a social system are possible—indeed, even desirable and necessary—the change agent should try to facilitate the activities of groups with different perspectives. This is a common approach followed by change agencies with substantial resources.

CONSIDERATIONS FOR SELECTING FACILITATIVE STRATEGIES

Awareness of the Change Target

Facilitating strategies do not work simply by virtue of assistance being made available to the client system. Awareness among the client group of the availability of help must exist in sufficient detail and clarity that potential clients know exactly what is available and where and how assistance may be obtained. Thus public relations and other information dissemination activities are particularly important elements of programs pursuing facilitative strategies. Community participation in immunization programs in rural areas of Central America has been shown to be strongly affected by the degree and nature of preprogram information campaigns (Lin et al., 1971). Advertisers have long been aware of the need to provide very specific information such as telephone numbers, and mailing and street addresses where goods and services can be obtained. This is particularly important during the introduction of new products and services. The "let your fingers do the walking" theme promoting the use of the Yellow Pages telephone book is one of the most well-known commercials employing a total facilitative strategy.

The various means by which awareness of the existence of a needed innovation takes place and knowledge concerning its acquisition is disseminated are potentially as numerous as the number of different communication channels (and their combinations) that are available in any given situation. Determinants of the actual set of communication channels to be used for establishing awareness about the existence and acquisition of a needed change are varied. They include (1) the nature of the particular change or innovation, (2) the desired content of the information, (3) the characteristics of the change target, (4) the decision-making stage of the target system (a) as a whole or (b) of its key segments, (5) the feasible communication channels which exist, and so forth.

One way of creating within the target group an awareness of specific problems or creating a particular perspective with respect to an acknowledged problem is to formally involve members of the target system in the decision-making process of change agencies. This also appears to produce greater commitment to change within the target group. Various organizations are using or experimenting with client advisory councils. The United States government is playing a particularly active role in increasing citizen participation in social action agency deliberations (Mogulof, 1969).

Four objectives of client system participation in the deliberations of social service programs have been identified with respect to poor people (O'Donnell and Chilman, 1969). These objectives can be stated more generally with equal application to target groups defined in other ways.

1. Participation lessens feelings of alienation from change agencies;
2. Participation enhances client system feelings of control over decisions affecting it;
3. Communication among different groups within the target group is stimulated and facilitated; and
4. Participants become somewhat more socialized into the change agency thinking and operation.

The problem-solving component of the survey feedback problem-solving strategy is also a good example of a facilitative strategy. Data are collected from organizational members to identify organizational issues or problems that the organization needs to work on. The data are fed back to intact work groups who then diagnose the problems and come up with action strategies for solving the problem. This involvement of organizational members in the decision-making process has several advantages: it increases their commitment to following through on solutions to problems; it has the potential for generating more varied information to use in the problem-solving effort; and by working with intact work groups, this activity tends to reinforce the organization's structure and communication channels.

Degree of Commitment

A variety of client commitments, many of considerable magnitude, may be involved in the seeking of help and the establishment of a specific helping relationship. Provisions must be made for identifying and facilitating these commitments. Many of these decisions or commitments are noted below (Landy, 1965):

1. The help-seeker must feel that he is handicapped with regard to himself, family, friends or society;
2. He must face the probability that these relevant others will know of his disability and question his role and achievement capacity, thus rendering him culturally disadvantaged;
3. He must be willing to admit to the helper that he may have failed as a person and is incapable of handling and solving his problems as the culture demands and expects;
4. He must be willing to surrender some autonomy and place himself in the dependent client relationship, relying on the ability of the helper to give the help he needs;
5. He must decide to ask for assistance;
6. He must make a number of economic decisions, each of which not only has consequences for the way he views his own capability, but which influence his path to the helper;
7. He must decide whether to risk possible threat to existing relationships by opening them to deep probing.

The degree to which facilitation should take place is difficult to determine. At least it can be assumed that what appears to be maximum facilitation effort is not always optimal. For example, dispensing contraceptives free of charge and presumably removing cost as a barrier to acceptance does not appear to produce as much consumption of contraceptives as charging a nominal fee (Harvey, 1974). This appears to be true even among the economically disadvantaged members of developing societies. Thus, from a programmatic standpoint, delivery of contraceptives is facilitated by charging at least a nominal fee rather than no fee.

Various techniques for facilitating service utilization have been tried with different degrees of success. Mobile X-ray units providing free service have been used to reach people for whom travel to stationary health units or financial considerations represent psychological and financial obstacles. There is a very real possibility that the basic problem underlying low use of service may be a lack of motivation to address a problem. Change agents in various contexts have argued that under such circumstances facilitative strategies are especially important: the lower the client's motivation or commitment to solve a problem, the easier it must be made for him to use a service not now used. Thus the use of telephones to report medical symptoms has proven facilitative in some cases for some types of people: "Generally speaking, those higher in education, occupation, income, and social class were relatively more likely to use the telephone for reporting symptoms and relatively less likely to use face-to-face contacts than were those lower on these socio-demographic variables" (Pope, Yoshioka, and Greenlick, 1971, p. 161).

Cernada, Lee, and Lin (1974) report considerable success in the use of the telephone for establishing contact with people who might otherwise not be contacted or reached with family planning information. In the first nine months of operation, the Taipei pilot project called "Contraceptive Guidance Special Line" had received well over 5000 phone inquiries, and the Seoul project (Family Planning Counseling Service) received almost 2300 calls after 17 months. Importantly, it was found that over 50% of the callers in each city used a public or office phone, suggesting the possibility that the absence of a private phone was not an obstacle as had been feared. In both cities the availability of the telephone service was widely publicized. The authors note that news releases about the programs seemed especially helpful in stimulating phone calls. Again this indicates that facilitative strategies often must be linked with informational strategies, particularly for persons with low motivation or little commitment to the idea of a particular change.

The question of who is committed to the change or innovation is also important. Bean et al. (1975), in their study of the implementation of operations research/management science activities, found that a key variable predicting successful implementation was top-level management support. Apparently, top-level management support conveyed to the rest of the organization that it was "worth their own efforts" to adopt the change. Bean et al. (1975) also found that the way the leader exhibited this commitment was also related to successful implementation. Their data suggest that ". . . leaders who spend a high percentage of their time accommodating OR/MS models and outputs to the organization's operating environment and a small percentage of their time in selling and administration will be leaders of successful groups (1975, p. 95)." Clark (1972) has also demonstrated the importance of top-level sanction for change. Sanction for a change in managerial philosophy was not established in the project he studied. As a result, when members of the change target system saw that, even though the words of top-level managers often implied approval of a change, the actions of top-level managers did not, they withdrew their support for the change.

Change agent commitment to the change is also an important factor in inducing commitment among potential adopters of an innovation. For example, canvassers in an Indian vasectomy campaign were particularly convincing when they were able to show their own vasectomy scar to the potential vasectomy client (Rogers, 1973). Similarly, clients are apt to become change agents or facilitators under some circumstances such as drug use. One survey of middle-class illicit drug dealers found dealers typically to have been drug users prior to becoming dealers and, as dealers, are more likely to initiate nonusers when the drugs are the so-called "riskier" drugs as opposed to low-risk drugs (e.g., marijuana) (Atkyns and Hanneman, 1974).

Perceived Need for Change

Making available a service in, say, mental health, or a commodity such as a new strain of rice does not guarantee its utilization. There must be a perceived need for the advocated change. There must be a belief within the target system as a whole or among key persons in that system that the need can be met without undue social, psychological, or financial cost. Theodore (1968), speaking with reference to a health setting, notes that, apart from the existence of a physiological or psychological problem, the client must (1) perceive the presence of these problems, (2) be willing to remedy problems through appropriate health care services, and (3) have the ability to transform perceived needs into overt demand for medical assistance.

It appears, too, that when it is the clients who initially recognize the problem or need they are more likely to continue working to satisfy that need than when the need is initially diagnosed by a change agent (Kadushin, 1958–59). This suggests that an important task for the change agent is to facilitate client recognition of problems or needs by providing the necessary tools to diagnose a problem. Implicit in this is the assumption that the need can be satisfied using current technology, independently of whether the need satisfying innovation presently exists. A major trade association recently concluded a study that investigated the causes of the commercial failure of several new industrial products. In a substantial minority of the cases, it was discovered that many potential adopters simply did not believe that the current state of technology in the areas of concern could produce a satisfactory solution to their particular need, and hence the potential adopters did not believe claims made by salesmen.

The existence of a perceived need does not guarantee the utilization of services or commodities designed by external parties to satisfy the need. The client or target system must establish a connection between his need and the advocated solution for satisfying that need. Establishing this connection is often the responsibility of the change agent; in any event, it should never be assumed that the client will automatically "see" the relevance of an innovation to a particular need.

When there is a disagreement among members of a client system about the appropriate solution to a commonly agreed on and collectively felt need, it may be desirable to offer multiple solutions. For example, in family planning, the most successful programs appear to be those which offer a variety of contraceptive technologies to satisfy different preferences (e.g., male- vs. female-oriented contraceptives). This form of facilitative strategy might be called *solution diversificaion.*

There may also exist situations in which disagreement exists about the felt need. Here, too, solution diversification may be an appropriate strategy. Often the expressions of felt need are very diverse and closely identified with professional backgrounds. Thus some professionals perceive malnutrition primarily as a result of too large a population; others may view the problem largely as a result of ignorance of proper child feeding practices, underproduction in agriculture, and so forth. In many cases major social problems and needs are in fact determined by a multiplicity of factors. One response to this in the health care area has been to integrate health services, although this has met with considerable opposition from specialists.

Capacity of the Client to Accept Change

The capacity of an individual or social system to change is a function of many, often interrelated factors. One factor might be broadly labeled the mental set of the change target. Does the individual possess the necessary attitudes and skills required by the change? Are the existing social norms or formal policies of the organization consistent with the advocated change? The chapter on resistance to change identifies several capability constraints. Logan (1973) describes a situation that is not uncommon in many parts of the world where mental sets embodying humoral medicine as a folk belief may restrict the capacity of persons or groups to accept modern medicine. "When physicians prescribe medicines or dietary regimens that conflict with a patient's belief in the humoral concept, the successful treatment of that patient can be adversely affected" (p. 385). The adverse effect occurs when the physician's prescribed treatment is not in direct opposite temperature quality to the perceived illness. For example, among some groups chicken pox is classified as a hot disease, and the physician-prescribed treatment of aspirin and vitamins is also classified as hot. In this instance, physician-prescribed treatment will probably be rejected. Similarly, penicillin is a hot treatment among some groups and will likely be rejected as treatment for hot symptoms or illnesses. Logan suggests that this dilemma may be avoided by selecting effective clinical treatments whose temperature quality is opposite that of the patient's problem. Where effective treatments with an opposite temperature quality are not available, placebos having opposite temperature quality might be jointly prescribed to "neutralize" the basic prescription.

A second factor affecting the capacity to change concerns environmental restraints and opportunities. This involves both the physical environment and the social environment. The magnitude and frequency of natural disasters in Bangladesh, for example, are serious incapacitating factors in that country's overall development (Chen, 1973). The social nature of educational settings as learning environments can also affect learning among disadvantaged minority groups (Coleman et al., 1966).

Available social and financial resources are very important factors. The absence of needed skills and social institutions necessary to initiate or support change lessens substantially the capability of the change target to alter its behavior. Income is an obvious factor affecting the capacity to utilize innovations, although the role of financial resources should not be overestimated (McKinlay, 1972; Monteiro, 1973). For example, some institutions require of their membership a medical examination before providing insurance, and often institutions will require participation in health care programs while at the same time providing substantial income. In such instances it is a nonfinancial organizational policy that accounts for the use of medical services as much as the higher incomes these organizations pay their employees.

One problem arises when facilitating action is incomplete. This can occur when the change agency does not facilitate the entire process with which it is concerned and the client does not have the resources to pick up where the change agency left off. An example is a large hospital constructed in a West African country by an external donor. The external donor, however, made no provision for staffing, directing, and continued financing of the hospital after its construction. Moreover, the recipient country lacked the resources to provide these needed services (Lowenthal, 1973). Thus the capacity or capability to provide a service existed but the means for its effective delivery did not. Van Willigren (1973) has identified several attributes of resources that affect the success of facilitating strategies employed by a change agency. Several prescriptive statements are derivable from the understanding Van Willigren provides concerning the functioning of these attributes:

1. Change will be more likely if the resource-providing institution is located within the client system. Later in this chapter community organizations are discussed as a partial mode of implementing this strategy.

2. Long-run and more pervasive social change is more apt to be achieved if resources are applied to the community rather than the individual (when such alternative choices exist). Certain "economies of scale" may be realized when this approach is employed.

3. The more general the goals to which a resource is committed, the more likely it is to be used effectively. Presumably, the more general goals permit a wider latitude of choice by the change target in selecting problems to which the resources can be applied. However, several of our interviewees commented that diffuseness of goals often leads to dissipation of funds and makes evaluation research difficult.

4. Tying a resource to a specific time period inhibits community participation in a change program and thus reduces the effectiveness of resources committed to the program. As in item 3, the concept of optimality is a key consideration. Establishing a specific timetable can cause misuses of resources if the timetable is so short that adequate assessment of alternative ways of implementing a program is not possible. On the other hand, a timetable that is far in the future or very diffuse in its openendedness may cause a program to move toward desired goals very slowly as a result of exploring too many options at various decision points.

5. The creation of new roles within a client system is desirable if existing roles are inadequate to utilize a needed resource. For example, it has been advocated that special roles be developed to serve as linkers between the marketing management of a country and the various components of its family planning program. The person in the linker role would be able to identify exactly what marketing management expertise would help specific aspects of the family planning program. The linker would also be able to identify individual organizations or persons able to supply management expertise.

Capacity of the Client to Sustain Change

Facilitative strategies are relevant when the client or target group lacks the resources or capability for continued implementation of a decision. It is likely that efforts will be necessary to facilitate continuation of the change for at least a limited period while the target system develops its own financial, human, and other resources. Long-term subsidies of change programs, particularly in government to government relationships, are not uncommon.

Efforts to introduce new OR/MS technologies into organizations provide a good example of the problems that a client system can have in sustaining a change (Bean et al., 1975). Once a new technique is presented to a client system, the key problem then becomes how to integrate the new process into ongoing organizational activities. Often the change agent leaves the client system at the point at which the change is presented to the system. It is erroneously assumed that the organization can now simply take the change and implement it itself. This is an example of the technocratic bias discussed in Chapter 1. This oversight misjudges the complexities of getting the change accepted. Also, organizations often need help after implementation to debug problems that arise.

The experience of a large metropolitan police department illustrates the above problem. The department was introducing a new computer-assisted dispatching system with the help of a consulting firm. However, no provisions were made to provide periodic help sessions with the dispatchers as they encountered problems using the new system after the initial implementation stage. Complete utilization of the new system was thus hampered because personnel did not have the necessary training and experience to work through the "bugs" in the system.

Resources Available to the Change Agent

In most instances, change agents have many competing demands made on their resources. Many social change agencies will not enter into an agreement to undertake a change program if the continuation of that program will require continued resource expenditures long after the initial implementation of the change. For example, social marketing programs in the family planning area have been subsidized by major funding agencies, with the explicit understanding that the programs would be self-supporting within a specified period of time, beyond which further subsidy would not be forthcoming.

Segmentation of Target System and Decision-Making Stage

It was noted earlier that some members of a target group adopt change more quickly than other members, even when all members of the target group are exposed to the advocated change more or less at the same time. It was also indicated that it is desirable in such circumstances to segment the target or client system using differential response to innovations as a segmentation dimension. The implication of this is that early adopters might optimally require different facilitative strategies than later adopters. Thus, over time, the mix of facilitative strategies used might vary as different potential acceptors become open to change (perhaps as a result of educational and persuasive strategies). For example, regional differences in the acceptance of abortion services have been noted despite roughly comparable needs for abortion services as defined by persons outside the target system. In this instance, it makes sense to allocate presumably scarce resources for establishing abortion clinics in those regions of a country or sections of a city most open at the present time to the use of such services. At the same time, persuasive strategies might be pursued in the more resistant areas.

Moreover, it is possible in this example that the early adopters would be—in fact have been—willing to use the services of a specialized clinic, whereas later adopters, when ready for acceptance, would be more attracted to the services of a well-known and well-established hospital. In general, then, what the change agent does to facilitate change may vary as different segments of the client system become open to change at different points in time.

Magnitude of the Change

The larger the magnitude of the intended change, the more important it is to undertake facilitative efforts, although facilitation strategies alone may be the least effective type of strategy unless very strong motivation and commitment are present in the target system. In marketing contexts requiring a significant change in purchase behavior, it is common for the vendor to offer small sample sizes of a product, or allow a demonstration for a specific and limited time period in the case of capital equipment. Thus a major change is turned into a small change for a limited period, with the intention of getting the buyer used to the product and thus more open to a full-scale acceptance. When the change involves a financial outlay felt by the buyer to be sizeable, special introductory prices or "cents-off" coupons are offered when launching new products. Thus lower prices reduce the magnitude of the financial aspect of change for a limited period of time.

Anticipated Level of Resistance

When resistance is anticipated, facilitative strategies intended for the target group as a whole will be relatively ineffectual. The rationale here is that facilitative strategies assume that the target system recognizes its problem and is therefore open to external assistance. There is a certain assumed openness to change on the part of the target system. Facilitative strategies do not include specific techniques for creating this openness. This is something that persuasive and reeducative strategies focus on more directly. Facilitative efforts might be helpful if directed toward those (probably few) who are inclined to change despite group and/or cultural pressures toward maintenance of the status quo.

Nature of the Change

Difficulty in using or implementing something new is a major barrier to successful change. The use of many consumer and industrial goods has been greatly enhanced by automating certain aspects of their operation. The simplification of many complex technological innovations in the medical area has permitted their wider distribution in developing nations in which maintenance of climate-sensitive equipment is difficult. Simplification and automation are two facilitative strategies. Various "outreach" programs are facilitative in nature. Efforts to broaden the delivery of family planning services to remote areas using retired military personnel, the use of satellites for broadcasting educational programs overseas, or the offering of English language courses very early in the morning prior to the time nonEnglish speaking United States residents leave for work are other examples of facilitative strategies in response to changes that at one time had restricted availability.

Time Requirements

In many instances, a client system is ready to implement change, and once facilitation is provided the change occurs relatively quickly. Facilitative strategies are frequently most appropriate when an openness to change exists, whereas power, educational, and persuasive strategies are more appropriate when openness to change does not exist. When time is a crucial factor and openness to change is not present, strategies other than facilitation are indicated.

Change Objectives

Facilitative strategies are appropriate when the change agency is simply concerned with making easier whatever it is that some individuals or groups are already committed to do or are currently engaged in doing. When the objective is more concerned with altering behavior among persons or groups not predisposed toward the advocated change, facilitative strategies are not helpful. The philosophy that simply making goods and services available will itself produce change does not work in practice.

ILLUSTRATIONS OF FACILITATIVE STRATEGIES

Let us turn to several examples of facilitative change agencies and strategies. Perhaps one of the most prominent facilitating organizations in the United States is the Ford Foundation. The role of the Ford Foundation varies greatly depending on the particular situation. In some instances, such as the Indian Education case detailed below, the agency takes the initiative and provides extensive financial, educational, and research resources. Alternatively, the Ford Foundation may simply provide financial support for an autonomous organization, such as a Community Development Corporation (CDC), and the supported organization in turn acts an an instituion and impetus for change in the community. Let us look at these specific illustrations in more detail.

Indian Education*

On reservations and in nonreservation schools, Indian children typically do quite poorly and have a high dropout rate. This situation was identified as a problem by the social development section of the Ford Foundation. The agency decided to begin addressing the problem by focusing on nonreservation school treatment of and programs for Indians. Specifically, the agency personnel selected a school system in New Mexico that was felt to offer the easiest access and greatest visibility as a starting point for creating change.

The nonreservation schools provided a unique situation. Since Indians do not pay the usual property tax that supports public education, a special method for funding the education of Indian children in nonreservation schools exists. This is the Johnson O'Mally fund, which is administered by the Bureau of Indian Affairs and compensates nonreservation schools for Indian students. The use of these funds by each school is to be approved by the parents of Indian students. In fact, in the district described in this case, no audit had ever been conducted and no one appeared to know what was being done with the money.

Two change strategies appeared feasible. The issue could be fought by the agency through the court system. Although it was likely that such a court case would be successful, it would be a lengthy process and would not involve the Indians themselves in the change process. The alternative was to work with the Indian parents to bring the Johnson O'Mally funds under their control and ultimately improve Indian education. The agency elected to pursue the latter alternative.

A two-step approach was used to facilitate the Indian parents obtaining control over the Johnson O'Mally funds. First, agency personnel conducted extensive research on the school budget to formulate a case for the parents to present to the school and in court, if necessary. When the research was complete, descriptive materials were prepared in a form that untrained persons could read and understand. The second step involved working with the Indian parents, teaching them to read school budgets, and training them to confront school officials.

The facilitative strategy was successful. In New Mexico it was discovered that $2 million out of $3 million in the fund was being misspent. The state

---

* The case was taken from an interview conducted by Carol Scott with a member of the staff of the Ford Foundation. Alice Tybout authored this case and the one following.

conceded the case without going to court, and the money is presently under the control of the Indian parents. Because the parents have been trained and educated, it is unlikely that the problem will arise again. The effort to improve Indian education has expanded to other states. However, many states, having observed the New Mexico results, have voluntarily begun reform.

Several unplanned or second-order consequences of the strategy have also been observed. The victory has brought a new feeling of self-esteem to the Indian community. It has helped to overcome feelings of helplessness and powerlessness and modified the opinion that deviousness is the only effective strategy for change. Armed with this success and a new self-esteem, the original project community has gone on to attack other issues.

DISCUSSION. This example illustrates the use of a facilitative strategy by the Ford Foundation. The agency facilitated a particular strategy, confrontation of school officials by Indian parents over use of the Johnson O'Mally fund, by providing financial, educational, and research resources. In this instance, there appears to have been an awareness of the problem and a perceived need for change. However, prior to intervention by the facilitating agent, the Indian parents lacked the resources and skills to exercise the legal power they possessed. As observed earlier, the absence of needed skills and social institutions necessary to initiate and support change lessens substantially the likelihood of the change target altering its behavior.

The lack of resources among the Indian parents and the consequent inability to effect change in their children's education made it unlikely that any change would have occurred without the facilitating agency taking substantial initiative. The Ford Foundation considered two alternative methods of intervention in this case: (1) direct use of a legal power strategy and (2) facilitation of the exercise of legal power by the Indian parents by providing financial, educational, and research resources. Where the former strategy probably would have won concessions in court, it would not have involved the Indian parents.

Involvement of the parents was considered important for several reasons. First, involving the parents as the actual change agents increased their commitment to the change effort. Also, there are a number of benefits generally found to accompany participation, including reduction of alienation, enhancement of feelings of control over decisions affecting the group, increased communication between groups (i.e., target, agent, facilitator), and increased understanding of the change effort.

The increased commitment and understanding resulting from the parents' involvement heightens the probability of maintaining the change in the absence of the facilitating agency. Because the Indian parents have been trained to read school budgets, confront school officials, etc., they now have the ability to insure that the change persists. Furthermore, these newly acquired skills may enable Indian parents to tackle other issues and increase their feelings of self-worth. Thus the Ford Foundation helped increase the effective power of Indian parents over their children's education, and this power may generalize to their relationships with other bureaucratic organizations.

In the Indian education case, the facilitating action by the agency was complete; the Ford Foundation provided all the critical resources. Had the agency simply performed the research and then handed the reports to the Indian parents without any effort to train them to read the budgets and confront school officials, its effort might have failed. It is important for a facilitating agency to assist in all areas in which the change agent has insufficient resources; otherwise the effort may fail, and the contribution of the facilitating agency will be wasted.

In summary, the Ford Foundation acted as a facilitator of change by providing financial, educational, and research resources to the Indian parents. This strategy was selected by the agency rather than a direct power strategy because it would involve the parents and consequently would be more likely to result in long-term change after the agency left the situation. It also had positive second-order consequences such as increased self-esteem for the Indians. A facilitating strategy was successful because there existed an openness to change among the parents.

The Indian parents used the resources provided by the agency to employ a legal power strategy. Although Indian parents had potential legal power as specified by the Johnson O'Mally fund, they lacked actual power because they did not possess the resources necessary to exercise that power. The Ford Foundation facilitated the exercise of legal power by the Indian parents over the schools. A facilitative strategy was not appropriate with respect to the school district itself, since the district did not appear to be open to change. However, it should be noted that the agency could also have facilitated the use of a persuasive or educative strategy by the Indian parents, had such a strategy been appropriate.

Community Development Corporations (CDC)

Community development corporations illustrate facilitative strategies and agencies at several levels. Often such agencies have supporters such as the Ford Foundation whose primary input is that of facilitating the organization and its operations by providing financial resources. The CDC itself may, in turn, facilitate a wide range of change efforts in the community by providing financial, educational, and research resources to change agents, as well as undertaking some direct change efforts on its own. Let us look at some brief descriptions of two large CDCs that have received support from the Ford Foundation.

Zion Non-Profit Charitable Trust

The Zion organization, under the leadership of Reverend Leon Sullivan, conducts a variety of social and economic programs in North Philadelphia. In January 1964, Reverend Sullivan opened the first Opportunities Industrialization Center (OIC) where Black youths could learn industrial skills. Now a national program, OIC has trained over 40,000 unskilled people in 95 cities in the last six years, and the bulk of its $40 million annual budget comes from federal grants. In June 1962, Reverend Sullivan initiated the "10-36" plan, a program through which his parishioners would invest $10 per month for 36 months to fund community educational and charitable activities and to create a capital base for local housing and economic development. The "10-36" plan now includes 6000 investors and is the foundation of the Zion program.

Over the years, Zion has engaged in a number of economic development activities. A $1 million, 96-unit garden apartment complex was financed through the Industrial Valley Bank and the Federal National Mortgage Association (FNMA). The 76,000 square foot Progress Plaza Shopping Center was developed at a cost of $4 million for land and construction, with the participation of Philadelphia's First Pennsylvania Bank and Trust Co., the Metropolitan Life Insurance Company, and the Ford Foundation.

Early in 1968, Reverend Sullivan persuaded General Electric to provide technical assistance to help Zion develop an electronics and mechanical parts plant. Based on a $91,813 labor-training contract from the United States Department of Labor and a loan of $680,000 from the First Pennsylvania Bank and Trust Company, Zion launched Progress Aerospace Enterprises, Inc. At its inception, the new company was managed by personnel from General Electric's aerospace plant at Valley Forge, Pennsylvania. Now, after an intensive nationwide search for top-flight management, the firm has its own interracial team of managers, some recruited from the aerospace industries of California, who supervise a labor force of 200 people.

Progress Aerospace Enterprises, Inc. is located on a new three-acre industrial park purchased by Zion in northwest Philadelphia. The park site, which will eventually house several enterprises, is being renovated by Zion's Progress Construction Company. Progress Construction Company also holds a contract in a joint venture on a $5 million office building. Zion is currently planning future large-scale programs that will comprise commercial construction, including new shopping centers in and around Philadelphia, and a number of low-and moderate-income housing projects. The combined Zion enterprises currently employ approximately 350 people and produce gross annual revenues of over $3 million.

Zion's wide-ranging social programs are conducted through the Zion Non-Profit Charitable Trust. The Trust, with a budget of $3.5 million, is governed by a six-member board and has 144 employees. It operates a number of locally based programs including a tutoring and financial assistance program for high school students, a day care center accommodating 140 children, loan packaging, management assistance, and training for minority businessmen, and a general counseling service for residents interested in continuihg or resuming their college education.

Grants from the Ford Foundation totaling $3,025,360 are helping to strengthen Zion's economic development activities in commercial and housing development.

South East Alabama Self-Help Association

The South East Alabama Self-Help Association (SEASHA) grew out of a community education program sponsored by Tuskegee Institute in 1965. The small original staff was responsible for basic community organizing and providing various kinds of aid to the rural poverty population of its target counties in obtaining social services from unresponsive local agencies, assisting residents in retaining their land, and in securing adequate water systems. The association was conceived as a multipurpose community development corporation to improve social and economic opportunities in 12 southeast Alabama counties.

Some of these counties, five of which are located in the Alabama Black belt, are the most destitute in the nation. Their back roads are without gravel, sewer and water facilities are minimal, some 28.7% of all Black families in the area earn less than $1000 per year, and 40% of the entire population, White as well as Black, have less than an eighth-grade education.

Chartered in 1967 and directed by a board made up of residents from each county, SEASHA has attempted to act as an ombudsman on behalf of 8000 Black families, many of whom find their problems compounded by frequently hostile local agencies and institutions. By 1968, SEASHA's program objectives began to broaden beyond these functions, partly in response to local need and partly as a result of the availability of relatively large ($480,000) funding from the Office of Economic Opportunity. SEASHA is today defined by its executive director John Brown as a "rural economic development corporation." Since 1969, the SEASHA staff, currently numbering 28, has assisted 90 local individually owned businesses in loan packaging, management controls, and accounting systems. SEASHA has also established a credit union with 1300 members, total assets of almost $100,000, and 270 loans outstanding, and a 52-family feeder pig cooperative. A principal objective of the credit union and the cooperative has been to establish credit worthiness of Black families and farmers with local banks. SEASHA, whose membership through its county affiliates has grown to 6200 people, continues to provide assistance to its constituents in securing basic social services such as welfare, aid to dependent children, and social security.

SEASHA's future plans for development include a 300-acre housing and industrial development in Bullock County, Alabama; it has already acquired 100 acres of prime land for this and related rural housing development.

Ford Foundation support to SEASHA has totaled $575,000.

DISCUSSION. The CDC system may be viewed at several levels. First, there are private and public agencies such as the Ford Foundation and the former Office of Economic Opportunity that may facilitate the operation of the CDC itself through financial support and thereby facilitate a wide range of change strategies. In view of the fact that the CDCs are designed to be autonomous community organizations, outside agencies such as the Ford Foundation do not take the initiative in their development, nor do they typically provide the educational and research services that were provided in the Indian education case.

.The basic strategy of a funding agency such as the Ford Foundation is to find viable organizations and support them with financial and technical assistance. Criteria employed by the Ford Foundation in selecting a corporation for support include (1) adaptable management style, (2) variety of services offered, (3) must be nonprofit, (4) locally controlled, (5) connections or clout with local government, and (6) credibility and cooperation of the community.

The CDC itself acts as a resource and an institution for change in the community. It may provide resources that serve to facilitate change in the community, or it may undertake direct change efforts. For example, Zion provides tutoring and financial aid for high school students, day care, job training, loan packaging, management assistance for minority businessmen, and numerous other services. SEACHA provides a similar range of offerings. The day care service may facilitate change by freeing mothers to work on community issues. Management assistance for minority business may facilitate the change efforts of groups concerned with increasing the number of Black entrepreneurs in the area and raising the probability of their long-term success.

CDCs have also undertaken massive change efforts of their own in the area of economic development activities. Zion trained Black youths in its opportunities for industrialization center to increase their employment opportunities. Both Zion and SEASHA have undertaken massive housing and shopping development projects. These projects will have a broad range of implications for the community's standard of living and for its morale.

In summary, the typical change strategy employed by CDCs entails providing the community with resources ranging from financial to educational, that can be employed to achieve both short- and long-term change. The resources may be employed to facilitate persuasive, educative, or power strategies as deemed appropriate in a particular situation.

This chapter can be summed up by the following principles:

1. Facilitative strategies can be used when the client system recognizes a problem, agrees that remedial action is necessary, is open to external assistance, and is willing to engage in self-help.

2. A facilitative approach must be coupled with a program of creating awareness among the target groups of the availability of assistance.

3. Facilitative strategies making it very easy to change may be necessary to compensate for a low motivation to change: great ease compensates for low motivation.

4. The facilitating of change through solution diversification is desirable when members of a client system desire different ways of satisfying a common need.

5. A change will be more likely if the resource-providing institution is located within the client system.

6. Long-run, persuasive social change is more apt to be achieved if resources are applied to the community rather than the individual.

7. The more general the goals to which a resource is committed, the more likely that it will be used effectively.

8. Tying a resource to a specific time period inhibits community participation in a change program and thus reduces the effectiveness of resources committed to the program.

9. The creation of new roles within a client system is desirable if existing roles are inadequate to utilize a needed resource.

10. Facilitative strategies such as the provision of funds or capabilities are

necessary when the client system lacks these resources to continue a change.

11. The change agent should assess the client's ability to sustain change itself, and its own ability to provide continued assistance if the client does not have the ability to sustain change.

12. Different subgroups within the client system may require different facilitative strategies at any given point in time.

13. The larger the magnitude of the intended change, the more important it is to undertake facilitative efforts.

14. The greater the resistance to change, the less effective facilitative strategies will be.

15. Certain attributes of the change object such as complexity, accessibility, and divisibility may require offsetting facilitative efforts.

16. When change must occur quickly and an openness to change does not exist, a facilitative approach is unlikely to be effective.

17. When the change objective involves altering a firmly held attitude or firmly entrenched behavior, a facilitative strategy alone is unlikely to be helpful.

Reeducative Strategies

Planned change has been defined in terms of reeducation; thus it is not surprising that reeducation is itself a strategy for change. A reeducational strategy is one whereby the relatively unbiased presentation of fact is intended to provide a rational justification for action. It assumes that humans are rational beings capable of discerning fact and adjusting their behavior accordingly when facts are presented to them. The prefix "re" in reeducative is used because this strategy may involve the unlearning or unfreezing of something prior to the learning of the new attitude or behavior.

Reeducative strategies are felt by many change agents to be the most neutral of all strategies. Ostensibly, this approach does not present biased information or information not clearly rooted in objective fact. Unlike facilitative strategies, reeducative strategies do not always point to a particular means or channel for implementing intentions to engage in a new or different behavior.

CONSIDERATIONS FOR SELECTING REEDUCATIVE STRATEGIES

Timetable

Reeducative strategies are feasible, other things being equal, when time is not a pressing factor. An example of this is the effort of the Soviet Union in 1929 to increase the "need for achievement" in its society by introducing, in the lower grades of elementary school, textbooks and lecture themes conducive to increasing achievement needs. Each year, as the highest level grade of the initial group moved to the next grade level, the subject matter at that level was altered to conform to need achievement norms and messages.

Reeducational strategies are sometimes used in advance of the actual availability of an innovation. This occurs most frequently when it is necessary to identify for potential adopters the connection between their needs and wants on the one hand and the advocated change on the other hand.

Reeducational strategies are also appropriate when power is used to bring about rapid behavioral change. In this case, reeducational strategies are intended to establish, over time and subsequent to the power strategy, attitudes or beliefs that are congruent with the forced behavioral change.

The creation of educational institutions is one educational strategy that can have a long time frame. An example of this is the Ford Foundation support of the National Institute of Management Development in the United Arab Republic and similar institutions in several other developing nations. Although the Institute became relatively well established as judged in terms of other institution building efforts, it nevertheless required a number of years and very imaginative planning to become an effective, nationally supported and operated training agency. The case reported below, partially edited from taped interviews, reflects the concern of the change agencies with the time factor.

In the late 1950s and early 1960s, the Ford Foundation had a modest technical assistance program in Egypt. Its resources were limited, and it could not support large projects of a continuing nature. Instead, it chose to focus on discrete activities emphasizing development of institutions in several key sectors. Management development was one such sector.

THE PROBLEM.   The United Arab Republic, like other underdeveloped countries, lacked trained personnel. It also had few institutions directly relevant and contributing to rapid economic development. President Nasser had initiated a program of development that emphasized nationalized industries and public sector enterprises. However, it was very difficult to find middle- and senior-level executives to run these enterprises. When the nationalization program was begun, a large number of expatriate executives left Egypt, creating a vacuum that had to be filled by trained U.A.R. nationals.

President Nasser recognized the serious need for trained managerial personnel. The traditional educational institutions were commerce-, not management-, oriented. They also were highly academic and uninvolved with the practical problems of industry.

The Ford Foundation representative in Cairo initiated discussions directly with the President's office, and offered the Foundation's resources and expertise in the development of middle- and senior-level managers. The Egyptians were very receptive to this idea, and a plan for institution development was developed jointly.

ALTERNATIVES.   In the early 1960s, Egypt had five universities, two of which had schools of commerce. Therefore, the planners had a choice of working within one of these schools and transforming it to meet the managerial need, or, alternatively, to build an entirely separate and new institution.

In evaluating the alternatives, there were two basic considerations—time and conflict. Reform of an educational system is generally difficult and time consuming. The existing schools of commerce operated within the larger university system, and the scope for changing a subpart of the system was limited. The planners wanted relatively rapid results, and the use of existing institutions seemed to be a lengthy process. There was also a question as to how dynamic and flexible the management training program could be if it were embedded within a large institutional structure. However, there were problems of a new institution posing a threat to established institutions, leading to opposition from these institutions.

THE STRATEGY.   The planners chose to build a new institution. Fortunately, President Nasser's interest in this institution became very strong and very visible, and hostility and opposition from various interest groups was minimal. A decision was made to place the proposed institute directly under the Office of the President and not the Ministry of Education. This decision was initiated in part by a desire to keep the institution outside the regular control of government bureaucracy, thereby enabling more rapid implementation.

The plan involved implementation of the agreement immediately upon the availability of physical facilities. This required two operations simultaneously:

1. To have foreign faculty provide the training requirements in the first few years;
2. To have Egyptian nationals receive advanced training in the relevant disciplines in order to take over the institution gradually and completely in the next few years.

The Ford Foundation's role was to select the foreign faculty, support their teaching activities in the U.A.R. for 2–3 years, and to support the education of Egyptians in the United States. The specific components of the plan, such as design of educational curriculum and administration of the institution, were initially developed by the foreign faculty members.

In order to keep the new institute highly visible, support was generated from the top echelons of the government, attractive salaries were offered to draw personnel of high calibre, and a problem orientation was emphasized to build credibility with industry. Unlike traditional educational institutions, NIMD was designed as a growth-oriented, flexible system. The leadership was decentralized, and authority and responsibility were widely delegated. The faculty members recruited for the institute were allowed opportunities for professional growth, including travels abroad and participation in a wide spectrum of activities.

THE RESULTS. NIMD today is not only a viable but a very prestigious institution in the U.A.R. It has acquired a reputation for quality. Its activities have expanded from training seminars for middle- and senior-level managers to longer-term educational programs for junior personnel. Through its consulting wing, it has been able to involve itself with problems of several major industries in the country.

Over a period of 3–4 years, NIMD was able to increase the Egyptian contribution to its own design, planning and administration. Development of curriculum, selection of faculty, etc., are now totally within Egyptian control, and the dependence on assistance from external agencies has been considerably reduced. The NIMD is considered to be flexible and open to change and experimentation.

Awareness of Change Target

Reeducative strategies can be effective in providing the foundation for future action without identifying the specific action desired by change agents (Jain and Gooch, 1972; Jain and Sinding, 1972). For example, a number of pro-abortion groups in the United States and abroad followed a strategy of presenting facts to the public about the consequences of illegal abortions and their frequency without also making suggestions concerning actions the public could take. It was believed that, at the time these groups first became active, an open advocacy position would cause many people to "tune out the harsh facts," as one of our interviewees put it. Subsequently, as the general public became better informed about illegal abortions, it became easier, and indeed desirable, from the standpoint of these action agencies, to recommend specific citizen action. Thus the more controversial and/or threatening the advocated change is, the more desirable it is to present only the basic facts about a social problem, to be followed later, after the basic information has been received, by specific courses of behavior. The advocated partisan action is then more apt to be seen as a logical step given the factual information. Some premature negative reaction is avoided.

Degree of Commitment

Generally speaking, the stronger the degree of commitment (e.g., attitudinal and behavioral compliance) required, the less effective reeducational strategies will be, at least relative to persuasion. This is more likely to be true in situations in which (1) factual ambiguity exists, (2) the advocated change is highly ego involving, (3) active opposition is manifest, and (4) the change targets do not possess or do not feel they possess the skills or information necessary to assess for themselves information provided by change agents.

Commitment may be expressed in a variety of ways such as financial contributions, the donation of services and commodities, signing a petition, displaying a poster in a window, providing personal testimony concerning the change, as well as compliance with the basic idea of the advocated change. It is not clear which particular form of commitment various educational strategies are most uniquely suited to.

The issue of commitment was very important in the organizational development program described later in this chapter. The reeducational strategy involved survey feedback and collective problem solving. This particular program required, at least initially, considerable time on the part of the teachers involved. In extensive discussions with teachers and school principals, a major concern was uncovered involving the degree of commitment to the project on the part of key personnel in the superintendent's office. Teachers and principals were very reluctant to commit themselves to the project prior to any firm signs of commitment by the district superintendent and his staff.

Some change agents have argued that reeducational strategies are particularly appropriate in obtaining financial support for change programs and in establishing long-term commitment to such programs. This, they argue, is especially true when arguments against the advocated change can be expected. Apparently, reeductional strategies are effective in immunizing people against appeals to resist change or revert back to the previous situation.

Perceived Need for Change

A distinction must be made between problem recognition and felt or perceived need for change. A client system may experience a difficulty that it wants remedied, but it may not know exactly what the problem is, that is, problem diagnosis is difficult even though symptoms are obvious. On the other hand, a problem may be clearly defined but not perceived to be of such a magnitude that serious concern or need for change exists. Often the change agent is confronted with the task of educating and persuading a client that a problem exists, that it has particular causes, and that there is a strong need to remedy the problem because of the serious consequences of letting the problem go unresolved. Inherent in this situation are several obstacles that reeducation can be used to overcome. Such strategies may be effective in connecting causes with symptoms, creating awareness that a problem exists by indicating how much better off the client could be, and demonstrating that it is indeed (if it is) possible to remedy the problem.

The survey feedback problem-solving change strategy is a good example of how reeducative strategies operate in this regard. In their study in school systems, Coughlan et al. (1972) found that the survey feedback problem-solving strategy brought teachers together to work on school problems. By analyzing data collected through questionnaires, teachers began to work on problems and generate solutions. The data provided participants the opportunity to understand problems and thus begin to resolve them. The study found that the survey feedback collective decision-making strategy helped the organization to become more proactive and gain control over its problems.

Not all situations are characterized by change agents helping to establish awareness of problems, their consequences, and solutions. Many so-called clients speak of their difficulties in convincing change agencies that certain difficulties exist, that they are caused by particular factors, that the difficulties are serious ones warranting solutions, and that reasonable solutions are possible. This is particularly the case when the change agent has control over scarce resources such as money, human skills, and commodities. Client systems seeking resources often amass huge quantities of statistics, undertake pilot or demonstration projects, and use testimonies (both solicited and unsolicited) to convey the essential facts supporting their perception of the situation. Private organizations often commission position papers prepared by academics and others to help justify their position with respect to possible regulations under consideration by governmental agencie;.

Generally speaking, the greater the degree of consensus between client and change agent concerning the existence of a problem, its character, and the need for and nature of remedial action, the lower the need for reeducational strategies. The greater the discrepancy between client and change agent in any of these dimensions, the more desirable it is to use reeductional strategies. This does not mean exclusive use of reeducational strategies; persuasive strategies are also likely to be required.

Some social change agencies will become involved in change only when a felt need for change already exists within the target system. One such agency is World Education. One less-developed country interested in increasing literacy levels and "critical thinking" among its uneducated enlisted the assistance of World Education. Through means that need not be detailed here, World Education determined that genuine felt needs for increasing literacy and critical thinking existed at both the governmental level and at the level of the target group. A change agent was assigned by World Education to work with its government personnel. This change agent described her working assumptions in this situation:

1. Adults are reluctant to sit in a classroom to learn to read and write.
2. The areas covered must be functional to them in everyday life—the concepts must be put in use every day.

3. People are initially overwhelmed by large books—which appear too big a task for them.
4. People teach as they were taught—most teachers prefer a lecture method, since it is more secure to the teacher to say to students, "do as I say."
5. Most governments do not have good background training in an area and expect too much too soon from programs.
6. There is a question of fierce pride in most countries—they do not want to be "taught" by outsiders completely.

This change agent then proceeded to assist the developing country to undertake a base line survey, a pilot test, and an evaluation of the pilot test or program. Typically, World Education will not assume prime responsibility for implementing a full-scale change program, although it will assist in devising curriculum materials and so forth.

World Education has minimum resources and relies heavily on local governments to obtain and mobilize funds and people. It also tries to work with labor unions, local universities, and local interest groups to better integrate programs. In the case mentioned here, the government was also able to extract funds from local banks to fund programs. World Education also makes extensive use of existing educational delivery systems to minimize any major structural change. World Education's primary resources to implement change seem to lie in a good reputation of innovative educational techniques that can be devised and rearranged to meet specific needs. They use American facilities (like the Michigan Research Center) to keep abreast of current educational developments.

In the case related by the change agent, the country had four specific goals:

1. Attain higher literacy rates in villages.
2. Increase basic skills of writing and arithmetic.
3. Increase technical expertise of local people—specifically, the base line survey showed eight district market segments in this country (corn farming, wheat farming, etc.). Therefore, eight different curriculums had to be set up (although each segment could use all eight curriculums if desired).
4. Facilitate change in attitudes through "critical thinking" process. Getting people to think about the benefits of having a small family, etc.

To facilitate the reaching of these objectives, World Education presents the government with a synopsis of various approaches used in other countries with similar problems. The government then makes decisions, although World Education will assist in providing innovative materials and so forth. For example, to avoid problems of too large a book, World Education developed a loose-leaf book whereby the person begins with an empty binder and day by day increases the number of pages—each page covers a different subject area (one concept, one page). Thus village people are not overwhelmed. Each concept is discussed in a context specific to family life problems. The person is then tested not only on the concept but on how it relates to his situation. The stories (concepts) themselves are based on village life and are actually written by writers from the villages. The people can identify with characters and project themselves into all the new concept situations.

Although this particular program was judged successful, a number of problems arose in the changes:

1. Teachers had difficulty accepting the nondoctrinaire approach to teaching.
2. Rural people had the passive attitude that decision making on the whole should be left to others ("You are a university man—tell me what to do") which inhibited "critical thinking."
3. In their zealousness to get the program going (coupled with World Education's policy of noninterference), the host government planners had allowed too little time for educational materials development. For example, they had not allowed for a comprehensive pilot test of materials.
4. Those people who set objectives were not always in close communication with programmers (who implemented and evaluated program) who were in turn not always closely linked with professional writers, who interpreted objectives and put them in stories as they perceived them.
5. Research tools were sophisticated (multiple choice, etc.), whereas peasant farmers were not. They could not understand multiple choice questions, and thus results were meaningless. (These tools were not pretested, again because of time pressure exerted by the government.)

Capacity of the Client System to Accept Change

Reeducational strategies are necessary when effective use of the advocated change requires skills and knowledge the client system does not possess or when there is a lack of understanding concerning the nature of the change (Zaltman and Pinson, 1974). One early strategy used by computer manufacturers to gain acceptance of their products was to provide free training in computer technology to potential and actual customers, thereby removing one barrier to change. Similarly, simple descriptions of how computers worked removed still other barriers erected as a result of an overestimation among potential clients of technically complex industrial and consumer goods. The use of comic books and commercial ads has been an effective means of increasing the usage of contraceptives and enriched foods through graphic illustration of their use. Considerable technical training among farmers in developing countries was necessary before advanced agricultural technology could be implemented. In one program in Thailand, it was determined that functional literacy was required before new farming techniques could be introduced and utilized. The approach used was rather unique. Each farmer-student built his own book at his rate of progress. Each page was a picture with two to three words identifying the prominent features of each picture. Thus a picture of a person working in a rice field would have the word "farmer" written next to the picture of the person. The farmer-student would select the pictures from which he wanted to learn in the order he wished, and those were placed in his own loose-leaf book. Only when he demonstrated mastery of each set of words on the selected picture would he be allowed to make another selection. An interesting feature of this particular technique is that each picture was taken locally; thus the persons displayed were likely to be individuals the farmer knew, the tractor one he may have used or ridden on, the farmland property he could recognize, and so forth.

Studies assessing the effect of management development efforts have indicated some of the characteristics of both individuals and organizations that have led to successful change. Those individuals more satisfied with their job have a more favorable reaction to management development activities (House and Tosi 1963) as do managers who are more risk taking (Baumgartel and Jeanpierre 1972). At the organizational level, when participants perceive that there is top-level support for the training, that the training is relevant to one's job, and that there is going to be an opportunity to apply the training, individuals are most positively oriented to the management development efforts (Carroll and Nash, 1970; Zaltman, Florio, and Sikorski, 1977). Baumgartel and Jeanpierre (1972) found five attributes of organizations related to the successful adoption of management development activities. Management development activities are likely to be successful when: (1) higher management is considerate of lower management, (2) the organization stimulates and approves of innovation and experimentation, (3) the organization is anxious for executives to make use of knowledge gained in management courses, (4) free and open communication occurs among the management group, and (5) top management is willing to spend money for training (Baumgartel and Jeanpierre, 1972, p. 690).

Capacity of the Client System to Sustain Change

Reeducative strategies often work only slowly. If an agency does not possess the resources to sustain a long-term involvement, a reeducative strategy alone is not indicated. This is also true if the agency needs a quick success to attract financial and other resources to help sustain its cause. One of the major barriers to the continuation of organizational development (OD) programs in educational and other institutions is the absence of trainers in OD skills after the initial introduction of such programs. Natural turnover among trained OD leaders recreates the situation (i.e., an absence of OD leaders) that existed prior to the initial training of leaders. The increasing absence of trained leaders from leadership positions eventually causes the program to break down.

Resources Available to the Change Agent

Reeducation to bring about major social change may require very long-term commitment of financial and human resources.

Often, mass media time is a free resource given as a public service by broadcasting stations, although, not uncommonly, the day and the time of day chosen for the broadcast is one at which audience viewing or listening is lowest. A campaign sponsored in part by the United States Agency for International Development and conducted by the former Organization of Central American States (ODECA) relied partially on donated time and partially on purchased time. The donated time was nearly always the least desirable time, whereas the purchased time was usually prime time. Once USAID support stopped, ODECA was financially unable to continue the campaign long enough to achieve any real behavioral change in child-feeding practices. It had considerably heightened sensitivity to problems in child feeding and achieved modest success in increasing accurate knowledge about appropriate child-feeding practices. However, the lack of resources prevented further education efforts designed to convert this new knowledge into enduring change in selected child-feeding practices.

Segmentation of the Target System

The target system is not always a uniform group. Some members may feel negatively disposed toward an innovation or change, others favorably disposed, and still others unaware or uninterested in the existence of the advocated change. Thus several strategies may be necessary more or less at the same time, each directed at a different segment of the target system. There is the possible danger that one strategy may spill over to an unintended segment and cause ill will or backlash. Reeducational strategies are the least likely to cause dysfunctional consequences among segments not intended as the audience for such strategies.

Among the possible ways of segmenting a target system, such as geographically, on the basis of benefit received, demographically, and so forth, is the criterion of role difference. It is necessary to distinguish between the user of the advocated change, the decision maker, the person who implements decisions, the influencer, and the gatekeeper. One change program in education involving curriculum innovation at the senior high school level successfully diagnosed different information needs and circumstances surrounding the introduction of the innovation. The change agent determined that the gatekeeper and user both required a reeducational strategy, although the information provided to each role differed: communication directed at the gatekeeper primarily stressed the demonstrated effectiveness of the curriculum material and only secondarily addressed the issue of its simple use. The degrees of emphasis were reversed in the communications directed to the teacher, who was considered the user in this change effort. Somewhat more persuasive strategies were directed at the decision makers and influencers.

Decision-Making Stage

Reeducative strategies involve, to a greater degree than other strategies, informing the target group about the nature of the problem. Many change agents use reeducative strategies for creating awareness about problems and then follow up with more persuasive strategies concerning particular problem resolution efforts. This in turn may be followed by facilitative strategies to make easier the transition of intentions to change to actual change. In several cases reported to us, power strategies have been used as a last resort if the persuasive strategy-facilitative strategy sequence is not effective.

Reeducative strategies are sometimes used even after the desired change is brought about. Where the change is difficult to discontinue, reeducative strategies may still be used to provide adopters with information to reduce dissonance which, if unchecked, could create negative word-of-mouth communication. Where the adopted change is reversible, that is, is readily discontinued, reeducational strategies provide supportive functions for the adopter, particularly when he does not experience any significant social support for his actions. Agricultural extension agents, salesmen for industrial goods, disseminators of educational innovations, and change agents in several other contexts have all reported the need for reeducative strategies as follow-up subsequent to adoption.

Magnitude of the Change

Reeducative strategies are appropriate when the advocated social change requires extensive information and skills among members of the target group before it can be successfully introduced. Generally reeducation strategies alone are insufficient for accomplishing large-scale change in the short run and where motivation for learning is low. However, reeducation strategies are often desirable as a means of establishing a foundation for more "active" change strategies in the long run and for influencing, in the short run, a small number of people who may be motivated to change but simply lack appropriate information. The early users may serve as demonstrators for those persons who are the objects of a long-term educational program.

Anticipated Resistance to Change

When resistance is prevalent and based on inaccurate information, educational programs are highly desirable. The exact nature of the reeducational strategies will be influenced by the source(s) of misinformation as discussed in Chapter 3. In general, the higher the anticipated level of resistance, the more necessary it is to initiate educational programs well in advance of the actual introduction of the change.

Resistance to change may develop only after the change has been introduced. The longer resistance is allowed to persist and grow before remedial educational programs are launched the more difficult it is to combat resistance. For this reason, program planning should anticipate possible sources of resistance and have counteractive contingency educational programs prepared prior to initiating change.

Nature of the Change

Reeducational strategies are essential when the change involved is a radical departure from past practices and high uncertainty exists within the target system concerning its ability to successfully perform the new practices. This uncertainty is particularly likely when the change is perceived, correctly or incorrectly, to be complex in its implementation. If the relative advantage is unclear or if there is a skepticism that the purported advantage is really achievable, it is desirable to undertake an educational campaign.

Objectives of Change

Reeducational strategies are desirable when the intention of a program is to heighten target group awareness of a problem and a possible solution. Reeducational strategies in themselves are inadequate in bringing about behavioral change unless a strong felt need and a strong motivation to satisfy that need exists. Reeducational strategies generally do not heighten the need or motivation to change. This is still compatible with the ability of educational programs to increase the salience of a problem or issue as well as knowledge about ways to remedy a problem.

An interesting example of a change strategy that was coercive for one group and reeducative for another is provided by the truth-in-lending (TIL) legislation. It was a coercive action from the standpoint of lending institutions and reeducative in intent for consumers. This particular strategy appears to have failed in its objectives. It involves a mismatching of objectives with strategy.

TRUTH-IN-LENDING (TIL). The TIL legislation was intended to increase consumer awareness of interest or credit charges and rates concerning consumer goods. However, this was not the ultimate goal. The ultimate goal was to stimulate or increase cost-effective comparative shopping for interest rates, and perhaps affect decisions about whether to enter credit transactions at all. Thus the goals and strategy of TIL were (1) to increase awareness of certain information, (2) to produce more effective decisions by consumers in selecting from among competing products, and (3) to provide information that would help consumers decide if a particular product should be purchased at all.

The Federal Reserve Board has been evaluating the impact of the legislation. Their most empirical studies are those conducted prior to the enactment of the legislation and 15 months after the law went into effect. Unfortunately, the methodologies used and the reporting of the data contain flaws that are sufficiently serious to render both studies nearly useless, particularly for before-and-after measures of effectiveness. Most important, the surveys to date have not measured rate awareness among consumers at the time the transaction occurred, which is the critical point at which TIL is to have a direct impact on buying behavior. Apparently, no consumer surveys by the Federal Reserve Board have been undertaken since 1970. The FTC conducted a survey of creditors in 1971 that indicated that only 14% of the creditors surveyed were not in substantial compliance with the law. The FRB reported in 1973 that more than 36% of the creditors surveyed did not comply with the oral disclosure requirements. This figure rose to nearly 50% in 1974, indicating that compliance seemed to be declining.

Angell (1971) attempted to ascertain the impact on consumers in a survey on bank loan officers 9 months after the TIL legislation was passed. Relying on bank officials for such data is a questionable approach, and therefore the following findings of Angell about consumers are presented with a note of caution:

1. Eighty-two percent of all officials indicated that borrowers exhibited no interest in credit information.
2. Thirty-six percent of the officials reported that borrowers understand that rates are unchanged but simply restated.
3. Forty-six percent of the officials reported that borrowers had to be informed that legislation did not change rates but only the manner in which rates were stated.

Day and Brandt (1974) attempted to examine the effect of information disclosure on knowledge of annual percentage rates and dollar finance charges and on credit purchase behavior over time. This study, too, has methodological flaws, but the authors' summary conclusion is of interest: "What is clear, however, is that it is not enough to simply provide consumers with more information. This is simply the first step in a major educational task of getting consumers to understand the information, and persuading them to use it" (p. 31). An important finding mentioned by Day and Brandt supporting this statement is that mere exposure to information does not lead to learning.

Probably the most thorough review of interest rate awareness in connection with TIL is an article by law professor William C. Whitford (1973), who concludes that:

> Despite the difficulties in interpreting their significance, the surveys pretty clearly suggest that in the first year or so of truth-in-lending there was only modest improvement in consumer awareness of prevailing annual percentage rates. It is more difficult to measure improvement in consumer knowledge of specific annual percentage rates at the time a transaction is concluded, but the available evidence implies that any improvement has also been at best modest. Moreover, any improvement in awareness has been concentrated in upper income groups.

Whitford also cites evidence that consumers misused information disclosure in accordance with TIL. The annual percentage rate was used by consumers in such a manner as to indicate a finance charge roughly twice the proper amount.

TIL also appears to have had little, if any, impact on comparative interest rate shopping (Day and Brandt, 1972, cited in Whitford, 1973). The Day and Brandt study, as reported in their 1972 monograph, "was unable to determine conclusively whether the truth-in-lending helped stimulate whatever credit shopping behavior was found or aided credit shoppers in reaching more intelligent and cost-effective credit decisions."

Control Possibilities

Reeducational strategies involve the least degree of control by change agents over clients. Heavy emphasis is placed on the assumption that man is rational and will himself initiate appropriate action once he has digested the relevant facts. This ignores the problem of barriers to action and the problem that what constitutes rational behavior from the perspective of the client or change target is often not viewed by the change agent as rational client behavior.

ILLUSTRATIONS OF REEDUCATIVE STRATEGIES*

Let us turn now to a discussion of three cases involving reeducative strategies. A reeducative strategy has been defined as one in which the relatively unbiased presentation of fact is intended to provide a rational justification for action. It is the word unbiased that distinguishes this strategy from persuasive ones. However, in practice, the line between biased and unbiased presentations may be a fine one. In fact, it has been argued that it is impossible to develop an unbiased presentation.

Reeducative strategies assume that individuals are rational beings, capable of discerning fact and adjusting their behavior to respond to such information. A reeducative change strategy that used survey feedback as a basis for diagnosing and remedying problems is presented below. The case illustrates the importance of having relatively objective information as the basis for problem-solving efforts among peers and between those in superior-subordinate roles.

The Organizational Development Program

This organizational development program involved a reeducative change strategy tested in a Chicago area school district. This research was supported in part by a grant from the National Institute of Education. The program was designed to provide administrators and faculty with information about their problems, allowing them to diagnose and ultimately remedy those problems. The procedure entailed four stages. First, survey data on work attitudes were collected from faculty and principals. Questionnaire

---

* Alice Tybout reported these cases on the basis of interviews with various change agents.

items focused on issues such as task relations between school and district personnel, special services, professional and nonprofessional work loads, materials and equipment, student needs, etc. These data were given to the school personnel to provide an objective basis for problem and need identification.

Next, task-oriented problem-solving meetings were held by principals and teachers for problem analysis and solution generation. These groups were encouraged to generate and analyze a number of alternatives prior to selecting a final solution. Once a solution was determined, a timetable was developed for its implementation.

Following the problem-solving meetings, a program was undertaken to implement the suggested solutions. Teams of both faculty and administrators and special documents were used to facilitate the communication, sanctioning, and implementation of recommended changes.

Finally, the program included a self-monitoring feature that allowed each group of teachers and principals to evaluate the success of the program.

RESULTS.   The results of the research indicate that the program made its greatest impact by improving communication within the schools. In the opinion of both teachers and principals, the project improved faculty morale, cohesiveness, and problem-solving capabilities. Statistical analysis of results demonstrates that teachers' attitudes toward their work environment improved significantly. The greatest improvements were found in teachers' attitudes toward (1) performance and development (their opportunities for professional growth), (2) their voice in the educational program of their schools, and (3) their relations with their principals. The program also resulted in improved teacher attitudes toward the educational effectiveness of their schools, the procedures used for evaluating students, and school-community relations. The program did not have a very great effect on teachers' attitudes toward such things as buildings and facilities.

The OD program was very successful in improving the problem-solving and change capabilities of the participating schools. Faculty-initiated changes occured in most of the schools involved in the previous program: a teacher appraisal and evaluation program was developed and implemented; a teacher-administration committee developed and monitored a modular scheduling system; teachers assumed responsibility for making assignments regarding playground, lunch, bus duty, etc.; student evaluation procedures were improved; parent-teacher conferences were restructured and extended. According to principals and program leaders, the program also facilitated the implementation of administratively-initiated changes, for example, individualized reading.

DISCUSSION. The above case of the organizational development program is an example of successful utilization of a reeducative strategy. The researchers identified a need for better problem-solving and change capabilities, as well as greater work-related satisfaction among faculty and administrators. Since one of the primary objectives was the development of change target capabilities for problem solving, a reeducational strategy was most appropriate. Furthermore, Argyris has stated that reeducative strategies are appropriate when inconsistencies exist in the target's behavior that, when made salient, are likely to cause discomfort and induce change. The lack of time constraints and the benefits of increased commitment resulting from the change target participating in all phases of the change process also favored the use of a reeducational strategy.

The approach taken in this case clearly meets the criterion for an educative or reeducative strategy. The information given to the change target was the target members' own uninterpreted attitudinal responses to a questionnaire. The change agents did not specify specific problems or solutions, but instead allowed the members of the change target to study the data and arrive at their own interpretation of problem areas and solutions. The change agents merely provided the structure and method for identifying and solving problems.

Several observations about the use of a reeducative change strategy can be made. First, as mentioned earlier, the reeducative process is often a lengthy one and may require long-term involvement on the part of the change agent. The organizational development program took several years to complete, and its long-term impact is still being monitored. Second, in spite of the effort of the change agent to remain neutral and allow the target to identify and solve its own problems, the change target may attempt to force the change agent to provide the problems and solutions. The very presence of outside researchers who structure the situation may lead the target members to rely on the change agent for questions and answers rather than develop their own. If the change agent gave in to these efforts, it would defeat the major purpose of the strategy, that is, to increase the problem-solving ability of the target. Remaining neutral and uninvolved is one of the more difficult, important, and perhaps impossible tasks for the change agent employing a reeducative strategy.

Finally, a major benefit of a reeducative strategy is the more complete understanding of the change by the target and therefore the greater ability to sustain the change after the change agent has withdrawn.

Next, we discuss an instance in which a reeducative strategy was essential to successful social change, but where such a strategy was not employed and the change effort ultimately failed.

Rural Water Supply Program *

In the late 1950s, the rural water supply in El Salvador was irregular and unsatisfactory. The need for an improved system was acute and widely recognized. To meet this need, the Ministry of Community Development decided to undertake the building of water supply systems in rural villages. The problem then became how to design a water system, communicate the government's intention to build it to the villages, and build and successfully implement the system.

Two alternative courses of action were considered by the Ministry. Villages to receive the water system could be chosen according to some criteria (limited resources prohibited supplying the system to all villages), and the government could then send in experts and the necessary materials to design and build a system that would be technically feasible and best in each village. Alternatively, they could seek greater involvement in the project from residents of each village and design a water system that conformed to their desires, subject only to accessibility to water. It was felt that involvement of the villagers would increase their commitment to the adoption of the new system, and therefore the second alternative was chosen. Technical assistance and material resources were provided by the Ministry.

Specifically, the effort began with advertisements and announcements placed in newspapers and on the radio to communicate to the villagers the government's intention to build the water supply systems. Interested villagers were asked to request assistance. Engineers and sanitarians were sent to approximately half of the villages requesting the system; the remaining villages could not be served because of limited resources. These engineers and sanitarians explained the program in detail, solicited community support by having villagers vote on the system, and finally enlisted villagers' assistance in locating the wells and communicating village requirements to the technical team. When the planning was complete, the water supply system was built with technical and material resources donated by the Ministry. Villagers were hired by the government to perform the necessary manual labor. No education of villagers regarding the system and its maintenance was undertaken.

RESULTS.   Two years after the systems were completed, 75% were inoperative because of poor maintenance by local residents. Five years later, only 5% were operative.

---

* This case was developed on the basis of an interview conducted by Ruby Roy with an official of USAID.

DISCUSSION. When the program was intiated, it was designed to fill a very important need in the villages. The need was widely recognized, and the solution to it was technically feasible. The problem was defined essentially in terms of implementation, with emphasis on the design and construction of the water supply system. By involving the villagers through an active vote on the system, as well as in site selection, the program designers had assumed that they met the local needs.

However, little emphasis was placed on educating the villagers to maintain the system. Repairs were needed continuously for which expertise, labor, and materials had to be provided by the villagers themselves. They were not educated on the need for maintenance. There was insufficient community involvement, and the villagers depended entirely on the Ministry for resources and expertise. Behavioral commitment to the system that might have been generated through the use of voluntary (unpaid) labor and village resources was not established. Such a commitment might have ensured sustained interest in and maintenance of the program. (A similar situation occurred in Laos, where latrines were accidentally built on the grounds of a Buddist shrine. The latrines built on village property quickly deteriorated, while those built on the property of the shrine were very well maintained by the villagers.) However, the Ministry learned from these mistakes and avoided these problems in their Mobile Rural Health Program.

Mobile Rural Health Program *

The villagers in El Salvador suffered from diseases and poor health. Sanitation and hygiene facilities were highly unsatisfactory and access to medical expertise was very limited. The Ministry of Community Development therefore embarked on a Rural Health Program in 1963.

The program objectives were essentially twofold: to create awareness of health needs among rural dwellers and to provide health services. Both the curative and preventive aspects of health care were incorporated into the program. From their past experience in the Rural Water Supply program, the government officials in El Salvador had become convinced of the need to educate and involve the community in the program to a maximum degree.

THE PROGRAM. The approach selected was a mobile health service program. A mobile health team was drawn up consisting of a doctor, a

* This case was developed on the basis of an interview conducted by Ruby Roy with an official of USAID.

nurse, and a sanitarian. The team visited each community weekly; illnesses were diagnosed and treated in the medical outpost, and medicines were dispensed. More serious cases were removed to neighboring hospitals.

The salaries and traveling expenses of the medical team were paid by the government. Medicines and other requirements were also provided by the government authorities. The villagers were charged a nominal fee (to enhance commitment) for the medical services; the fee was decided on by a committee elected by the villagers themselves. The revenue generated from these charges was available to the village committee to spend as they desired. This group incentive was intended in part to overcome resistance by individuals.

In order to have a medical team visit a community, the villagers had to provide a building to serve as a medical outpost. Daily maintenance and supervision of the outpost was made the responsibility of the elected committee.

PROGRAM IMPLEMENTATION. A primary concern of the Ministry was creating sufficient community involvement and understanding about the program so that the difficulties encountered with the rural water system could be avoided. The following steps were taken in an intensive effort to educate and involve the villagers. First, teams of health educators visited rural communities and discussed the mobile health program with community leaders. Next, community meetings were arranged at which movies on health and sanitation were shown and talks were given by doctors, nurses, and community leaders.

Following these meetings, villagers were asked to publicly indicate approval of the program and elect a committee to develop a project proposal. This elected committee prepared a proposal detailing the procedures for (1) the selection of the outpost site, (2) the generation of resources from the local community for the construction of the health and administrative facilities, and (3) the levying of charges for services rendered by the medical team to the villagers.

After the proposal was approved, group discussions on health were initiated by health educators to generate awareness of health needs and sources of prevention and treatment. Sanitarians visited each house in the village to advise and educate on matters of hygiene and sanitation. When the outpost and other facilities were made available, medical teams were routed through these villages, and the committees were made responsible for the collection and administration of funds and other resources.

Finally, posters and monthly bulletins were used to inform the committee members on matters pertaining to the use and administration of health facilities. For special campaigns such as immunization programs, media such as comic books were used to educate the villagers.

RESULTS. The program was very successful. Nearly all the villages visited by the health teams responded very favorably to the mobile health program. Where suitable sites were not already available, villages built their outposts, donating money and their own labor. It is estimated that over 60% of the funds required for building these facilities were generated by the villagers themselves.

Once the medical teams began their weekly services, the villagers made satisfactory use of the facilities. On the average, about $100 per month was received as fees. Several village committees used these funds to finance an auxiliary nurse stationed permanently in the village; others used the funds to build water supply systems.

DISCUSSION. What distinguishes the rural water system case from the mobile health program case? Many of the circumstances were similar in the two cases. In both instances there was sufficient time to allow use of a reeducative strategy. Furthermore, conditions that might favor a persuasive rather than reeducative strategy, such as factual ambiguity, high ego involvement, and active opposition, were not present. Finally, reeducative strategies were called for because effective use of the changes in both instances required skills and knowledge that the change target did not possess.

The differences in the two instances are twofold. First, the Ministry simply did not recognize the need for education in the rural water supply case. The Ministry was aware of the need for some form of commitment on the part of the change target; however, the commitment they sought was relatively passive and did not ensure understanding of the workings and maintenance of the rural water system. This oversight was avoided in the mobile health program at least partially because of the diagnosis of the failure in the rural water supply program. Second, there was a basic difference in the extent of the Ministry's involvement in the two cases. The rural water supply system was a "one-shot" change program for which external resources were deployed only once. The mobile health program provided continuous service and therefore may have facilitated greater opportunity for community education and involvement. Thus a very important point is the need for change agents to emphasize the *continuation* of an innovation by a change target after the initial implementation or adoption. This also means that change agencies should devise reward systems that motivate their agents to seek continuation of an innovation by change targets.

This chapter can be summarized by the following principles:

1. Reeducative strategies are feasible, other things being equal, when change does not have to be immediate.

2. Reeducative strategies can be effective in providing the foundation for future action by establishing an awareness of a need (general or specific) for change. It may be desirable not to mention a specific change if it is potentially controversial until a clear need has been established.

3. The stronger the degree of commitment a change requires to be effective, the less impactful reeducational strategies alone will be.

4. Reeducational strategies are effective in immunizing people against appeals to resist change or to revert back to the previous situation.

5. Reeducational strategies can be effective in (1) connecting causes with symptoms, (2) creating awareness of a problem, and (3) establishing that a known problem can be resolved.

6. Reeducational strategies are necessary when the use of the advocated change requires skills and knowledge the client system does not possess.

7. When a change agency does not possess the resources to sustain a needed long-term involvement, a reeducative strategy alone is not indicated.

8. Reeducational strategies are particularly useful at the awareness stage of the adoption process.

9. Generally, reeducational strategies alone are insufficient for accomplishing large-scale change in the short run, particularly where motivation to change is low relative to the magnitude of the change required.

10. The higher the anticipated level of resistance, the more necessary it is to initiate educational programs well in advance of the actual introduction of the change.

11. Reeducational strategies are essential when the change involves a radical departure from past practices.

12. Reeducational strategies are feasible when little control over the client is necessary and the rationale for change is clearly presented in terms of the client's perspective.

Persuasive Strategies

Persuasive strategies are "strategies which attempt to bring about change partly through bias in the manner in which a message is structured and presented. They attempt to create change by reasoning, urging, and inducement. Persuasive strategies can be based in rational appeal and can reflect facts accurately or be totally false" (Zaltman et al., 1972). Persuasive strategies are commonplace. Most advertising by profit and nonprofit agencies encountered so frequently is essentially persuasive in nature. Similarly, much everyday interpersonal communication among friends, business associates, and others has a strong persuasive content. Quite often, neither the communicator nor receiver is explicitly aware of his use of or involvement with a persuasive strategy or tactic. This is discussed in more detail later. The discussion immediately following focuses on the various criteria to consider concerning the effective use of a persuasive strategy.

CONSIDERATIONS FOR SELECTING PERSUASIVE STRATEGIES

Awareness of the Change Target

If it is desirable for a change to be introduced in a low-key manner, perhaps because of its potential for stimulating unwanted conflict, an explicit persuasive strategy should not be used.* If a persuasive strategy can be developed that is not obvious, it indeed may be very effective in creating an openness to change or at least a receptivity to information about change. Again, a good example of latent persuasive messages comes from the family planning area. Father Cornelius Lagerway, a Roman Catholic priest working in the Philippines, prepared a commercial entertainment film whose main story was unrelated to family planning. Numerous well-planned latent messages advocating family planning were structured into the film but not to a degree to which a relatively unsophisticated, youthful audience would perceive the family planning messages to be the main concern of the film. The film used famous Filipino actors and actresses who were assigned to specific roles on the basis of careful research into the popular images of the players. The specific themes and their development were well researched and pretested in the best marketing research tradition. The film played in commercial movie houses and was financially successful. There appeared to be no public reaction to the latent family planning themes. Unfortunately,

---

* Conflict and resistance may be desirable (as indicated elsewhere in this book). In this case, a persuasive strategy would be employed openly to the extent necessary to reach the optimal levels of conflict and resistance.

no information is presently available about the impact of the family planning appeals on audiences. There is an ethical issue here that the reader may have already sensed. The issue concerns the appropriateness of attracting an audience on the basis of one theme, the major manifest plot of the film, without alerting them in advance to the latent content and keeping the latent content at a level at which it would be received by the audience but not so salient as to (immediately at least) evoke any defenses people may have about accepting family planning as a positive practice.

Degree of Commitment

The lower the degree of commitment to change among the target system, the greater the need for persuasive strategies. Commitment can be enhanced by stressing, realistically or unrealistically, either the possible benefits of change and/or the costs of not changing. Programs are underway in the family planning area that give rewards to couples for practicing contraception and punish families who have more than three children by increasing the proportion of the medical expenses that must be paid by the family (IGCU, 1973). Financial incentives have had some limited success in keeping gangs as a whole and their members individually from engaging in illegal activities.

Persuasion can be used to increase both attitudinal and behavioral commitment. Structured group discussion techniques to reinforce the content of a radio broadcast or lecture have been used effectively in a variety of settings to change and strengthen attitudes and behavior. The People's Republic of China is probably the major practitioner linking mass media and interpersonal channels in media forums. "Approximately 60 percent of the adult Chinese population, over 250 million people, regularly participate in study groups where print material is read and discussed" (Rogers, 1973, p. 281). Rogers cites a number of studies that lead him to conclude that mass media channels are most effective, at least with regard to family planning activities, when they are linked with interpersonal communication channels such as media forums. Although the evidence is somewhat sketchy, it appears that radio forums are at least as effective as any other media forum such as newspapers, magazines, etc. One interesting way of enhancing commitment is the use of coupons to attract buyers for a new product. At first, a "cents-off" coupon, that is, a coupon that entitles the bearer to (say) a twelve-cent discount for a new frozen item, is attached to a frequently purchased product such as bread. The consumer may then use the coupon to buy the new product at a discount. This establishes a first-time use. Later, when the consumer buys the frozen item, he finds another coupon attached, this time to the product itself. This second coupon, which may entitle the buyer to an eight-cent discount, is intended to persuade buyers to purchase the product a second time, thus taking a perhaps hesitant first step toward establishing a commitment (preference) to this particular product (this is known as brand loyalty). The cents-off coupon may be terminated or lowered again depending on the estimated time required to establish brand loyalty for this product. (For a concise discussion of consumer and sales incentives, see Turner, Jr., 1973, and Stein and Miller, 1973).

Felt Needs

A reeducative/persuasive strategy is indicated when a problem either is not recognized, that is, specific symptoms are not associated with specific causes, or not assigned a high priority for remedial action. Persuasive strategies have been found effective in helping to influence first-time mothers to breastfeed their children and in getting mothers to breastfeed their second child after bottle feeding their first child (Lee, 1975). Persuasive strategies have had some effect on increasing the salience of driver safety and presumably the need for safer driving (Bryk, 1973). On the other hand, persuasive strategies are found ineffective in increasing the salience of mining safety among coal mine supervisors (Bryk and Schupack, 1972).

Often, a felt need exists along with an appropriate means of satisfying that need, but the person experiencing the need is unable to connect the need with the solution. The person or group may be unaware of the solution or perhaps unable to see how the solution would apply to its problem. In the latter case, the change agent must perform the creative function of connecting two items previously unrelated in the mind of the target person or group. This generally calls for persuasive strategies, although reeducative strategies would be appropriate as well.

Capacity of the Client System to Accept Change

If the client does not have the capacity—financial or otherwise—to accept change, a persuasive strategy is not useful. In many community development programs, for example, there are virtually no real resources within the community, and a persuasive strategy would have no positive effect. Persuasive strategies may be used, however, to convince an organization to reallocate resources from one context to another where the need for change may be especially great. (The reader is urged to consult Zimbardo, 1972, p. 6, for a good discussion of how a target system's capacity to change can be enlarged through judicious use of persuasive strategies.)

Capacity of the Client System to Sustain Change

The use of a persuasive strategy to convince the client system to sustain a change is appropriate or at least feasible when the client system has the capacity to sustain change but is reluctant to allocate resources or capabilities for the continuation of the change. However, when the capacity for continuation is not present, persuasive strategies are not appropriate. For example, many new products of both a consumer and industrial nature require "product champions." Product champions are persons or groups who actively work to gain a commitment by the firm to add the product to their present product line. The product may or may not have been developed initially by the firm. Product champions must work hard and persuasively to have a share of limited resources committed to their project. It is not uncommon for a product champion to be switched to other duties, with no other staff members assuming the role of product champion for the product or service involved. Thus the product or service typically will be brushed aside by other goods or services with strong positive support and which are competing for the same resources.

Resources Available to the Change Agent

The fewer the resources available to the change agent that are attractive and valued by the target system, the more he must rely on nonpower strategies (see Scott and Scott, 1971, and King and McGinnies, 1972). The client-valued resources are a source of power over the client or target system. Also, the fewer the resources a change agent has with which to initiate and sustain a change, the more quickly he must act to bring about change. This suggests the use of persuasive strategies, as is evident in the discussion of time as a criterion in strategy selection.

One type of resource that is an exception to the rule-of-thumb presented in the preceding paragraph is knowledge. One of the authors was accused of unethical behavior when he suggested the use of lottery tickets as incentives to improve and maintain voluntary attendance in a nutrition education class in a Central American country. The upset reactor to the suggestion claimed that such action took advantage of known human frailties. Without debating the ethics of the use of knowledge about selected behaviors or whether the purchase of lottery tickets is a frailty, the decision to use lottery tickets as an incentive did indeed represent a persuasive tactic that was unconnected functionally with nutrition behavior. In general, the more knowledge a change agent has about a target system, the more persuasive he can be.

Segmentation of the Target System

Not all members of the target system will be in the same stage at the same time. Additionally, different members of the target system will be differentially disposed to the advocated change. Finally, different people, regardless of their decision-making stage and disposition to accept a given change, may be differentially responsive to different kinds of strategies. Thus it is desirable to determine whether members of the target system can be clustered or segmented into subgroups that are internally homogeneous with respect to an important criterion and yet differ from one another in terms of that same criterion. It has been demonstrated, for example, that teachers can be segmented in terms of their general openness to change, and that persuasive strategies are more effective for those teachers who are less open to change (Kasulis, 1975).

There is an important problem in segmenting strategies if the target system cannot be readily isolated in terms of the persuasive treatment. A community medical clinic in Athens, Georgia used several fear appeals in local media to encourage area residents to be examined for venereal disease. A modest increase occurred in the number of persons seeking such examinations. Most interesting, however, was the fact that there was a decrease in the kinds of people who typically did not use the clinic. What apparently occurred was a differential response to the fear appeal. It was effective with one group, but caused another group to discount the severity of the problem as a defense against the threatening messages. Unfortunately, the campaign could not prevent the negative responders from being exposed to the fear appeals.

Decision-Making Stage

If the target system is well into the decision-making process, perhaps at the evaluation stage or legitimation stage, a persuasive strategy is appropriate. In this instance, an attempt would be made deliberately to bias the information on which the evaluation is based or to use visible and highly credible or legitimate sources of information to disseminate information with either a persuasive or nonpersuasive structure and content (Lee, 1975). Commercial advertising often uses this approach: reeducative strategies at the awareness stage, persuasive strategies at the evaluative stage, and facilitative strategies at the adoption stage. This is very evident in the practice of comparative advertising in which alternative brands are identified clearly in whatever product test is used in the commercial. There have been a few instances in which one firm has demonstrated in a commercial how its product is superior to specific competitive brands in a test of water-absorbing capability, whereas, in another commercial, one of the competitive firms is able to demonstrate how its product is superior in the same way to other brands, including that of the first firm mentioned.

Magnitude of Change

The larger the change involved, the more likely it is to be perceived as risky and disruptive. Also, the larger the change involved, the greater the likelihood of more persons being affected and hence the larger the number of persons or groups who may actively resist the change. As perceived risk and other factors increase in importance, so does the need for persuasive strategies. The very magnitude of a significant change may prevent any trial adoption either on a limited scale or for a limited time period. Thus the use of a rational or reeducative strategy such as a demonstration in the target system is not possible.

Incentives are an example of a persuasive tactic in which something of value is given to an adopter and/or change agent to engage in prescribed behaviors. Incentives may take the form of money or specific other commodities and services such as clothing, kitchenware, medical services, education, and so forth. This tactic is considered persuasive because it involves a deliberate restructuring of the total change offering without enhancing the performance of the change.

Incentives of various forms have been suggested as a means of helping to alter nutrition practices among low-income and low-education consumer groups who lack the motivation to undertake such a major change (Zaltman, 1975). Incentives may be classified in the following manner. First, there are adopter versus diffuser incentives. Adopter incentives involve rewarding the consumer who adopts the desired change, for example, adults who enroll in (adopt) a nutrition education program (the change). Diffuser incentives are provided to persons who recruit attendees for the program. Diffusers are essentially salespersons. A second classification of incentives is immediate versus deferred incentives. For example, should incentives be given only at the time adults enroll for the program? Should they be given for each meeting attended at the conclusion of each meeting? Or, should the incentives be given only after the nutrient education program is completed? Another distinction concerns individual versus group incentives. Should enrollees be given their incentive, say, at the end of the program based only on their own attendance, or should the incentives provided be dependent on the group's overall attendance? A variation is when the group is provided with an incentive shared by all versus an incentive that can be used privately by any one individual without making it simultaneously available to all group members. A fourth distinction is between positive and negative incentives. Should people be rewarded for performing an activity and penalized if they do not? For example, extra food stamps could be provided to persons using food stamps if they provide evidence of attendance at nutrient education classes, and they might receive less than the normal ration of food stamps if they do not participate in a locally held nutrient education class. Incentives also vary in monetary and nonmonetary terms. Should money be given or should some nonfinancial commodity such as food, health care books, health magazine subscriptions, kitchen ware, or nonfood or nonhealth-related products or services?

The relative merits of each incentive vary greatly according to the consumer's social group and general social setting, the nature of the advocated change, and many other factors. The basic idea or concept of incentives, which basically go beyond relying on the target group's self-interest in improving their well being, is controversial but quite possibly highly effective.

Time

When a change must be instituted quickly, even though the target group is reluctant to change, a power approach may be necessary. However, when such an approach cannot be maintained for very long, persuasive strategies are indicated. Persuasive strategies would be used to convince the target group as quickly as possible that a real need exists and that it is essential to maintain the change voluntarily. This holds regardless of whether power strategies are used. Persuasive strategies may be desirable when resistance to the advocated change is expressed and a demonstration of the change is not feasible within the period in which it must be implemented.

Nature of the Change

A persuasive strategy is indicated when the change or innovation is risky, not amenable to limited or small-scale trial, is technically complex, has a strong potential impact on interpersonal relations, has no clear or highly salient relative advantage, and originates from a poorly regarded source. Persuasive tactics are necessary to compensate for these disadvantages. The tactics should be designed to counter these innovation attributes. For example, simplicity of use is stressed for technically complex equipment; improvement of interpersonal relations is stressed for changes that could alter existing social interaction, as is often the case in organizational development programs, and so forth. Care should be exercised not to underestimate the problems or exaggerate the advantages of a change. If expectations are not met, unfavorable word-of-mouth communication can occur very quickly and overshadow some very substantial real benefits associated with the change.

Anticipated Resistance to Change

Persuasive strategies are potentially very effective in combatting resistance to change. The use of two-sided arguments by a change advocate, for example, may be an effective persuasive tactic. Presenting the best arguments for change first and last and placing the weakest arguments in the middle of a list can enhance retention of the best reasons to accept change. Present day comparative advertising that openly compares an advocated product with its competitor is an example of a persuasive strategy that uses a two-sided argument but involves a clear conclusion about which product is best.

Persuasive strategies may be counterproductive and create resistance. Marketers have found, for example, that too many claims in a commercial about the virtues of a product or service, even if accurate, can create a skeptical audience, an audience somewhat jaded perhaps because of the failure of other products to live up to their promises or the claims made about them. It is sometimes better to let some advantages come as a surprise to the consumer or change target. Thus the optimal claims for an innovation may be well below the maximum number of claims. It is less clear whether the optimal mix should involve all the best claims or only some subset of them.

It is necessary to distinguish between resistance apparent at the time a change is introduced and resistance that develops later as the change is in the process of implementation and routinization. The techniques cited two paragraphs above are particularly relevant for countering resistance that may be present at the time the change is to be introduced. Tactics that have been used to combat resistance developing after the change has been introduced include the discrediting of the sources of resistance and the use of testimonials by average or typical users of the change. Another technique involves the stressing of beneficial second-order consequences and desirable unanticipated consequences. Second-order consequences are those consequences resulting from intended effects of the change. For example, an intended effect of a series of group discussions between teachers and parents in a Los Angeles elementary school was to reduce considerable parent hostility related to language difficulties of the children and a new method of coping with these difficulties. The children were predominantly Black and teachers predominantly white. Thus two different language standards were used that produced conflict in the children and parental anger toward teachers. The teachers too were disturbed by the parents. The group sessions were effective; parental hostility to the teachers and the new language learning technique decreased considerably. Also, student learning increased, and general classroom behabior improved as a result of the lessening of tension between teachers and parents. These facts—parents approved and children behaved better—were used to justify continuation of the new technique even though some teachers felt that there were still better pedagogical alternatives.

ILLUSTRATIONS OF PERSUASIVE STRATEGIES

Two examples of persuasive strategies are presented. Persuasive strategies have been defined as strategies that attempt to bring about a particular change through bias in the manner in which a message is structured and presented. Such strategies may attempt to create change by reasoning, urging, and inducement, and they may be based on rational and/or emotional appeals.

Philippine Family Planning Campaign*

In an effort to increase the incidence of family planning among poor Filipinos, a persuasive campaign was developed in the form of a radio soap opera. Several different serials were developed and played on the radio over a period of months. The following is a brief description of one of these serials entitled "If You Really Care."

BACKGROUND. The story revolves around Madias, a fisherman, and his wife of eight years, Linda. The couple has six children and is having difficulty providing for all of them; however, they do not practice family planning. Madias comes from a family of 14 children and is greatly influenced by his mother, who opposes any form of birth control. The story unfolds in a five-part series.

FIRST INSTALLMENT. The story begins with a noisy family scene. Active children are playing and quarreling in the background, and a tired mother is heard complaining and scolding. The sequence emphasizes the problems of a large family, focusing primarily on the economic difficulties. When the mother, Linda, suggests family planning to avoid additional children, her husband argues against it, citing traditional arguments such as that family planning would be going against God's will. In the discussion, the arguments opposing family planning dominate. The installment ends with the listener being urged to tune in the next day to see what happens.

---

* This case was developed on the basis of a report summarizing the campaign written by Ruby Roy. The radio series was developed by Father Cornelius Lagerway.

SECOND INSTALLMENT. Linda's younger, single sister, Ada, comes to visit Linda from the city and brings with her a basket of fruit for Linda's family. The hungry behavior of the children in response to the fruit and Linda's obvious financial difficulties trigger a sisterly discussion of family planning. Ada provides support for Linda's interest in family planning and advances reasons why it is not a sin to adopt family planning. However, in a later encounter with her mother-in-law, Linda again meets with strong opposition to family planning.

Linda is presented as being so afraid of becoming pregnant that she cannot be responsive to her husband. The installment ends with Linda leaving her husband and taking her children to stay with her at her parents' home in the city.

THIRD INSTALLMENT. After Linda leaves him, the distraught Madias visits his mother. His mother criticizes Linda for not adhering to the traditional role of women and attributes Linda's behavior to her city upbringing and her stubbornness. Madias defends his wife, stating that she simply was trying to keep their present children, and she just did not want to have more children than they could afford. Linda is also troubled by leaving Madias and attempts to explain her behavior to Ada by stating that it appeared to be the only way to avoid becoming pregnant.

Then, Linda's mother-in-law visits Linda in the city to inform her that Madias is very sick. This provides Linda with a face-saving opportunity to return and nurse her husband back to health.

The scene ends with Linda fainting. It is made obvious that some time had elapsed since her return home, and she is back to performing her normal household duties.

FOURTH INSTALLMENT. The scene reopens with the fainted Linda being attended by her husband and her mother-in-law. She recovers from her fainting spell only to discover that she is pregnant. Deeply distressed by this news, Linda leaves her children with her mother-in-law and goes to the city to seek an abortion. Accompanied by Ada, Linda visits a quack midwife who had performed an abortion for her cousin. The midwife is portrayed as an evil person, a cackling old witch who shows little concern for Linda. Grave opposition is voiced by Ada to abortion in general and the midwife in particular; however, she cannot offer any alternative solutions. Despite her fright, Linda determinedly has an abortion. The scene ends with Linda bleeding and Ada screaming.

FIFTH INSTALLMENT. The scene opens with Linda sick in her parents' home and greatly troubled by her sinful act. Linda's father goes to visit Madias, who accompanies him back to the city. Her father tells

Madias that Linda has lost the child and that she does not want more children. He counters Madias' objections to family planning. Arguing that, while children are God's gifts, man's rational faculties have also been given by God and therefore employing them to limit family size is not against nature. He adds that killing life through an abortion is far more sinful than family planning.

Madias becomes angry at himself for not listening to Linda earlier. He and Linda decide to learn from their past mistakes and practice family planning so that they may give the children they already have a better life.

The series ends with the commentator stating that couples like Madias and Linda can be helped through family planning and that competent nurses and doctors can help couples space children according to their own conscience.

DISCUSSION. This case is an example of a persuasive strategy. The information is presented in a highly biased manner; it favors family planning and characterizes alternatives to family planning, such as large families or abortion, very negatively. Persuasive strategies exert greater pressure toward a particular attitude and/or behavior change than educative strategies and less pressure than power or coercive strategies.

An educative strategy would not have been adequate in this situation, since it might not have led to any action. The change agent desired adoption of a particular solution to the problem of large families.

A power strategy would also have been inappropriate for several reasons. First, for ethical reasons, it would have been unacceptable to force the target to adopt family planning. Second, power strategies would result in compliant behavior that would only be likely to be maintained under surveillance. In this case, surveillance would be impractical and extremely expensive. As a result, a persuasive strategy would seem to be most feasible.

Persuasive strategies provide the target with information and may attempt to manipulate various source and message cues to facilitate acceptance of a desired position or rejection of an undesirable one. Let us look at the utilization of source and message cues in the Philippine family planning case.

The first step in developing a persuasive communication is to identify the audience or target's initial attitude toward the issue. In this case, the initial attitude of the target, poor Filipino couples with several children and no use of family planning methods, was one of opposition to family planning primarily because of tradition and religious reasons (i.e., children are God's will). When the audience initially opposes the issue, members of the audience are likely to resist persuasion by counterarguing or generating arguments opposing the issue as they listen to the message. To overcome this negative attitude, a combination of source and message strategies may be employed that inhibit counterarguing. First, we look at the manipulation of source credibility.

When the target audience initially opposes the position advocated, a high credibility source may increase persuasion. If a highly credible source is employed, the target may find it more difficult to simply dismiss the message. Three dimensions of source credibility may be manipulated—attractiveness or similarity, trustworthiness, and expertise.

The characters in the message have been carefully selected to appear similar to the target audience. Linda, with her problems in caring for a large family on little ·money, may be viewed as similar to female members of the target audience. Studies have demonstrated that, when a source is similar to the audience and when such similarities are related to the issue at hand, similarity is likely to increase persuasion. Furthermore, the use of common everyday people as the characters may increase credibility, because such characters are likely to be viewed as trustworthy, or having nothing to gain by not telling the truth.

In addition, Linda's sister Ada, and Linda's father are presented as more knowledgeable or expert than other members of the family in the area of family planning. The expertise of a source has been demonstrated to influence persuasion; highly expert sources are more persuasive than ones low in expertise. However, Ada's believability as an expert may be undermined by her youth and the fact that she is single and thus may be considered to lack experience. An older, married sister might have been a more effective choice. While Linda's father may claim expertise by virtue of his age and experience, his credibility with the target may be undermined by the fact he is from the city, whereas the target audience may be largely rural. The impact of this factor would be a function of how rural people view city people.

Message content and execution may also be manipulated to overcome counterarguing by the target. The persuasive communication employs a number of fear appeals to enhance persuasion. Fear appeals have been found to enhance persuasion when presented by a highly credible source. The message suggests that Linda's fear of becoming pregnant and attendant unresponsive sexual behavior may drive her husband elsewhere for sex. Furthermore, Linda's fear actually does drive her to leave her husband. Fear is also introduced in the presentation of abortion as a highly negative alternative to family planning. The abortionist is presented as a cackling witch, and the abortion ends with Linda bleeding and Ada screaming. Abortion is also stated to be a far worse sin than family planning.

Finally, the execution of the communication is such that both positive and negative points about family planning are made. Research suggests that such two-sided communications should observe the following rules of thumb: (1) acknowledge negative arguments at the outset to inhibit counterarguing by the target, (2) refute negative arguments persuasively, and (3) only acknowledge negative arguments of which the audience is aware. This communication generally followed these guidelines. Negative arguments were stressed in the first two installments; these arguments were ones that the audience was likely to be familiar with, and the negative arguments were refuted compellingly in later installments. Below is a listing of both pro and con arguments regarding family planning that were presented in the communication.

CON

- It is against God's will to interfere.
- Man cannot do anything about the number of children he has, and he has no right to do anything about it.
- It is the wife's duty to accept whatever is given.

PRO

- It is impossible to take proper care of the children one already has and more children will only make things harder.
- Too many children make the wife too tired to attend to her husband.
- Fear of getting pregnant can drive a wife away from her husband.
- God is not opposed to family planning; he gave man the ability to reason.
- Man cannot have children by God's will alone, man also participates; God only gives what man asks for.

In general, the persuasive strategy appears to have effectively employed available source and message cues to inhibit counterarguing and enhance persuasion. However, some caution should be taken in using two-sided communications when the audience has little education. The presentation of both sides of an issue may be confusing for such an audience. Further, the major argument, that family planning is not sinful because God gave man the ability to reason and therefore expects man to utilize that ability, may be rather sophisticated for the target audience.

Let us now look at another illustration of a persuasive strategy. Persuasion was the primary strategy utilized by a national lobbying organization in its effort to pass a land use bill.

The Land Use Bill*

This case relates the efforts of the Environmental Policy Center (EPC) in attempting to pass a new piece of land use legislation.

* This case was developed on the basis of an interview conducted by Alice Tybout with a staff member of the Environmental Policy Center.

BACKGROUND OF THE BILL. Since the 1920s, all states have adopted "model zoning enabling acts" as the primary land use control device. These bills take the states' inherent power of planning and regulating land use and delegate it to the local governments. The problem with this method of regulating land use is that there is no review of local decisions and no corodination between local governments in spite of the fact that the decision of one local government might have implications for surrounding communities. For example, if an area needs a power plant but none of the local governments want it in their communities, how is a decision made about where to locate the plant? A land use bill that would provide for the resolution of such issues was developed and was being seriously considered by the 93rd Congress, when the EPC entered the picture. The bill proposed that states resume a limited role in land use decisions. In areas of more than local concern, ones that would have an impact outside of that local area, decisions should be made either by the state, or by the local governments, subject to state review. Four types of decisions are likely to fall into the category of "greater than local impact":

1. Decisions on critical environmental issues.
2. Decisions about large-scale development.
3. Decisions about the development of regional benefits.
4. Decisions about key facilities such as highways.

The bill left it up to the state to formulate the specific method for providing review of local land use decisions. However, all states wishing to receive federal funds for land use would have to comply with the bill.

The utilities and oil companies viewed the land use bill as an opportunity to remove siting decisions from local control and give them to the federal government. Much of the opposition to siting decisions by utilities and oil companies is local; therefore, these firms wanted to replace local control with less resistant federal control. The utilities and oil companies attempted to amend the land use bill to provide for a federal review or override. It was this move to amend the bill that the EPC opposed. There were two major reasons for their opposition: (1) local influence may put pressure on the companies to improve their design, protection of the environment, and so on, and (2) addressing local issues such as public utilities siting decisions are the heart of local environmental organizations. Removal of such issues would make it difficult for such agencies to build interest and membership.

In summary, the EPC wanted to get the land use bill that provided for state review on issues of more than local concern through the Congress. However, they wanted to prevent the utilities and oil companies from attaching a federal override to the bill.

STRATEGY. In 1972, a land use bill passed the Senate with administration support. However, in the House, Representative Aspinwall amended the Senate bill in such a way that the opposition of environmentalists to the amendments, as well as jurisdictional disputes with other committees, combined to defeat the bill. Later, in 1972, Aspinwall was defeated in a Colorado primary by an environmental candidate.

In 1973, Senator Jackson reintroduced the bill that had passed the previous year. The EPC worked closely with Jackson's committee and succeeded in passing a version of the bill that did not contain a federal override. However, this success was not achieved without a fight with both local opposition and the energy interest groups. Local opposition revolved around the Chamber of Commerce, which sent out mailers and wrote articles in trade journals arguing that the land use bill would interfere with private property rights. They succeeded in scaring local realtors, ranchers, and farmers about the bill. The EPC did not attempt to counter these arguments at that level (the size and budget of the EPC precluded such a strategy). Instead, they addressed the congressmen directly through expert testimony because they felt that, on issues of substance, experts have more impact than constituency pressure.

The EPC also had to fight the energy interests who were lobbying for a federal override on land use interests. These interests wanted to add a provision stating that "the state must take adequate consideration of national energy needs in siting energy facilities and in the event of a dispute the Secretary of the Interior would determine what the national need was and whether it had been met." Initially, Jackson and his staff favored the federal override. The EPC had to work to convince Jackson's staff not to include the override in the bill. The EPC used three strategies to eliminate the federal override:

1. The EPC analyzed the proposed override and concluded that it interfered with state's rights, did not give cities full participation and that, if there was to be a federal override, the entire planning process, not just siting decisions, should be opened and everyone concerned should be represented and participate in decision making. (The energy companies, of course, were strongly opposed to opening up the entire planning process.) The EPC sent its analysis of the override to several hundred key people in state and local government and environmental groups and encouraged these individuals to write to Jackson and his committee members.

2. The EPC also went to Jackson's staff directly with the following appeals:
   a. As a presidential candidate Jackson would lose part of his constituency if he attached the federal override.

b. The override was a bad way to do things because it only applied to siting decisions. If there was to be an override, other issues such as need for the facility should also be considered.

c. Finally, it was argued that there was no need for a federal override, since the problems that the bill was designed to overcome had been at the state level. Giving the states power should resolve the issue.

3. The EPC also appealed to other members of Jackson's committee. When discussing the issue with conservatives, they argued that the federal override violated states' rights. When speaking to environmentalists, they argued that the override was an unfair process.

The EPC was able to defeat the override before energy people knew they were being attacked and were able to react. The amendment was not resurrected in the House because not all the utilities were behind it, and the EPC was able to persuade energy lobbyists that it would also be killed in the House.

In the House, the subcommittee and full committee worked on the bill and reported it out, at which point it was tied up by the Rules Committee. The Rules Committee refused to report the bill out in February 1974. This was a shock, since backers thought that the bill represented something everyone could live with. However, the opponents, mainly the Chamber of Commerce, worked to get the administration to withdraw its support for the bill on the grounds that it interfered with private property rights. Their appeal to President Nixon was strengthened by the upcoming impeachment vote in which the President needed as much support as possible. The opponents of the bill were also aided by the fact that bill supporters were caught unaware. They were not expecting any opposition.

The problem became how to get the bill out of the Rules Committee. The EPC appealed to individual committee members. It also used newspaper editorials in the hometown papers of committee members that pointed to the outrageousness of nine Rules Committee members killing a bill that had been debated for three years, had passed the Senate twice, and had been voted out of the House Interior Committee.

The opposition may have erred by opposing the bill in the Rules Committee rather than trying to amend it on the floor. Defeating the bill in the Rules Committee gave bill proponents several advantages: (1) It provided them with a rallying point when they still had time to do something about it. Had the bill been defeated on the floor, it would have been dead for the session. (2) The proponents of the bill looked like the underdogs, whereas the opponents appeared to have used rather unfair, undemocratic tactics.

The Rules Committee consented to hold several additional days of hearings and then take a new vote on the bill. The Chamber of Commerce sent mailings to local groups asking them to oppose the bill and request field hearings that would effectively kill the bill for 1974. The EPC countered by pointing out that there had been 140 witnesses in the Senate and 130 in the House. How much more could be gained by going to the field? Both sides presented witnesses in the several days of Washington hearings. The opponents attempted to impress the committee with a large number of witnesses. The EPC used high-credibility, big-name witnesses such as governors. The EPC also attempted to directly influence Rules Committee members through their constituents.

DISCUSSION. The land use bill case illustrates the use of persuasive strategies by parties on both sides of the issue—the Chamber of Commerce and the Environmental Policy Center. In this instance, educative strategies would have been inappropriate for either side, because such strategies would not necessarily have resulted in the behavior desired. The overt use of an overt power strategy would also be inappropriate in this context for ethical reasons. It is acceptable for lobbyists to persuade but not for them to openly coerce. However, it should be observed that threats of lost votes or support may be implicit in many of the persuasive arguments made. In fact, this case provides a good illustration of the point that persuasive and power strategies are simply different areas on a continuum of pressure strategies, and are not discrete categories.

Both the Chamber of Commerce and the Environmental Policy Center manipulated source and message cues to enhance acceptance of their respective positions. The Chamber of Commerce focused on establishing the appearance of a great deal of constituency opposition to the land use bill. The rationale may have been to manipulate trustworthiness and similarity by conveying the impression to the Congressman that the "little guy" he represents opposes the bill.

The Environmental Policy Center attempted to counter the Chamber of Commerce efforts by presenting expert testimony on behalf of their position to the Congressman directly. The EPC felt that, on substantive issues, experts have more impact than constituency pressure.

Both parties also manipulated their message content and employed various appeals for different target segments. The Chamber of Commerce appealed to realtors, ranchers, and farmers to oppose the bill on the grounds that it would interfere with personal property rights.

The Environmental Policy Center appealed to state and local government officials, requesting that they oppose the federal override on the grounds that it interfered with states' rights and did not give cities full partnership. Jackson's staff was asked to oppose the federal override on the grounds that Jackson would lose some constituency support in his bid for the presidency if he favored the override. Conservative members of Jackson's committee were told that a federal override would violate states' rights, whereas environmentalists were appealed to on the grounds that it was an unfair process.

This chapter can be summarized by the following principles:

1. Persuasive strategies are indicated when a problem is not recognized or not considered particularly important, or when a particular solution to a problem is not perceived to be potentially effective.
2. Persuasive strategies are desirable when the client is not committed to change.
3. Persuasive strategies are desirable when it is necessary to induce a client system to reallocate its resources from one program or activity to the activity advocated by the change agent.
4. Persuasive strategies are not feasible when the client system has no access to resources to sustain a change effort.
5. Persuasive strategies are often necessary when the change agent has no direct control over the client system through the manipulation of resources of value to the client system.
6. Persuasive strategies are particularly appropriate when the client is at the evaluation or legitimation stages of the adoption process.
7. When a persuasive effort is appropriate to one subgroup but not to another, care must be exercised to prevent persuasive content from reaching the subgroup that would respond negatively to such messages.
8. Persuasive strategies are appropriate where the magnitude of change is great and is perceived to be risky and socially disruptive.
9. The greater the time constraints and the lower the ability to use power, the more desirable or necessary it may be to use persuasive strategies.
10. Persuasive strategies are indicated when the change cannot be implemented on a trial basis, is difficult to understand, and has no very visible relative advantage.
11. Persuasive strategies are especially effective in combating resistance to change, although the strategies used to combat resistance to initiating change may not be those used to combat resistance after the change has been implemented.

Power Strategies

Unlike persuasion, power strategies involve the use of coercion to obtain the target's compliance. This coercion takes the form of manipulation or threat of manipulation of the target's outcomes.

The ability to exercise power is based on an obligatory relationship between the change agent and change target. This means that the target is dependent on the change agent for satisfaction of its goals. The strength of power is related to the degree of dependency, which in turn is a function of several factors: (1) the goals controlled by the change agent and the target's motivational investment in those goals, (2) the availability of alternatives to satisfy the target's goals, and (3) the cost of alternative modes of goal attainment. Thus power and the ability to use power strategies increase as a function of the goals controlled by the change agent and the target's motivational investment in these goals. The availability of alternative sources of goal attainment decreases the power of the change agent by reducing the target's dependence. The exercise of power has a cost to its user. First, there is the cost of rewarding the target if it complies with the request or punishing the target if it does not. Second, there may be additional costs in the form of retaliation on the part of the target. In situations in which the target also enjoys some measure of power over the change agent (the change agent is also dependent on the target for attainment of his goals), it may, in turn, exercise its power, causing the change agent to incur damages.

Thus power may be defined as the ability to change the probability of the target performing some behavior by manipulation or threat of manipulation of the target's outcomes. Thus a power strategy entails the use of coercion to alter the probability of a target performing some behavior desired by a change agent. The use of such power may result in both direct and retaliatory costs to the user.

CONSIDERATIONS FOR SELECTING POWER STRATEGIES

Degree of Commitment

A dimension of change that warrants explicit consideration is the degree of commitment required. This variable was suggested in our earlier discussion of compliance. Power strategies typically result in compliance, which implies a low level of commitment to the change. Forced compliance requires surveillance to maintain the change. If the objective is to ultimately achieve a self-sustaining change, educative and/or persuasive strategies are probably more appropriate.

Furthermore, even if surveillance is feasible, power strategies may lead to suboptimal implementation of the change by the target. Power strategies, in essence, do something to the target rather than with the target. This handling of the target may invoke a response that effectively undermines the change. The target may comply in a manner that meets the standard set but creates other problems. For example, production workers who feel management has set unreasonable quotas for their work may turn out more but defective products or may overproduce to the extent that a backlog occurs at the next stage in the production process.

One suggestion clearly involving coercion or power is a proposed modification of the food stamp program to improve nutrition behavior among selected groups with a low level of commitment to changing their nutrition practices (Zaltman, 1975). Currently food stamps can be used to purchase any consumable food product. In the extreme, a family could spend all its allotted food stamps on soft drinks, colas, and licorice sticks. A modified food stamp program, it is argued, would eliminate certain food items such as candy and related snack foods from the pool of food products that could be purchased with food stamps. Alternatively, food stamps could be coded so that a certain portion must be used to buy meats; others could be coded to be usable only with fresh vegetables, dairy products, and so forth.

Perceived Need to Change

The lower the perceived or felt need for change within a target group, the greater the need for a power-oriented strategy. The federal requirement that all automobiles have seat belts with mechanisms that make it difficult to avoid their use is an example of the use of legislation to enforce a practice many individuals did not believe necessary. The head restraint and seat belt regulations are good examples of the exercise of power over one agency, the automobile industry, to achieve change in a target group, that is, automobile drivers. The safety belt legislation is also an example of a power strategy whose impact on its ultimate target group was very limited because of a low level of commitment among drivers and an associated low level of perceived need to change (Robertson, 1975). We will now examine the head restraint and seat belt legislation.

O'Neill (1972) assessed the impact of the U.S. Department of Transportation regulation requiring head restraints in cars after 1968. Using data on automobile insurance claims for whiplash injuries, he found a significant decrease in the incidence of such injuries. This study is noted here because it is a completely "passive" strategy, that is, it requires no effort by the person protected, nor does the person protected have the opportunity to avoid its use. This is in contrast to the use of safety belts discussed later.

Robertson and Haddon (1974) evaluated the impact of the buzzer-light reminder system. The basic question they posed was whether people having cars with buzzers wore seat belts more than those without the buzzer-light system. The basic strategy employed by the government was a facilitative strategy that made seat belts available to everyone. No significant differences in seat belt use were found between those having the system (18%

used seat belts) and those not having the system (16% used seat belts). Robertson (1974a) also observed seat belt use in six urban areas in late 1973 and early 1974 in order to compare strategies that attempt to increase safety belt use. It should be recalled that 1974 automobiles were required to have an interlock system. This new legislation essentially was coercive. This system was designed to minimize circumventing of the seat belt use regulation by making it difficult to disconnect the interlock system. There was indeed a difference between the usage of the buzzer-light system at the end of 1973 (28%) and the usage of the interlock system in early 1974 (59%). What is most interesting, however, is that only 59% of the people having an interlock system used it despite the effort required to render it inoperative. Robertson (1975) notes, however, that dissatisfaction with the interlock system was not associated with observed belt use or nonuse. The interlock system was eventually abandoned at least in part because of the strong public feeling against the system.

Sweetser (1967) found no relationship between perceived probability of being in a serious traffic accident and possession and use of seat belts. This raises the issue of how effective scare or fear appeals might be. Robertson (1974b), pursuing this issue further, observed belt use over a nine-month period of a television campaign advertising seat belt use. The intent of the study was to determine if a large television campaign based on previous research findings would lead to increased seat belt use. He compared use rates of a group using cable television and receiving seat belt messages with a control group of cable television viewers not receiving the messages. The television campaign stressed disability-disfigurement themes adapted to suit different viewing audiences. Appeals could be characterized as stressing social obligations and fear. The program was judged a failure: There was "no measured effect whatsoever on safety belt use."

Capacity of the Client System to Accept Change

A power strategy will be ineffective if the client group does not have the requisite resources for adopting the change and the change agency cannot provide them. One school board in a large city in the southeastern United States issued a ruling that a particular reading and speech education approach be adopted in their elementary schools. (This decision was related to a controversy concerning language demands placed on Black ghetto children by white teachers.) The advocated approach involved the use of video equipment, which none of the schools owned. When supplementary budget requests were made by the principals to acquire the equipment, the school board discovered that they had no funds to purchase the equipment for all schools. For political reasons they chose to not purchase any equipment, even for a few schools in which it was most needed.

Capacity of the Client System to Sustain Change

A power strategy may be necessary if the client or target group has limited resources and is generally unwilling to allocate available resources to the continued implementation of a change. For example, a particular health education program in nutrition was initiated in an East Asian country with funds from an international humanitarian assistance agency. The external funds were sufficient to initiate and sustain the program for a four-month period and were granted with the understanding that the ministry of health would continue the program for the additional three months needed to complete the full design of the educational program. However, when the external support terminated, the responsible department within the ministry discontinued the program because of limited funds and higher priorities assigned to other activities. The nutrition education program began promptly when a high ministry official issued a formal directive to the health education director instructing her to restart the program. We are told that the high ministry official in turn was responding to informal but unequivocal pressure from the external funding agency. Thus the health education department was forced to reallocate funds from other projects to sustain a project not initiated with its own resources. It appears, from information provided by the source of this example, that the health education department director did know at the outset that the external agency expected the program to be continued beyond the first four months. What is not clear is whether or how the director expected to support the project for the final three months.

Resources Available to the Change Agent

As observed in our discussion of the definition of power, power requires that the change agent have some control over resources valued by the target. This provides the change agent with potential rewards and punishments necessary to induce compliance. The absence of such resources precludes the use of power strategies.

Additionally, the change agent must have the ability to exercise the power potential he possesses. His ability to exercise power may be restricted by the prohibitive costs of doing so. For example, he may not have the financial resources to monitor the target's behavior. Alternatively, the resources could be used more productively elsewhere. Or, exercise of power may cost the change agent the friendship of the target, which in itself may be a valued resource. Finally, the use of power strategies may be prohibited by legal or normative sanctions for behavior in a given situation. The power strategies available to members of a channel of distribution are regulated by the Justice Department and the FTC as well as trade associations.

A good example of a power strategy was a campaign to reduce time lost from duty by air force personnel who were in alcohol-related auto accidents. The strategy employed was twofold: (1) an information campaign (an educational strategy) was used that labeled as "sick" the behavior associated with heavy drinking and driving while under the influence of alcohol, and (2) a more power-oriented approach involving automatic administrative review of service records was undertaken to consider giving a dishonorable discharge from service to airmen who lost duty time because of injury in an alcohol-related traffic accident involving private vehicles. Barmark and Payne (1969) evaluated this campaign and concluded that there was a significant reduction in time lost resulting from such circumstances. The researchers attributed this directly to the power approach rather than the reeducative approach. The Air Force, the change agent in this example, exercised power through its control over the type of discharge a serviceman receives. The type of discharge, for example, honorable, dishonorable, etc., has a major impact on the serviceman's civilian professional life and is thus of great concern. A favorable discharge is the serviceman's valued object and the resource the Air Force can control.

Segmentation of the Target System

A power strategy can be effective when various individuals or groups may be favorably disposed to change but one person or group is acting as a bottleneck or barrier. The United States Supreme Court school desegregation ruling in 1954 was a partisan power intervention in a situation in which different groups in society had very different preferences about segregation in public schools. Prior to that ruling, school boards primarily in southern communities were using power to segment their student constituency. In this instance, partisan power was used to create or maintain segmentation within the target system. Viewed from this perspective, the Supreme Court ruling was an effort to desegment a target group.

Decision-Making Stage of the Target System

The selection of strategy is also influenced by the decision-making stage of the target. Power strategies are most appropriate with innovations in which the target experiences no felt need for a change and time constraints prohibit a lengthy education process. Power strategies are also appropriate when decision making is complete and the outcome is not favorable. Other strategies such as education and persuasion have been employed to influence the initial decision; however, after a decision is made, the most likely strategy is power. The outbreaks of violence and rioting among low-income Blacks in the 1960s illustrate this point. All other means had failed to achieve change; consequently, the only feasible strategy that remained was a power strategy. Legitimate avenues for change were also blocked. Furthermore, the only base of power available to the Blacks was their ability to disrupt the status quo either with peaceful sit-ins or violence (e.g., Watts).

This is a typical situation for low-power groups. Legitimate channels are unresponsive; that is, decisions are made without regard to the educative and persuasive appeals of the group. Strategies that dramatize their plight and bring groups with greater power to the bargaining table are needed.

Magnitude of the Change

The magnitude of the change desired is also an important variable. Larger, more costly changes (from the target's perspective) will require greater inducements to achieve change. Thus the change agent would need to possess increasing degrees of power over the target to effect larger changes. Furthermore, when major changes are required, they are more likely to require reeducation and persuasion to be successful in the long run. If a large change is the ultimate goal, a combination or sequential use of change strategies may be appropriate. Power strategies may be most effective in such situations if the change can be subdivided into small parcels and the client can then be moved toward the ultimate goal by compliance with small requests that require correspondingly smaller inducements.

The greater the magnitude of the change initiated, the more likely it is to have a significant impact on the target system, and hence the greater the number of different people or organizations affected by the change. As a consequence of more people or groups being affected, there will be a stronger likelihood that resistance will develop. The use of power strategies is one way of overcoming resistance fairly rapidly. This does not necessarily mean that the sources of resistance for everyone will disappear, but simply that visible symptoms may be suppressed.

Anticipated Resistance to Change

If a high level of resistance to the change is anticipated, the use of power to induce change quickly may be appropriate. It was observed by Huntington (1968) that:

> The most successful and rapid racial desegregation in the United States, it has been observed, frequently occurred where those in power introduced abrupt, firm, and irreversible policies without much prior preparation.

Such policies brought about effective changes in behavior without attempting to alter attitudes and values. Changes in the latter, however, are likely to follow changes in behavior. A more gradual approach to desegregation did not on the other hand increase the likelihood of its acceptance by those in the community opposed to integration.

Opportunity and time for preparation of the public for change is not necessarily related to "effectiveness" and "smoothness" of change. An interval of time for change not only may be used for positive preparation, but may also be used as an opportunity to mobilize overt resistance to change.

Nature of the Change

Rogers and Shoemaker (1971) have discussed five attributes of innovations that influence their rate of diffusion: relative advantage, complexity, compatibility, communicability, and divisibility. As complexity increases, the rate of diffusion decreases. As compatibility, divisibility, communicability, and the relative advantage of the innovation increase, the rate of diffusion increases. These same factors also influence the selection of an appropriate change strategy.

A highly complex innovation or change would require education before adequate adoption could occur. A change that was communicable, compatible with the targets' life-style, and offered a demonstrable relative advantage over other products would lend itself to persuasive strategies.

An innovation which the change agent believes will be found to be positive and which is divisible or can be adopted on a small scale before dramatic investment is required may lend itself to a power strategy to force initial trial. Power strategies are also likely and appropriate when the change or innovation scores poorly on the dimensions that facilitate adoption, because in the absence of coercion, adoption will move slowly.

Lin and Zaltman (1973) and Zaltman, Duncan, and Holbek (1973) have identified several other dimensions of innovations. The impact on interpersonal relations is an important attribute of an innovation. As indicated earlier, the greater the impact on each personal relationship and the larger the number of relationships involved, the more likely it is that resistance will develop. Here the use of power may be necessary to overcome sources of resistance located in interpersonal relationships. The less susceptible to modification an innovation is, the greater the need for a power strategy to force the target group to adapt itself to the advocated change.

Time

Time is a dimension of objectives that is particularly important. There are two aspects of time which affect the nature of the change effort: (1) the amount of time available to achieve the change and (2) the time horizon over which the change is to be sustained. A change agent operating under great pressure to achieve change quickly will find power strategies most expedient. This is a result of the nature of the change; no understanding or attitudinal support is required, and thus change may be achieved quickly.

However, there exists a tradeoff between the speed with which change is achieved and its maintenance. A change achieved with a power strategy will only persist in the short term or as long as the target believes he or she is under surveillance by the change agent. If surveillance can be maintained efficiently, as in the HEW example below, long-term change will result.

Objectives of Change

Power strategies are appropriate when the change agent's purpose is to obtain concessions or compliance with a request from the change target. It has been noted that a party is unlikely to make substantive concessions voluntarily (Walton, 1965).

Compliance leads to adoption of the induced behavior by the target because the target expects to gain specific rewards and avoid punishments by conforming. Consequently, compliance is only an appropriate objective when either short-term behavior change is needed or when the change agent can maintain surveillance of the change target's behavior and administer rewards and punishments as appropriate.

ILLUSTRATIONS OF POWER STRATEGIES

A power strategy has been defined as the use of coercion to change the probability of a target performing some behavior desired by a change agent. Two examples are presented here.

Common Cause*

The first example summarizes the efforts of Common Cause (CC), a national and state lobbying organization, in its attempt to pass a package of bills directed at legislative reform in California.

BACKGROUND. California has a large and active chapter of Common Cause. In 1973, the chapter had a membership of approximately 30,000, a budget of over $100,000, four offices, and ten staff members. The California

---

* This case was developed by Alice Tybout on the basis of an interview with Thomas Belford of Common Cause.

chapter set as its objective in 1973 the passage of legislation that would reform campaign finance, ethics, lobbying, and open meetings. It was felt that although California ranked first or second on these issues of legislative reform in a Common Cause study, their statutes were far from ideal. They were simply the best of a bad lot. However, in light of the state's relatively high ranking in the area of legislative reform, legislators and their constituents were generally complacent and did not regard the situation as problematic. Thus the first task of CC was to convince the public of the need for legislative reform through documentation of problems with the current legislation and introduction of new legislation. Public support would give CC a base of power from which to present its platform to legislators. When public recognition of the problem was achieved, CC began lobbying both directly to individual legislators and through the media for passage of its bills. Its lobbying efforts (basically persuasive) met with little success among legislators who still did not see the situation as critical and had little desire to regulate themselves voluntarily. CC decided that a new strategy would be necessary to achieve passage.

THE STRATEGY.   The members of Common Cause decided to use the power available to them via California's initiative process. Under this process, anyone who can collect 350,000 signatures of registered voters in the state of California can have his issue placed on the ballot and decided in the next election. CC informed the legislature that, if action were not taken on the proposed legislation, the initiative process would be used to pass an even stronger version of the bills. After making this threat, CC simultaneously made one last effort to pass the bills in the legislature and began collecting signatures to qualify the initiative. It succeeded in passing the campaign financing and ethics bills, but not the lobbying or open meetings bills. In view of the failure of the lobbying reform and open meetings bills, CC decided to carry out its threat to use the initiative process on the June 1974 ballot, and it was passed by the voters. Passage of the initiative was an important victory not only in terms of the specific legislation, but also because it established the organization's ability to carry out its threat, thereby increasing the power of future threats.

DISCUSSION.   The Common Cause example illustrates a combination of strategies. The organization began with an effort to create awareness of the problem and need for change by employing educative and persuasive strategies. It documented problems with the existing legislation and then moved to persuade the public and legislators to push for the passage of the new legislation. Only after efforts to persuade the California legislature to pass the reform bills had been exhausted was a power strategy employed.

Power strategies involve the use of coercion to obtain the target's compliance. This coercion takes the form of manipulation or threat of manipulation of the target's outcomes. Common Cause threatened to manipulate the legislator's outcomes by passing more stringent legislation via the initiative process.

The employment of a power strategy is often a last resort after alternative strategies have been exhausted. A primary reason for reluctant use of a power strategy is that such a strategy has high costs for both the target and the user. The cost to the target in this case is the possibility of more stringent legislation via the initiative process. The costs to Common Cause may be both direct—man hours and money necessary to carry out the initiative—and indirect—alienation of legislators whose support may be desired in the future.

Let us analyze how appropriate the use of a power strategy was in this instance. Power strategies are appropriate when the change agent's purpose is to obtain concessions or compliance with a request from the change target. The compliance obtained is a function of the target's desire to obtain rewards and/or avoid punishment; as such, the compliant behavior is not likely to be self-sustaining. It will probably only persist as long as the power user monitors the target's behavior. Therefore, power strategies are most appropriate when short-term compliance is desired or where monitoring the target's behavior and awarding rewards and punishments over the long run is feasible. The use of a power strategy by Common Cause was appropriate because the objective was short-term compliance, that is, voting for the reform legislation.

The degree of commitment required on the part of the change target for successful change must also be considered. Compliance, which is the usual result of a power strategy, typically implies a low level of commitment. As mentioned above, compliance requires surveillance to maintain the change. In the Common Cause example, the legislators are likely to have little commitment to the recently passed reform legislation; however, a high degree of commitment may not be essential. Once the legislation is passed, the legal system assumes responsibility for monitoring compliance with the new laws, and therefore legislators are likely to be forced to comply even after CC has withdrawn from the situation. However, although legislators may be forced to comply with the new legislation, their lack of commitment to it may result in behavior that effectively undermines the intent of the legislation while still technically complying. If an educative or persuasive strategy could have been successful in obtaining passage, greater commitment and consequently more enthusiastic compliance might have resulted.

We have observed that the change agent's ability to exercise power over the target is dependent on his having control over some resources valued by the target and consequently having the potential to reward and punish the target. The ability to reward and punish stems from the agent's control over various bases of power. In this example, Common Cause used the legitimate base of power available to it via the initiative process. Although the initiative process is potentially available to any person or group, its use would be prohibited in many instances by the money and man hours necessary to qualify the initiative. Therefore, it is important to note that CC's own economic and labor resources were essential to its ability to employ the legitimate power base.

A power strategy is also appropriate when the anticipated level of resistance is high. Because power strategies are relatively swift, their use is often recommended when the passage of time would simply allow opponents to mobilize resistance. In the CC case, continued pursual of persuasive strategies would have allowed opponents (such as big business, which opposed disclosure of campaign financing and registering of lobbyists) time to redouble their efforts in opposition and therefore seemed unlikely to be successful.

Finally, the appropriateness of a power strategy varies with the stage in the decision-making process. In this instance, the power strategy was employed when decision making was complete, the legislature had passed only part of the package of reform legislation, and the outcome was not satisfactory. CC wanted all the legislation passed. This was an appropriate time to employ a power strategy, since the opportunity to use other strategies had passed (i.e., the bills had been voted down).

It should be observed that CC was only partially successful when it threatened to employ a power strategy; two of the four desired bills passed. To obtain complete success, CC had to actually carry out the threat and therefore incur greater costs than it would have if the mere threat had been successful. However, because CC was able to carry out the threat successfully, future threats of a similar nature may be more successful.

It is useful to look at another example of a successful power strategy.

Protection of Minors in HEW Research*

The U.S. Department of Health, Education and Welfare is involved in numerous research projects including medical research involving minors. Historically, parental or guardian consent has been sufficient to allow the child to participate in such research. However, HEW felt that, although parents have authority to determine the activity and lives of their children, this authority does not extend beyond things that are done in the best

* This case was developed by Alice Tybout on the basis of an interview with Joel Mangel, a lawyer with HEW.

interests of the children. Typically, medical research differs from medical treatment in that it may focus on some objectives not clearly in the child's best interests; therefore, parental consent is not adequate. In light of this, HEW developed an internal policy to protect the interests of minors in research, typically medical research involving transplants. The policy states that a lawyer must be appointed for any child participating in research and that the lawyer will submit arguments on the child's behalf to court review. A child may only participate if court approval is granted.

In implementing the review procedures, HEW met with some resistance from the doctors involved in research. These doctors saw the procedure as an intrusion and an unbearable burden. Although doctors resisted implementing the policy, the agency used its power of refusing to approve operations to force compliance.

In 1974, 10 or 12 cases were reviewed. All cases involved bone marrow transplants in which the healthy sibling of an ill child was needed as a donor. The review procedure attempted to protect the interests of the healthy child who would be forced to undergo some pain and suffering to serve as a donor. After several cases had been reviewed, the reluctant doctors agreed that the procedure was beneficial and workable.

DISCUSSION.   In this example, HEW coerced the doctors by threatening not to approve their requests for operations unless they complied with the new procedure. This ability to exercise power over the doctors was possible because an obligatory relationship existed between the doctors and the agency. The doctors were dependent on HEW to satisfy their goal of performing research. The obligatory relationship may be two-sided; HEW may also be dependent on the doctors to satisfy the agency's research goals. If a mutually obligatory relationship exists, retaliation by the doctors to the exercise of power by HEW is possible. For example, the doctors may withdraw from the institution and conduct their research elsewhere. For this reason, HEW would want to be cautious in its use of power.

The viability of a power strategy should also be considered in light of the objectives, the resources available to the change agent, and the anticipated level of resistance. The primary objective in this instance was to obtain long-term compliance on the part of the doctors. Although a power strategy is unlikely to lead to commitment and self-sustaining change, long-term compliance can be achieved in this case because it is feasible to monitor the change target's behavior. Furthermore, it may be appropriate to use power to achieve behavior change. Attitude change following behavior change is likely when the trial is a positive or at least not a negative experience. A trial did lead to attitude change in this case. This attitude change reduces the probability of negative consequences such as retaliation by the doctors.

After several cases had been processed, the opposing doctors acknowledged that the procedure was workable and beneficial. A secondary objective was to implement the change relatively rapidly, and the use of a power strategy allows achievement of this objective.

The potential power of HEW over the doctors stems from its legitimate authority over all research sponsored by HEW. The ability to exercise this potential power is dependent on HEW having the resources necessary to do so. The necessary resources might include a procedure for monitoring all operations involving minors, sufficient funds and personnel to carry out this procedure, etc.

Finally, the anticipated level of resistance is a factor to be considered when selecting a change strategy. It appears that the doctors opposed the procedure, and persuasive or educative strategies were not sufficient to overcome this resistance; therefore, a power strategy was appropriate.

This chapter can be summarized by the following principles:

1. Although power strategies may be desirable when commitment by the client system is low, they are also unlikely to increase commitment.
2. The lower the perceived or felt need for change among a client system, the greater the need for a power-oriented strategy.
3. A power strategy will be ineffective if the client system does not have the requisite resources to accept change and the change agency cannot provide them. A power strategy may be effective, on the other hand, in getting a client system to reallocate resources to initiate and sustain change.
4. Power strategies are desirable when a protracted adoption–decision-making process is likely but change must be immediate.
5. Power strategies can be effective in overcoming resistance or in creating change rapidly before resistance can be mobilized.
6. The less susceptible to modification a change is, the greater the need for a power strategy to force changes within the client system. Power strategies may also be useful in securing a trial use of the change.

The Use of Multiple Strategies

This chapter presents two cases that use a variety of strategies. The joint use of two or more strategies is common in social change, particularly when the target group can be segmented and the advocated change is susceptible to modification to become more consonant with the varying needs and concerns of the different segments. Different strategies may be used simultaneously, in sequence, or both, as illustrated in Figure 8.1.

The introduction of a new product or service may begin primarily with re-educative strategy aimed at those persons most prone to adopting innovations early, that is, innovators and early adopters. As the innovation diffuses among these groups, a more persuasive strategy may be appropriate for the more difficult or less change-oriented groups, such as the late majority and

| Innovators | Early Adopters | Early Majority | Late Majority | Laggards |

Reeducative Strategy   ← Increasing Education   Increasing Persuasion →   Persuasive Strategy

← ——————————— Facilitative Strategy ——————————— →

FIGURE 8.1   Simultaneous and Sequential Use of Different Strategies

laggard groups. Independent of which strategy and/or group is of focal concern at any one time, it may be desirable to be pursuing a facilitative strategy, making the innovation more readily available, easier to use, and so forth, for every group. What might be most facilitative for one group may not be most facilitative for another group.

Earlier chapters discussed various criteria to consider when selecting a particular strategy. However, other factors must be considered when using more than one strategy. It is necessary to program the mix and use of strategies. For example, one issue is whether to follow a strategy of least resistance or of greatest resistance. Should the change agent focus first on those individuals or groups easiest to change or concentrate resources initially on those most difficult to change? On one hand, it is important that people who are models for others adopt early, but on the other hand, they may be likely to adopt early anyway in the absence of any significant efforts by the change agent. Appealing first and primarily to early adopter groups for whom educational strategies may be most appropriate would be a strategy of least resistance. Appealing first to late adopters for whom persuasive strategies might be most appropriate would be a strategy of greatest resistance.

Care must be exercised that one strategy does not cancel out the effects of another strategy used earlier, one being used simultaneously, or one to be used later. For example, a power strategy may evoke ill will that could render subsequent educational strategies ineffective. The director of a large New York hospital established a consumer advisory council despite the protests and expressed misgivings displayed by many of his staff. The council was established with very little advance legitimation seeking on the part of the hospital director. He subsequently sought to convince through rational argument those who were resisting the innovation and who, according to the director, were rendering it ineffective. His efforts to use a reeducational strategy met with little success. Interviews with resistant staffers suggested to the authors that the initial use of a power strategy created sufficient ill will that any efforts by anyone to gain active acceptance of the council were doomed to failure. The council continued to exist because of pressure from government agencies to have such councils, but its utility was very limited for this particular hospital.

Another situation little studied in the social change literature but very familiar to consumer researchers occurs when two or more change agents from different and perhaps competing change agencies are directing different strategies to the same target groups. Faith healers have been observed in Bangladesh, for example, to mingle in crowds attending an outdoor health education meeting in a regional marketplace. The government persons who were conducting the outdoor sessions on hygiene and general health were using a straightforward reeducational strategy. The faith healers were using a persuasive strategy and a facilitative strategy, since they had their health restorative products available. The government personnel heightened the salience of health matters, and faith healers converted this heightened awareness into action (a purchase) through persuasive face-to-face communication. This of course was entirely unintended and unwanted by the agents conducting the outdoor session. The phenomenon was interesting to observe. Most people were directing their attention to the government; however, as the faith healers moved among the crowd, little knots of people would cluster around them. In this illustration, the government's agents, contrary to their intentions, were helping faith healers, who were using a different strategy, to sell competing ways of dealing with health problems.

Two cases are presented to illustrate how a single change agency can use multiple strategies.

THE INTERNATIONAL LONGSHOREMEN AND WAREHOUSEMEN'S UNION*

Historically, labor unions have been active social change agents. The following is an illustration of a change effort by the International Longshoremen's and Warehousemen's Union (ILWU), as described by Mr. Patrick Tobin, a Washington representative for the organization.

Background

Enacted in the 1920s, the Longshore and Labor Workers Compensation Act governs compensation benefits paid by employers to injured workers. It is a no-fault law in which compensation is not dependent on proof of negligence of either party. The act provides for compensation equal to

---

* This case was developed on the basis of an interview conducted by Ruby Roy with Patrick Tobin, a Washington representative of ILUW.

two thirds of the injured workers' wages; however, employer organizations attached an upper limit of $70 per week to the provision. Additionally, compensation for permanently disabled workers was limited to $24,000.

A 1946 Supreme Court ruling that equalized longshoremen with sailors and shipyard workers extended the longshoremen's compensation rights to include the right to free legal suit against an employer in the courts, thus enabling them to obtain larger settlements than provided for under the act. This ruling was known as the "seaworthiness doctrine."

The Problem

By 1972, the Compensation Act had remained unamended for nearly 30 years. As a result, it was inadequate to meet the needs of the workers. The average weekly salary of a longshoreman had grown to approximately $200, and the maximum benefit rate of $70,000 did not implement the concept of two-thirds. Additionally, the "seaworthiness doctrine" was being exploited by attorneys, and neither the employers nor employees were benefiting from it. Frequently, one law firm represented both the defendant and the plaintiff and claimed 33–40% of the worker's court settlement. Consequently, the ILWU decided to attempt to amend the Act in 1972. Model legislation was written to achieve the following specific objectives:

1. Removal of the dollar limit on benefits. Benefits would be equal to two-thirds of a worker's wages.
2. Improved administration of the Act by the Department of Labor.
3. Free choice of doctors for the injured worker. Historically, the employer made his doctor available to the injured employee; however, this doctor represented the employer and may not have been the best doctor for the worker.
4. Employment of lawyers by the federal government to reduce exploitation of both employees and employers.
5. Removal of the limit on compensation for permanent injury or death. Allowing compensation to be determined on an individual case basis.
6. Extension of coverage of Act from ship to dock.

Constraints

In pursuing the above objectives, the ILWU faced several constraints. First, rank-and-file members of the ILWU and the ILA (counterpart Eastern and Gulf Organization) were opposed to the giving up of the "seaworthiness doctrine" or any other benefits currently in force, including the right to hire their own attorney. Second, if employers agreed to accept the provision of the model act, it would effectively increase their insurance expense 85%. It seemed unlikely that they would voluntarily accept such high costs unless all compensation benefits were brought within the strict purview of the act, thus eliminating the "seaworthiness doctrine" and high court settlement. Finally, the industry was highly localized in the coastal states and therefore was only of interest to a small minority of legislators in both houses of Congress.

The Strategy

There were essentially three change targets: the union officials, the employers, and the legislators. A separate strategy was developed for each.

SECURING APPROVAL OF THE UNIONS. To convince the union officials, the ILA, and the ILWU Washington representatives justified giving up the "seaworthiness doctrine" on the basis that increasing the benefit rates and retaining the doctrine were essentially contradictory. Increasing the benefit rates eliminated the need for the doctrine. Officials were asked to wire Washington in support of the amendment. Some sent wires opposing the amendment; however, the majority supported it.

SECURING EMPLOYERS' AGREEMENT. The ILWU had two choices with respect to employer organizations. They could either secure employers' support before attempting to pass the legislation, ensuring relatively smooth passage, or they could proceed alone and fight the employers in a vote in each of the Houses. They chose the former strategy and took the following steps to obtain support:

1. To gain acceptance of the primary objectives—benefit rate, maximum compensation for permanent injury, and choice of doctors—the unions used threats. The unions threatened to use collective bargaining to obtain their demands in the absence of employer support, and, further, they threatened to expand the context of their demands if forced to resort to collective bargaining.
2. To gain acceptance of the employers on the issue of legal fees, the unions presented the costs and benefits of the present versus the proposed system. The argument was that both the employers and employees were being exploited and resources were being sucked out of the industry.
3. Ultimately, to gain the full support of the employer organizations, the union representatives agreed to make the compensation scheme the only medium for settling injury cases, excepting very unusual circumstances.

SECURING LEGISLATORS' SUPPORT. The general strategy selected for the legislators entailed concentration on members of the committees in the House and Senate that would have jurisdiction over the amendment. More specifically, they chose to focus on the leaders of the Senate Committee, Senators Javits and Williams. Concentration on these Senators was justified because they were involved with issues relating to labor interests, and their expertise on the issue was widely recognized in the Senate. Consequently, their stand was perceived as crucial to obtaining the support of all the other Senators. Furthermore, it was assumed that the Senate vote would be decisive in obtaining approval of the House.

In their appeals, the unions presented relevant statistics and information from internal sources as well as a copy of the report on Standards for Compensation written by the highly authoritative President's Commission in 1970.

The Outcome

The union obtained its major objectives with the exception of employment of lawyers by the federal government. The support of Senators Javits and Williams was obtained and resulted in quick passage in the Senate. In the House, passage was slowed for several reasons. First, the House Counsel had copied the language of the bill erroneously, and it required a conference of the House and Senate members to work out the differences. Second, the confusion over the language of the bill delayed the process and provided time for opponents to marshall opposition. Admiralty lawyers sent wires to House members opposing the bill and hired a lobbyist to oppose giving up the "seaworthiness doctrine." Ultimately, the fate of the bill in the House rested with Hale Boggs, the majority whip, whose decision it was to present the bill to the floor for a vote. Representative Boggs relied on the leader of a local union in New Orleans on issues pertaining to the maritime industry, and this union leader's support of the bill had been mixed. Finally, on the last day of the session, Boggs spoke to the local leader in New Orleans, obtained his support, and presented the bill to the House, where it was passed.

Discussion

This case provides a detailed illustration of the use of multiple change strategies. In this case, the change agent, the ILWU, segmented the change target into three groups: union members, employers, and legislators, and then employed different strategies to obtain the support of each of these segments.

When approaching union members, the ILWU primarily used a persuasive strategy. A persuasive strategy was appropriate because a specific attitude change was the goal. The persuasive communications relied primarily on message content to persuade the target. The communications carefully detailed the benefits of the new legislation to the union members. The fact that particular benefits would accrue to the workers made the use of persuasion possible. In the absence of demonstrable benefits, a power strategy might have been necessary. This was the case with the change target, employers.

It should be noted that an objective in persuading the union rank and file to support the legislation was to use these supporters to then persuade the legislators by having them send letters and telegrams to Washington.

The second target was the employers. Employers had little internal motivation to support the legislation in view of the fact that it would raise their insurance payments as much as 85%. In view of the negative consequences, persuasive communications probably would have had little influence with employers; consequently, a power strategy was employed by the ILWU. The ILWU threatened to resort to collective bargaining to obtain its demands and in doing so expand the context of its demands in the absence of employer cooperation. Of course, if the employers refused to comply and the ILWU carried out its threat, the consequences might be even more costly than under the present legislation. On the other hand, because of the possibility of very large court settlements in addition to higher insurance premiums, the employers refused to comply unless the ILWU compromised and made the compensation scheme the only medium for settling injury cases. The ILWU agreed to such a compromise because it would experience costs if forced to carry out its threat. This illustrates a situation in which both the change agent and change target had some measure of power over each other; thus, when the ILWU exercised its power, the employers retaliated.

A second strategy, educative-persuasive, was utilized to gain employer support for the legal fees issue. The union organization presented employers with the costs and benefits of the proposed versus the present system. It is difficult to determine whether this strategy presented an unbiased picture of the facts and therefore was educative or whether the presentation was biased and persuasive.

The final target was the legislators themselves whose votes were necessary to pass the new legislation. Efforts began in the Senate, where the ILWU focused its efforts on persuading the leaders to support the legislation on the basis of union and employer support, and a Report on Standards for Compensation by the 1970 President's Commission. Union and employer support combined provided a high degree of credibility for the issue, since these were the two groups primarily influenced by the legislation. Additionally, these groups combined covered a significant number of voters and campaign contributions in the Senators' home states.

The two Senators were selected as primary targets for persuasion because they constituted the experts in issues relating to labor interests, and therefore their support was critical to obtaining the support of their colleagues. Thus the choice of the Senators stemmed from a desire to capture the support of a highly credible expert source. These sources would then be used to communicate to other Senators. Further, the decision was made to begin the change effort in the Senate, since the passage of the legislation in the Senate might increase credibility in the House.

In developing its position, the ILWU used the report by the President's Commission. This report constituted a high credibility source, since it was written by experts who had no motivation to bias their presentation and therefore were trustworthy.

In the House, the delay due to miscopying the bill allowed opponents time to marshall opposition. Delay in a change strategy was mentioned previously as being dangerous for this very reason. Ultimately, Representative Boggs made his decision based on the position of a source whom he considered both expert and trustworthy. Fortunately, and perhaps because of the ILWU's efforts to persuade union members, the source favored the legislation.

THE FIRST ERNAKULAM VASECTOMY CAMPAIGN, NOVEMBER 20–DECEMBER 20, 1970

This case concerns a massive vasectomy campaign in Kerala, India. Kerala is located at the extreme southwest corner of the Indian subcontinent. It is one of the most densely populated regions, with an annual growth rate of more than 2% per year. The population is overwhelmingly agricultural, with 90% of the land holdings smaller than 5 acres.

Ernakulam District is one of Kerala's administrative areas that accounted for a sizeable share of the state's population. The government of India had selected Ernakulam District along with 50 other Districts in the country for an intensive family planning program. These 51 Districts were selected on the basis of population density, and the expectation was that a successful population control program in these selected areas would contribute significantly to the total population control program.

The chief administrative officer of a district is the Collector. The Collector is also the Chairman of the District Family Planning Committee, which advised the Family Planning Bureau, the major administrative agency in this case. The District Collector in Ernakulam was a dynamic, change-oriented leader. Once his District was selected for the intensive program, he took the initiative in designing, operationalizing, and coordinating a mass camp approach to the family planning program. This approach was particularly suited to the characteristics of the district. Its rural population was scattered throughout remote, inaccessible areas. The logistics of delivering services to the villages were not easy to surmount. Instead, location of a camp at a single site offered many advantages. Service facilities could be concentrated; promotional efforts could be focused on a single location, and the logistics problem could be solved by moving the people to the camp.

Program Design

The Massive Vasectomy Camp was organized in Ernakulam for one month (November 20–December 20, 1970) after a one-day trial in which 746 vasectomies were performed. The essential feature of the camp was the integration and coordination of efforts that involved personnel from family planning and other government agencies, from the community, and from voluntary groups.

The design of the program consisted of six components. The first component was the service mix. Vasectomy had been selected as the sole method to be promoted. In addition to the actual operation, the service included a week's supply of injections and antibiotics, special tonics and vitamin tablets, prophylactics, and printed leaflets containing instructions for postoperative care.

The second component was the location. The Ernakulam Town Hall was selected as the site for the camp. It was a prominent, expansive building located in the center of the City of Cochin and on the National Highway. As the seat of the local township, it had high visibility and was readily associated with the government and authority.

Incentives, a third component of the program, were made available to individual acceptors and promotors. Both cash and commodity incentives were offered to individual acceptors. Cash incentives of Rs. 31/⁻ were offered: Rs. 21/⁻ was the incentive offered by Central Government, and an additional Rs. 10/⁻ was made available by the local government body that had jurisdiction over the specific acceptor. Commodity incentives included a week's free ration for the family, a CARE food packet, and a gift obtained from a lucky dip.

Promoter incentives consisted of cash and merit awards that were offered as incentives to individuals/groups who contributed significantly to the camp. Cash incentives were offered to field workers for maximum promotional effort, to Panchayats (local governing bodies) for sending the maximum number of persons, and to doctors for the maximum number of operations. Merit awards were offered for officials and nonofficials performing outstanding service.

Promotion, the fourth component, involved consideration of messages, symbols and media. The campaign was referred to as a crusade, involving the sacrifice of the participants and the public at the altar for the future welfare and prosperity of the nation. The symbol for the campaign was specially designed—it depicted a sad and careworn woman, visibly pregnant, and carrying an emaciated infant. It was an attempt to symbolize the desperate though latent demand of the Indian mothers for family limitation.

Every form of media available in the district was mobilized for the promotional campaign. (1) Public meetings were held in every local self-government area. (2) Microphone announcements, wall posters, banners, slides at local theatres, handbills, radio announcements, were distributed. (3) Press releases through both print media and All India Radio appealed for popular participation. (4) Special newspaper supplements were scheduled. (5) Variety entertainments and cultural performances on family planning with particular reference to the camp were organized. (6) House to house visits by family planning educators and village leaders reached every individual family.

A large signboard was also set up prominently in front of the camp to display the cumulative total number of operations performed.

The physical logistics of the entire campaign, the fifth component, had to consider two things: the layout of the camp and the transportation of the people. Inside the auditorium, 40 booths with operation tables and accessories were set up. The camp was designed to cope with up to 1000 operations. The flow of acceptors was designed to maximize throughput and orderliness. Arrangements for reception, registration, enquiry, pre- and postoperative waiting rooms, preparation, medical checkup, dispensing facilities for incentives, supplies, food, etc., were made to resemble a production line in a factory. The entire process from the registration to the final departure of a patient took only an hour.

For each day of the camp, two or three Panchayats or equivalent areas in the District were designated as targets. Government vehicles and private busses were specially requisitioned, and on the allotted days, the transport facilities were made available to the Panchayats. All intending acceptors from the same Panchayat were then taken to the camp and returned on the same day.

Administration, the final component, was considered comprehensively. Committees were set up at every level—District, Panchayat, Block, and Municipal. Each committee was drawn from local Family Planning staff and prominent citizens. Each committee formed subcommittees for different aspects of work, such as publicity, service, transport, accommodation, camp arrangements, food, remuneration, and so on. The responsibilities of the committees included generation of the cash incentive offered to acceptors from their jurisdiction, promoting the campaign, registering the participants, and fulfilling the quotas set up for each level.

Implementation of Program

The Collector wrote personal letters to prominent persons and institutions explaining the objectives of the Intensive District Program and soliciting their support. Their individuals and institutions included legislators, local government leaders, bureaucrats, school principals, doctors, businessmen, managers, pharmaceutical firms, women's organizations, etc. Members from the public were also included in the committees and various donations were obtained from firms and voluntary groups (e.g., free medicine, volunteers, etc.).

The District Committee decided on an average number of 500 operations per day; thus 15,000 vasectomy operations were expected to be performed at the conclusion of the one-month camp. Each Panchayat was allocated an equal quota of 150 acceptors; only the urban corporations and municipalities were given larger quotas.

Operations during the month-long camp were scheduled daily by Panchayats and municipalities. Each day was allotted to two to three Panchayats, and acceptors from these Panchayats were operated on the preassigned day. Intensive promotions were heightened 24 hours preceding the scheduled day for each Panchayat, and transport facilities were made available on schedule.

Two weeks before the camp, the promotional campaign was launched simultaneously all over the district. Public meetings were addressed by family planning workers, government and other prominent leaders, and acceptors who had already undergone vasectomy operations.

Family planning messages were painted on walls, hung on banners, printed on paper pamphlets, announced at street corner meetings, projected on movie screens, and published in newspapers.

As indicated earlier, a house-to-house education and persuasion strategy was implemented with the assistance of family planning and other voluntary personnel whose specific purpose was to break the resistance to family planning and motivate large numbers of people from each Panchayat to participate in the camp on the allotted days. Every conceivable opinion leader or figure of authority was mobilized for the promotional campaign—local medical practitioners, village revenue officials, Panchayat officials, teachers, women's organizations, and village elders.

The intensity of the campaign was heightened in each Panchayat 24 hours before its members were scheduled to visit the camp site. The entire "propaganda" machinery was concentrated on the selected Panchayats on the scheduled days.

In addition, selected segments of people were identified for special treatment. These segments were composed of lower caste groups, slum residents, and laborers. Special teams of family planning educators, doctors, officers and public men took the campaign message to these population segments.

Intentions to undergo the vasectomy operation were recorded at the village and block levels during the course of the campaign. The record then served as a control on the quota to be filled; those who had not committed themselves were focused on, whereas those who had committed themselves were provided with reinforcements.

During the course of the camp, a "walk-in" was organized through Ernakulam town. The participants were composed of acceptors who shouted slogans in favor of family planning.

The camp site was arranged to maximize the smooth flow of traffic and delivery of services, as well as to be itself a promotional medium. The physical layout of the camp was designed to follow the natural and orderly flow of traffic from reception to departure. A booth was set up for each Panchayat for reception and registration by staff members from these Panchayats.

Facilities for performing 40 vasectomy operations simultaneously were arranged. Forty booths with operation tables and accessories were set up. Supplies of medicines, linen, equipment, and instruments for performing 500 operations per day were arranged. Thirty medical officers and an equivalent number of nurses and nursing attendants, and the required number of ministerial staff, pharmacists, drivers, attendants, and barbers were mobilized and stationed at the camp.

In addition to the physical layout, the camp site was arranged to attract public attention and interest. The Town Hall and the surrounding premises were illuminated and decorated. An exhibition and entertainment program was arranged to attract crowds, and a large signboard functioned as a scoreboard and displayed the total number of operations performed daily. It served to build interest in the performance of the camp by both staff and members of the public.

Procedures to control the effectiveness of the camp were set up at each stage. For example, (1) during registration, individual persons were screened with respect to their age, wife's age, marital status, number of living children, and previous sterilization. Only those individuals with two or more children and a wife in the child-bearing age group were passed for the pre-operation medical checkup. Unmarried individuals and previously sterilized individuals were rejected. Individuals who had less than two children were advised to adopt some other method of contraception, and those who insisted on the operation were approved for sterilization after obtaining a special declaration from them. (2) A preoperation medical clinic was set up to check for physical conditions that were unsuitable to vasectomy operations. (3) A limit of 20 operations per doctor was set as the daily optimum number. The staff at the camp were provided with adequate rest rooms, periods of off-duty, and residential accommodations. (4) The sanitation and hygienic conditions of the camp were maintained by the Kerala Fire Force. (5) Postoperative injections and antibiotics were given to each sterilized individual as protection against infection and reaction.

Immediately after the operation, a week's supply of injections, antibiotics, supplies of special tonics, vitamin tablets, and prophylactics were distributed to each individual. Instructions on postoperative care were read and explained by the family planning staff during rest periods.

Arrangements were made to follow up each acceptor within three days of the operation and then periodically over the following months. Contact was to be made by members of the local family planning staff.

Results

A total of 15,005 sterilizations were performed during the one-month vasectomy camp. From a low of 117 operations on the first day, over 1000 operations were performed on the last two days. The total exceeded the annual target for the District. Even though such a large number of operations had been performed, infections and other complications were limited. Records indicate that less than 5% of those sterilized had minor complications or required hospitalization.

Evaluation

Some of the limitations of this particular camp have been highlighted by the District Collector. They include (1) use of medical supplies without prior testing of their strength and effectiveness. Use of some lotions led to skin reactions or allergies. (2) The quota of 20 operations per day by each doctor was not adhered to; some doctors performed more operations, very likely with less care and attention. (3) The choice of the month (time) for the camp had not considered religious and work patterns of the local population and may have reduced the number of people who would have responded to the campaign. (4) The promotional campaign would have been more effective if it had been implemented one month before camp. In the early days of the camp, less than 500 operations were performed, and it took a while to gather momentum. (5) The use of incentives attracted some ineligible people. Careful screening facilities (e.g., for semen examination) were not set up until the middle of the camp program. (6) Targets for each

Panchayat were set without reference to their population size and past family planning performance. (7) Although follow-up procedures were designed into the program, specific arrangements such as reference reply cards, detailing of family planning personnel, and use of mobile units, were not given the requisite emphasis. The follow-ups themselves were not monitored or evaluated by family planning personnel. (8) Increase in the promoter's fee (from Rs. 5 to Rs. 10) would have brought better results, and a reduction in the acceptor's fee (from Rs. 10 to Rs. 5) would not have affected the number of operations performed during the camp. (9) No mechanism for research and evaluation of the camp and all its facets was arranged.

The use of incentives has been variously labeled a persuasive strategy and a power strategy. Incentives are considered persuasive because they alter the perception of the change or innovation; that is, incentives offered to a group of eligible members accept the change may also create social pressure, forcing some individuals to adopt an unwanted change. Both adopter and diffuser or promoter incentives were used in the Ernakulam campaign. Thus persuasive strategies were used; if one accepts the argument that incentives are power strategies, particularly diffuser and group (in this case, local government bodies) incentives, then coercive or power tactics were also employed.

The promotional efforts also represent persuasion strategies. The symbol of a sad, careworn, pregnant woman with an emaciated infant would seem to qualify as a persuasive tactic. House-to-house visits by family planning educators explaining a vasectomy represent an educational tactic reinforced by special newspaper supplements and other mass media channels.

The location and physical layout of the camp and the transportation system established reflect considerable concern with facilitative tactics. The design of the camp and transportation system benefited substantially from a one-day trial of the entire program. This pretesting is just one indication of the sophistication and strong sense of applied social change displayed by the persons responsible for the program. Another illustration of a facilitative strategy was the concern with those who realistically could not be expected to reach the camp. Some villages in the easternmost district of Kerala were inaccessible to Ernakulam. A special one-week subcenter was established close to these villages, thus bringing the services to an essentially immobile group of potential acceptors.

The organizers of the Ernakulam program actively took into account many of the considerations discussed in the preceding chapter. Their identification of 66 groups for special attention reflects the segmentation consideration. Increasing the intensity of the campaign in each Panchayat 24 hours prior to the scheduled day for each Panchayat reflects concern for decision-making stages. The selection of a vasectomy as opposed to another contraceptive method reflects consideration of the inability of most of the main target group to sustain a change such as the adoption of condom or oral contraceptives, which require repeated decisions and probably unavoidable financial resources. The use of door-to-door visits and other intensive activity plus providing free transportation overcomes what was probably assessed as a low motivation to change. Door-to-door efforts plus incentives were used to help overcome resistance to change. The exhaustive and persuasive campaigns stressing the value of family planning reflects an assessment by the change agent that the felt or perceived need for change might be low. The location of the camp in the Ernakulam Town Hall, a prominent, expansive building that was decorated for the campaign, and the public display of a scoreboard displaying the total number of operations performed daily, reflect considerations about creating general awareness within the target group. Encouraging individuals to declare their intention to obtain a vasectomy reflects a concern for creating at least symbolic adoption and a partial commitment to the change, which increases the likelihood of actual change or behavioral adoption.

This chapter can be summarized by the following principles:

1. Social change should be presented to a client system as being in direct response to a strongly felt need that can be satisfied without undue social, psychological, or financial cost.

2. A change should not be introduced unless the client system has or can very readily acquire the necessary capacity—social and psychological—to accept it and sustain it.

3. Before undertaking a change effort, the change agent should forecast the resources needed to create change and match this forecast with available or expected resources.

4. The lower the degree of commitment to change in a client system, the greater the effort the change agent must expend to heighten commitment.

5. The change agent must carefully assess the nature of the advocated change and determine how it may be adopted to fit more closely to client system needs and behavior. It is also necessary to determine whether particular attributes of the change as perceived by the client require special promotional or introductory strategies.

6. The larger the magnitude of change required for the adoption of an innovation or change, the greater the need for inducement or incentives beyond so-called rationale justifications.

7. When designing an innovation and developing a program to gain its

adoption and diffusion, it is highly desirable to determine whether the client system can be subdivided or segmented into meaningful groups that might benefit from a differentiated innovation and/or implementation plan.

8. Diagnosis of potential cultural, social, and psychological sources of resistance to change should be undertaken early in the planning for social change.

9. Every effort should be made to identify potential undesirable consequences of particular strategies and to develop contingency plans for countering these consequences should they materialize.

10. The change agent should develop in his change plan the optimal timing sequences for various strategies and tactics.

11. The selection of strategies should consider the adoption–decision-making stage of relevant groups within the client system are in.

12. The change agent must determine whether the client system is at all aware of a potential change that could solve a problem or satisfy a need or are aware of a change possibility but unable to relate it to its problem.

13. The goals of a change effort should be stated very explicitly and operationally to facilitate implementation, evaluation, and control activities.

PART THREE

Actors in Change Processes

Chapters 9, 10, and 11 focus on change agents and the objects of their efforts—individuals and organizations. A large number of topics are discussed in Chapter 9 that illustrate the basic qualifications of change agents, generalizations about their successes, errors, motivations, etc. This chapter is intended to give the reader a feel for the dynamics of this important role. Chapter 10 highlights the considerations to which the change agent must be sensitive in attempting to secure change among or by individuals. This particular chapter introduces a new perspective on early adopters and raises several basic research questions that we feel are in need of further attention. Isolated individuals or individuals in informal groups are not the only · adopters of change. Formal organizations are also important targets for change agents. Chapter 11 is concerned with securing change in organizations. A special feature of Chapter 11 is the discussion of the environment as a major factor affecting organizational change. Another important feature is the treatment of "switching rules" to offset the dysfunctional consequences of a particular feature of an organization at particular stages of organizational change. The reader should consider Chapters 10 and 11 as related: the way organizations are structured and the changes that occur in organizations affect the way in which they (as change agents) influence their clients. Similarly, the characteristics of the change approach with regard to clients may influence the structure and innovativeness of organizations.

CHAPTER

NINE

The Change Agent

In earlier chapters, we focused on the different strategies of change and particularly the factors to consider in choosing from among these strategies. In this section we focus on the various actors in the change process. In this chapter, we study the change agent. In our discussion here we take a broad view of the change agent, to include both individuals within and outside the client's system who are attempting to create change in that system, whether it is sanctioned or not. Specifically, change agent is any individual or group operating to change the status quo in the client's system such that the individuals involved must relearn how to perform their roles.

This chapter discusses two broad aspects of the change agent role. The first part of this chapter discusses the characteristics and qualifications of change agents. Some of the common errors committed by change agents are also discussed. The second part focuses on the choices a change planner faces in structuring or designing the change agent role. Both parts draw primarily from the direct observations of the authors and the reported experiences of change agents. There appears to be little sound research that addresses the various ideas expressed in this chapter. The ideas do have firm rooting in participant-observer research, which we feel is a very sound methodology. However, many times the observers are not trained in this methodology. Nearly all ideas expressed have support in many different contexts, thus providing important convergent validity.

Social interaction is the basic commodity of social life and hence the basic medium through which social change occurs. Changes in social interaction may produce changes in social structure and social function. Planned change generally involves deliberate manipulation of social exchange relationships, often involving very different persons and groups. The use of change agents to induce changes in social interactions is one key method for altering social structure and function. Change agent behavior and related considerations such as recruitment, training, and motivation are discussed in this chapter. Change agents can be external or internal to the social system in which a change is desired. The desired change may also be classified as external or internal in origin. Because our concern is with the planned change, we do not focus on unwitting change agents, that is, persons who initiate change without a particular intention to do so, or even without awareness of their instrumentality as agents of change. Such persons should not be ignored, however, since they represent tools or mechanisms available to more deliberate change agents in facilitating their efforts to induce social change. These "nonagents" of change are important when they are opinion leaders for many persons and innovative with regard to the area of the advocated change. The person-to-person contact they provide is espe- cially important during the evaluation stage of the adoption process, since they can correct misinformation and may also be perceived as less biased than a formal change agent such as a salesman. They may be important as early adopters in demonstrating the use of a new idea, service, or practice. (Of course the opinions of "managements," however highly valued, may not always contribute to change. First, the content of their opinions may dis- courage a particular change. In this sense, they may be counter change agents. Second, their opinions might be outweighed by other factors that the change target must consider. Similarly, not all change agents are opinion leaders.

CHARACTERISTICS OF SUCCESSFUL CHANGE AGENTS

The interactional nature of social life gives the change agent a prominent role in planned social change. One of the basic functions performed by a change agent is to establish a link between a perceived need of a client system and a possible means of satisfying that need. Change agents may

Exhibit 1   Characteristics of Successful Change Agents*

---

1. The Change Agent Should Have These Attitudes and Values:

• Primary concern for the benefit of the ultimate user (usually students and communities in the case of education).
• Primary concern for the benefit of society as a whole.
• Respect for strongly held values of others.
• Belief that change should provide the greatest good to the greatest number.
• Belief that changees have a need and a right to understand why changes are being made (rationale) and to participate in choosing among alternative change means and ends.
• A strong sense of his own identity and his own power to help others.
• A strong concern for helping without hurting, for helping with minimum jeopardy to the long- or short-term well being of society as a whole and/or specific individuals within it.
• Respect for existing institutions as reflections of legitimate concerns of people for life space boundaries, security, and extension of identity beyond the solitary self.

2. The Change Agent Should Know These Things:

• That individuals, groups, and societies are open interrelating systems.
• How his other role fits into a larger social context of change.
• Alternative conceptions of his own role now and his potential role in the future.
• How others will see the role.
• The range of human needs, their interrelationships, and probable priority ranking at different stages in the life cycle.
• The resource universe and the means of access to it.
• The value bases of different subsystems in the macrosystem of education.
• The motivational bases of different subsystems in the macrosystem.

themselves be the means, as in the case of psychotherapy, or they may simply establish contact between the client system and the source of a need-satisfying product or service. In fact, the linker or change agent can perform many roles that are all subsumed under the larger role of change agent. These roles include diagnostician, information specialist and solution builder, evaluator, system monitor, innovation manager, and facilitator (Havelock and Havelock, 1973, p. 27). Change agents in any of these capacities are likely to be more effective if they (1) stimulate the user's problem-solving processes, (2) are sufficiently knowledgeable about the research and development processes that produce solutions so that they can

Exhibit 1   (Continued)

- Why people and systems change and resist change.
- How people and systems change and resist change.
- The knowledge, attitudes, and skills required of a change agent.
- The knowledge, attitudes, and skills required of an effective user of resources.

3. The Change Agent Should Possess These Skills:

- How to build and maintain change project relationships with others.
- How to bring people to a conception of their priority needs in relation to priority needs of others.
- How to resolve misunderstandings and conflicts.
- How to build value bridges.
- How to convey to others a feeling of power to bring about change.
- How to build collaborative teams for change.
- How to organize and execute successful change projects.
- How to convey to others the knowledge, values, and skills he possesses.
- How to bring people to a realization of their own resource-giving potential.
- How to expand people's openness to use of resources, internal and external.
- How to expand awareness of the resource universe.
- How to work collaboratively (synergystically) with other resource systems.
- How to relate effectively to powerful individuals and groups.
- How to relate effectively to individuals and groups who have a strong sense of powerlessness.
- How to make systemic diagnoses of client systems and how to generate self-diagnosis by clients.

---

* *Source: Havelock and Havelock, 1973, pp. 70–72.*

help stimulate these processes to function more consistently with client needs, (3) are able to foster communication and possibly collaboration between client systems and between change agencies, (4) are able to link particular users with an optimal number of change agencies, and link a particular change agency with an optimal number of users, (5) are willing to listen to new ideas with receptive but constructively critical ears, and (6) are able to introduce flexibility into the relationship between client and change agency in the event that adaptation of the need-satisfying product or service or the circumstances of its use becomes necessary.

This chapter focuses primarily on what change agents can do to bring about social change. However, the best change principles are unlikely to achieve their maximum effect if change agents themselves are inadequate interpersonally or in expertise. Thus very important considerations must be addressed prior to the application of social change technology. These considerations involve the recruitment and selecting of change agents, their training, their motivation, and their supervision. These topics are touched on only lightly here. There are several good sources for pursuing these topics in depth (Havelock and Havelock, 1973; Dodge, 1973; Stanton and Buskirk, revised edition, in press; and Davis and Webster, 1968).

A conference on Change Agent Training held at the University of Michigan suggested a number of characteristics change agents should have to be successful. Although the conference emphasized training as a means of instilling these characteristics, they are in fact obtained through a combination of successful recruitment, training, motivation, and supervision. The specific characteristics are presented in Exhibit 1. These characteristics are far from exhaustive. At the same time, it is a rare case in which a single change agent can be found who possesses as much as a small majority of these traits.

BASIC QUALIFICATIONS OF CHANGE AGENTS

Selecting change agents and agent leaders in the case of change teams is a crucial element in the process of organizing for change. The capabilities and performances of the team leader have often been the critical factors in the success or failure of many programs of planned change. In this section of this chapter, we focus on the necessary qualifications of the leader of change agent teams. Many of these qualifications are also desirable in individual members of the change team or the independent change agent.

Technical Qualifications

Perhaps the single most necessary trait the change agent team leader must possess is technical competence in the specific tasks of the change project. Dangers exist in bringing a person in from an unrelated or only marginally related field. Alternatively, there may be dangers inherent in selecting a generalist without the in-depth expertise that may be required occasionally. This is less a problem when the generalist recognizes his limitations and is willing to bring in the necessary experts as consultants. A candidate for a change team leadership position can have the necessary expertise "on paper" but still fail because he may not have kept up with current developments in his field. Also, lack of imagination when a novel problem arises may render useless the possession of considerable technical expertise or knowledge.

Another factor related to technological qualifications is the agent's ability to adapt and apply his skills to problems of a simple and mundane nature. This is the reverse of being underqualified. There is the danger of imposing complicated solutions on simple problems needing only simple solutions. This is sometimes a danger when external consultants are brought in to assist in solving simple problems. Consultants sometimes find it difficult to break a habit of applying complicated solutions to simple problems if most of their expertise is with complicated problems normally requiring complicated solutions.

A highly critical skill in many change situations involves institution-building capabilities. Does the change team leader have experience in developing institutional structures? For example, it is most important that a vacuum is not left when the change team leaves the client system. A durable, viable mechanism should be developed within the client system to take over the functions performed by the change team. Thus institution-building skills are important. The institutional mechanism replacing the change team may not, or perhaps should not, mirror the change team as a mechanism. This means that the team leader working in concert with representatives of the client system must possess the skill and imagination to develop comparatively new social institutions. These social institutions may range from an informal weekly luncheon meeting of those involved with, say, community development, to a permanent department or agency for community development employing a large staff of professionals.

Institution building is as difficult as it is important, and an unsuccessful first effort to build an institution can delay for years any attempt to develop an alternative institutional arrangement. Here, again, the abilities of change team leaders to recognize their limitations in institution-building skills is important. Similarly, a capability to develop complex institutions should not be allowed to create a complex structure where a simple one will do.

So-called "paper" credentials are sometimes important for establishing credibility. Thus it may be important for change agents to hold academic degrees, have occupational titles suggesting authority and expertise, and so forth.

Administrative Ability

Perhaps the most basic and elementary administrative requirement is the willingness to allocate time to relatively detailed matters. For example, the failure to check on whether the printing of training manuals for community development workers was on schedule resulted in the postponement of a one-month training program for nearly a year. The printing was three weeks behind schedule, and approximately ten months passed before a significant number of development workers could be assembled again at the same time.

This concern with detailed matters may at first give the impression of violating the cherished principle that a good administrator is one who delegates responsibility. One way of being attentive to detail is to delegate this task to a highly responsible subordinate. Often, however, such responsible people are simply not available.

Planning skill is another important capability a change team leader or program leader should possess. Particularly important is the ability to plan for the unexpected by allowing flexibility through alternative or contingency planning and the maintenance of reserve resources. Although the change team leader should be able to function well in a crisis, he should avoid crisis management as a characteristic mode of management. Planning is the process of deciding in the present what will be done in the future. When crises repeatedly arise and require major attention from the change team leader, there is good reason to believe that inadequate planning exists.

In selecting and training change team leaders, attention should be given to the element of personnel management. This is particularly important when the change team involves members of a client system that differs culturally. Reward, punishment, and motivation techniques and general supervising skills that are appropriate for change agents with one cultural background are not likely to be uniformally appropriate and effective with change team members with different backgrounds. This greatly complicates the personnel aspect of administering cross-cultural projects. A related consideration is the ability of the change team leader to develop the skills of the client system members of the team. This is an important element of institution building, since it is likely that the client system change team members will take over responsibility when the outside team members leave the project.

The ability to work well with others outside the change team is not only desirable, but may be crucial. Outsiders may include other agencies, both national and international, who are concerned with the same area in which change is desired, and unorganized members of the client system toward whom change efforts may be directed. It is important that different change agencies that share common goals do not embark on independent strategies or programs whose timing, content, and operationalization would cancel out each other's efforts. Good, informal relationships at all levels between members of different change teams can be very significant in achieving common goals. It is the responsibility of the change agent team leader to establish such a relationship. Unfortunately, rivalry often exists between teams, creating confusion and loss of respect from the client system Moreover, indifference by a team leader to persons and organizations outside the change team can create major barriers for others in the change team, no matter how well they are regarded by the client system.

Major organizations that many change managers must deal with are national and regional governments. Red tape problems tend to be disproportionately more numerous and troublesome with such organizations. Whether the governmental unit is the one for which the change team leader is working, or is part of the client system that is the object of change, the leader must be able to tolerate red tape. Moreover, and perhaps more importantly, he should be creative in overcoming red tape problems.

Interpersonal Relations

A very important trait for a change agent to possess is empathy. This refers to the ability of a person to identify with others, to share their perspectives and feelings vicariously. It is very important for the team leader to have such a capacity not only with regard to his own team members but also with others in the client system. An ability to empathize with his counterpart in the client system is especially important. In the case of his own team of change agents, empathy with their personal and professional difficulties improves morale and facilitates the resolution of problems.

Empathy with members of the client system can sensitize change agents to culturally delicate areas in which extra caution is necessary if an intervention involving those areas is under consideration. The insight empathy provides can also result in a better selection of strategies and a better way of implementing strategies. Also, when the client system realizes the change agent is empathic, its receptiveness to the change team generally is greatly increased.

Dealing with interpersonal issues is important for the change agent in dealing with colleagues as well as other members of the change target system. In change situations, participants face uncertainty, and strategies for dealing with these situations usually have not been covered by preestablished rules and procedures. Greater reliance must be placed on the informal network of relationships. Participants may also experience stress and anxiety as a result of the uncertainty experienced in the change situation. The ability of the change agent to communicate his concern and understanding of the client system's anxiety can help reduce its feeling of threat.

Dealing with interpersonal issues is also important because of the effects of good interpersonal relations on openness, risk taking, and trust, which are important for change and innovation. Argyris (1965) shows that, when individuals do not own up to their own behavior or are not open to the effects of their behavior on others, those around them are less likely to take risks and are more likely to conform in their behavior. This is likely to inhibit the change process. The study by Stephanson et al. (1971) showed that creative research and development labs were characterized by staffs that were willing to defend unpopular ideas and to take chances by being open and honest with fellow workers.

Job Orientation

MOTIVATION AND DRIVE. It appears that a major factor in the failure of change team leaders is a limited view of personal responsibility for achieving the change objectives. This is attributable, in part, to a lack of motivation and drive. All too often, ineffective change team leaders have been found playing a passive advisory function, giving limited information and even that only when requested to do so. Also associated with a lack of drive is the failure to give direction to team members.

A related problem characteristic of ineffective team leaders is a lack of initiative in going beyond the literal mandates of the change objective. Changes in one context often require changes in other contexts or utilization of resources available in other contexts. The initiative can be carried too far, of course, and create opposition or noncooperation within the other contexts. However, a good team leader will possess the necessary diplomacy and other skills to achieve what he wants in these other contexts.

It is not uncommon for a change agent or team of such agents to be assigned a task in which the problem is only vaguely understood by the sponsoring agency. Independently of the clarity of the problem and the degree of consensus as to its nature, there may be uncertainty regarding the appropriate responses or remedial strategies. The task of clarifying the situation in terms of defining the problem and determining the desired course of action should be performed by the change team. This requires initiative on the part of the team leader. It can be argued that, no matter how clearly defined the problems and courses of action are by parties other than the change team, the team leader should initiate his own problem diagnosis and strategy selection processes. The full team should also be involved in this effort. The change team leader should also take the initiative in discussing with others discrepancies between their evaluations and those of his team.

Another aspect regarding motivation is that the change agent should be clearly aware of his own motives. This is important in that it sensitizes the change agent to his own biases, which may act as moderators in dealing with the client system. Is the change agent really concerned about the welfare of the client system or does change agent activity satisfy his needs for power or control? If the latter motives are operating, this might cause the change agent to be more manipulative in dealing with the client system.

Also, if the change agent is too eager to show the client system how good he is, he may not allow the client system to develop its own capabilities for dealing with the problem situation. As a result, the client system becomes overly dependent on the interventionist. For example, an OR/MS change agent might be so overly concerned about proving the usefulness of his techniques that he does not allow the client system to develop those skills through its own trial and error learning so that it can effectively implement this innovation when the change-agent consultant is no longer present.

ACCEPTANCE OF CONSTRAINTS. One of the certainties of any social change situation is the presence of constraints on freedom of action. Constraints may come in a variety of forms: activities forbidden by company or government policy, inadequacies in the skills of team members, cultural and environmental barriers pertaining to the client system that prevent implementation of the most direct and logically most effective courses of change, limited financial and material resources, limited human resources in the client system, and so forth. Change agents generally must possess a well-developed ability to tolerate constraints. Moreover, creativity in adapting to constraints and working around them is necessary. It might be added that silent tolerance is often necessary for diplomatic reasons in the formal sense, or for other practical reasons, such as team members or clients not wanting a constant reminder that they are deficient in skills or other resources desirable for bringing a particular change into existence.

Probably one of the most important constraints that may face a change team and client system is the team or project leader's lack of enthusiasm for the change goal. This constraint becomes even more severe when the team leader does not agree with the policy underlying a change program. The team leader must seriously question his ability to carry out his obligation to implement that policy; by remaining in the role of team leader he is obligated to pursue the policy with the same skills and drive he would employ under circumstances in which he favored the advocated policy. This is extremely difficult to do.

DEVELOPMENT COMMITMENT. Development commitment refers to the dedication of a change agent to the development of client system skills and resources. This is sometimes expressed in terms of particular visible projects such as a new road, a sanitary system, the establishment of a grievance committee or a formal faculty change team within a school system. Such visible changes are more readily rewarded in terms of praise and respect than are less visible improvements in the client system. However, the needs of client systems more frequently seem to be changes of a less visible nature. In fact, even in the visible instances in which the change team personnel play the major roles, they might properly allow credit for such projects to go to members of the client system.

The need to focus on less visible social change projects or projects whose visibility will not surface until a distant period of time requires a commitment to development rather than demonstration. An even stronger commitment to development is necessary if client system members are given a disproportionately larger share of the credit for successful social change programs.

Leadership

POISE AND BACKBONE. Leadership in terms of executive abilities is an obvious requirement for a change team leader. In fact, were it not for the unfortunate frequency with which executive abilities are found lacking among change team leaders, it would not seem worth discussing because of its obviousness. One hallmark of a good leader is his ability not to act precipitously in the face of strong and unexpected resistance to his change program. Although instances of precipitous actions are not numerous, just one inappropriate response can effectively terminate a project. Not only must the team leader avoid precipitous action, he must react constructively when faced with a highly provocative situation. Related to this is the ability to accept reverses with outward poise. This is not to say that anger should never be expressed. On the contrary, a deliberate show of anger can be very effective in securing cooperation. The essential trait is the ability to know when and how to express anger and when to contain it.

Most social change programs encounter resistance, which in turn creates criticism and stress and demand for the cessation of or a major alteration in the change program. It is important, then, for the team leader to have the courage of his convictions. He must be willing to at least state his opinions in the face of controversy and, hopefully, be able and willing to defend them under attack. Firmness extends not only to disputes with the client system but also to those involving team personnel and the sponsoring agency if it is other than the client system.

POLITICAL FINESSE. The ability to promote a change that involves the active cooperation of others is another essential characteristic of the change team leader. Team leaders must be able to identify their major constituencies both at the client system level and at the sponsoring agency level. Once identified, he must be able to "package" or define the change program in a manner attractive to the various constituencies. This is a very difficult task not only because there may be numerous constituencies, but also because there may be conflict among the constituencies. The nature of the proposed change and the method for its implementation must somehow satisfy the basic concerns of most of these constituencies, in a manner that does not exacerbate the conflict. There is the very real danger that, if a change program is perceived to be in the interests of one group, another group having poor relations with the former group may automatically interpret the advocated change or the implemental machinery as being contrary to their own interests. Thus what is involved here is a sensitivity on the part of the team leader to undercurrents and tensions among persons and groups within the context of the change program. Sensitivity to undercurrents and tensions results in information helpful for determining the appropriate strategies and techniques to effect a change.

EMOTIONAL MATURITY. Obviously, the team leader should possess strong character in terms of personal integrity, particularly in matters that affect his well-being. Traits of honesty and ethical behavior are relevant here, as are standards of personal conduct. A related consideration is his own personal security or self-confidence. It is important for the change team leader to have that optimal level of security in his status and abilities that permits him to be open to suggestion and willing to act on advice as well as to admit to mistakes. Too much or too little self-confidence can be deleterious.

A summary of many of the issues discussed above are presented in question form in Exhibit 2.

GENERALIZATIONS ABOUT CHANGE AGENT SUCCESS

Rogers and Shoemaker (1971), in an extensive review of the literature, have developed a number of generalizations concerning change agent success. In this section, a number of these generalizations are presented and their implications for planned change discussed. The generalizations should be accepted only tentatively, and even then as applicable in limited contexts. However, they are suggestive of a number of applied actions that have been established as empirically important, thereby lending validity to the corresponding generalizations. Only those implications which have been established as empirically relevant are introduced here.

Generalization 1. "Change agent success is positively related to the extent of change agent effort" (p. 233). Perhaps the major factor that prevents this generalization from being foolishly obvious is that it is not at all clear that

*Exhibit 2   Basic Qualifications**

Technical Qualifications

- Can this candidate trouble-shoot and solve technical problems in the speciality that is the nub of this project?
- Is he too locked into high-powered and sophisticated approaches to relate to his counterparts on practical down-to-earth problems?
- Has he been active in the institutional as well as the strictly technical aspects of this kind of activity, and has he been effective in performing these kinds of functions?
- Will his credentials carry enough weight in this country to gain him the respect and high-level access the position requires?

Administrative Ability

- Does he take care of the administrative details that are part of his present job punctually and effortlessly, without slip-ups, flurries, or special reminders?
- Has he shown himself to be skillful in planning, and to be sufficiently well organized to stay on top of a number of ongoing activities at the same time?
- Has he shown himself to be relatively free of "blind spots" in judging the capabilities of his staff and assigning them suitable functions?
- Has he experience in working with the legal and quasilegal aspects of contract administration?

Interpersonal Relations

- Is he too self-centered or callous to be attentive to the needs of the people who will be dependent on him for assistance?
- Does he turn off the people who work with or for him by being aloof or autocratic, or otherwise trying in his working relations?
- Does he frequently offend others by ignoring common courtesies and conventions, or showing poor taste?

JOB ORIENTATION

Motivation and Drive

- Does he realize how much personal direction and push he will have to give to the project, and does he seem comfortable about playing so active a role?
- Is he an alert and reasonably aggressive self-starter?
- Can he be counted on to produce on an assignment, no matter how much time or energy may be required?

Acceptance of Constraints

- Is he willing to function in a quasiofficial role, as a part of the United States government structure?

Exhibit 2   (Continued)

Acceptance of Constraints (Continued)

- Does he fully understand the goals and mechanisms that have been established for this project, and is he willing to operate within them?
- Does he realize that the exercise of "freedom of speech" in this assignment must stop short of pronouncements that his hosts would find offensive?

Development Commitment

- Has he shown himself to be patient and skillful in developing more junior staff members?

EMOTIONAL MATURITY

Character

- Is he one of the people who can be counted on to support a decision that is for the good of the entire institution without getting caught up in its impact on him?
- Does he lack the stature and influence in his own home office that he will need to look out for the best interests of the project?
- Is he prone to trouble via liquor or women or other intemperate behavior?

Personal Security

- Does he take good advice when it is given, without regarding this as a threat or an affront?
- Does he admit to mistakes candidly, without indulging in elaborate excuses or rationalizations?
- Is he reasonably relaxed about his dignity and status, and about exacting his just due in amenities and respect?

LEADERSHIP

Poise
- Is he reasonably unflappable in high pressure or crisis situations?
- Has his staff found him to be patient and helpful when someone makes a mistake?
- Has he typically been a good sport about being overruled or turned down on something he considers important?

Backbone

- When he is right does he stick to his guns in a debate or controversy to the extent necessary to achieve the objectives?
- Does he take firm and timely action on personnel problems, without weaseling or passing the buck?
- Has he enough confidence in himself to deviate from prior agreements or instruction when there is an obvious need for modification?

Exhibit 2   (Continued)

Political Finesse

- Has he been popular and mobile in the organizations of which he has been a member?
- Is he usually fast in catching on to the "hidden agenda" in delicate negotiations or discussions?
- Has he shown himself to be a skillful psychologist and tactician in getting action on an idea or approach he wants to promote?

* *Source: Adapted from USAID, 1973.*

success is the result of change agent effort at least initially, but rather initial success stimulates change agents to greater activity which, in turn, results in still increased success. This further increase in success may be due more to the change agent's efforts than the earlier initial success. One implication of this latter interpretation of the generalization is that various other promotional aids such as monetary and nonmonetary incentives might be used initially to create a favorable initial condition of success that stimulates or reinforces change agent efforts. This is often done in commercial contexts in aiding salesmen by giving a cents off coupon to the buyer. The additional sales stimulated by the incentives provide the necessary momentum the salesman may require. This same approach has been used in programs involving vasectomy camps where canvassers were used in conjunction with incentives paid directly to the vasectomy adopter.

Whether a change agent is an initial or secondary cause of change, or both, it is appropriate that he understand the notion of critical threshold. That is, he should be instructed to expect a period in which efforts will not be particularly well rewarded in terms of numbers of innovation adopters, for example. Subsequently, however, other conditions being equal, effort expended will be well rewarded. To the extent that something of a bandwagon effect occurs, the change agent can use this phenomenon to stimulate greater acceptance of the advocated change. One change agent in a rice production improvement program summed up this technique as follows:

> After three frustrating weeks my first visible sign of success took place. [One] farmer adopted. I brought this fellow along with me at my personal expense in visiting several other farmers in the district. Frankly, he lied to the others about his successful experience because enough time hadn't passed for the [new practice] to demonstrate its superiority. Nevertheless they were quite impressed and adopted the practice. Subsequently I was able to mention these farmers to still others who quickly instituted the

practice. Had I not been able to refer to these others it is doubtful the program would have been the success it has. Of course some dishonesty on the part of the first farmer helped measurably. (Abueva and others, 1973)

Interestingly, the change agent did not report any efforts on his part to prevent the first farmer's liberty with the truth.

Generalization 2. "Change agent success is positively related to his client orientation, rather than to change agency orientation" (p. 237). Homans and others offer good theoretical foundations for this generalization. Here the importance of empathy is most evident. The client's orientation involves empathetic ability on the part of change agents. More importantly, perhaps, is the ability to communicate such empathy to clients, a trait possessed by most successful automobile salesmen who ostensibly side with the customer during the negotiation with the removed-from-view sales manager for a final car price. This generalization has implications for the selection of change agents. Agents should possess the necessary client traits whether they are linguistic skills that permit insight into the reasoning process or shared employment histories and industrial experience in the case of organizational development work. The generalization also suggests the use of simulation as a training tool to increase the speed of socialization or induction into the client system mores.

Many change agencies rotate personnel on a periodic basis to prevent too close an identification of change agents with the client system. What is in the interests of the client is not always compatible with the interests of the change agency. A change agent who becomes too closely identified with the client system may be adjudged by the change agency to be no longer capable of acting entirely in the interests of that agency. This is referred to by the somewhat disdainful expression, "going native."

Socialization is not a one-way process as often implied. One explanation that may be offered for generalization 2 is that the "client orientation" on the part of the change agent may be mirrored by an increasing change agent or change agency orientation on the part of the client. This results from the rapport established as the change agent becomes more credible in the eyes of the client. Theoretical support for this is provided by balance theory (Heider and others), which suggests that, as the client becomes increasingly attracted to the change agent, he will come to support or at least tolerate the belief system associated with the change agent. The same, of course, is true for the change agent, but in reverse.

Generalization 3. "Change agent success is positively related to the degree to which his program is compatible with clients' needs" (p. 238). An analysis of a long history of failure in planned change programs attests to the frequency with which the basic message—build programs on the basis of client needs—in this generalization is ignored. The change agent must identify the set of needs that relates to a social problem and select as the target need that which is most central to the behavior and belief systems of the client. An appropriate technique for doing this is to attach different needs to their corresponding motives in Maslow's (1954) hierarchy of motives. Tying a program to physiological and safety needs would tend to produce better program results than tying the same program to esteem and status or self-actualization needs. Similarly, Rokeach (1968) postulates a hierarchy of beliefs in terms of their centrality. Those beliefs which are highly central to the individual's ego (Sherif et al., 1965) and which are instrumental to the achievement of highly desired goals (Rosenberg, 1960) are the beliefs from which needs should be derived. The program of planned change should be linked as directly as possible to the most central of these needs. The program themes should incorporate the corresponding belief. The change agent should ask first, what are the most basic beliefs in the client system, and then ask what needs they serve. Barnett's discussion of various wants as the starting point for social change provides interesting theoretical background for this generalization as well as specifying particular types of wants that can be connected with change programs.

Marketing theory based on consumer research suggests that relevant needs may vary within the client system. Consequently, a given change project might require identification of the different needs of different groups. A project may, on one hand, appeal to the physiological needs of one group, and on the other hand, appeal to (or threaten) the status and power needs of another group within the client system.

Generalization 4. "Change agent success is positively related to his empathy with clients" (p. 239). The action implications of this generalization have been treated adequately in previous parts of this chapter. Theoretical foundations for this generalization can be found in Homer G. Barnett's (1953) discussion of basic innovative processes (pp. 181–225) and change advocates (pp. 291–328). Perhaps the most generally supportive theory is provided by George Homans (1974). The strongest empirical support can be found in Rogers' (1969) work on modernization. Additional evidence in a different context is provided by Evans' (1963) research on insurance salesmen–prospect interaction.

Generalization 5. "Change agent success is positively related to his homophilly with clients" (p. 242). Homophilly appears to be important with regard to social status, education, and cosmopoliteness. The implications of homophilly were also discussed earlier. It is likely that different types of persons will exist within the client system; hence the need to have different types of change agents corresponding to different sets of client system members. Again, exchange theory as articulated by Homans (1974) and Blau (1964) provides the theoretical umbrella for this generalization.

Generalization 6. "Change agent success is positively related to the extent that he works through opinion leaders" (p. 243). Opinion leaders are special types of elites whose influence can be enormous. They may function as facilitators of and barriers to successful change programs. Moreover, they may or may not be explicitly aware of their role in the influence process. Influentials such as opinion leaders play a central role in much social science theory, most notably in Merton's (1968, pp. 441–474) discussion of patterns of influence. Merton's observations concerning localite and cosmo-polite leadership lend credibility to this generalization. Further support is found in the works of political scientists such as Robert Dahl (1966). Clearly, it is necessary to identify opinion leaders and, through cooptation and other methods, to involve them in the change process.

Opinion leaders must be used deliberately and creatively. By deliberately, we mean that the change agent or agency should specify for itself the particular task that the opinion leaders are to perform: (1) What opinions are the opinion leaders supposed to reinforce and/or change? (2) Whose opinions are they to influence? (3) Are the opinion leaders to act their roles in a formal or informal manner? (4) At what stages in the particular change process is their involvement most important? Creative use of opinion leaders and other influentials involves altering the content of opinion and informa-tion they pass on, developing opinion leadership where none exists, and using opinion leaders to discourage acceptance of undesirable alternatives to the advocated change.

Generalization 7. "Change agent success is positively related to his cred-ibility in the eyes of his clients" (p. 245). Credibility has a number of dimen-sions, and the change agent must determine which type(s) of credibility is most appropriate for his setting and stress his possession of that type of cre-dibility. First there is truth credibility—is the change agent providing advice or services where he honestly believes they are appropriate? Second, there is technical credibility—does the change agent possess the requisite informa-tion and skills to ascertain that the change advocated will in fact work? A third dimension concerns motivational credibility—does the change agent have ulterior motives for securing a change that could cause him to act in a manner inconsistent with the interests of the client system? These and other dimensions of credibility are not mutually exclusive, although they may be differentially important in different situations.

Generalization 8. "A change agent's success is positively related to his efforts in increasing his client's ability to evaluate innovations" (p. 247). This generalization has broad-based theoretical support in the literature on clinical psychology and psychiatric treatment models. It is important that the change agent create an openness and willingness to evaluate innovations. Coupled with this is the change agent's provision of the client system with the skills and other resources to evaluate innovations. This has ranged from providing client system scientists and leaders with training in the design and conduct of social and agricultural field experiments to the provision of funds to hire experts to evaluate innovations. The survey feedback technique used in organizational development work is another useful means for evaluating new programs.

CHANGE AGENT ERRORS

The change agent may commit a variety of errors that can limit the effectiveness of his program or team (Kahneman and Schild, 1966). One of the major sources of trouble lies in the area of deficient planning. The discussion here focuses on some of the more commonly encountered planning deficiencies exhibited by change team leaders or individual change agents.

Premature Commitment to Change Strategies and Tactics

One error a change agent can make is to commit himself too soon in the program to a particular set of change strategies and tactics. This problem is more severe when this commitment is made publically, making it personally difficult to alter his strategies. The problem is compounded when significant other persons are not consulted prior to making a public statement. One newly appointed regional change team leader in a community development project in Thailand called a staff meeting his third day on the job and described in detail the various approaches to be followed in achieving certain objectives in the communities under his supervision. Strong opposition developed among both host country and United States staff personnel because they were not consulted. Interestingly, the opposition developed even though the staff personnel were in basic agreement with most of what the regional leader or director was advocating. Approximately one month of very vigorous conciliatory activity was needed on the part of a senior staff advisor who realized what the problem was. The opposition was not expressed openly and hostilely but rather surfaced in the form of a large volume of "clarification" requests, slow implementation of activities, and letters or reports to the regional advisor describing client system resistance to his strategies despite a successful past use of some of the very same strategies in the exact same communities.

Committing oneself too early to a particular stratagem may be the product of poor advance planning. Advance planning should involve the identification of the relevant groups affected by the change team's efforts, their interdependencies, and their need to feel involved in both goal-setting and strategy-design processes. Typically, the importance of cooptation is overlooked because the change agent may underestimate the value of the contribution members of the client system can make in helping structure and focus the change process. Also, pressure for quick action may dictate bypassing the cooptation process, which tends to be time consuming. There is still another dimension of this issue. Asking for advice seems to dictate that some of it be followed. One United Nations nutrition expert, when asked why more use· was not made of host country nutritionists replied, "When they are asked for advice they expect it to be used and all sorts of criticism follows when I don't use it." When pressed further, this same nutritionist indicated that she felt the host country experts were seeking her job, suggesting the possibility of personal threat or insecurity as a factor inhibiting more effective use of local resources in situations of this nature.

Failure to Get Client Participation

Two sometimes mistaken assumptions a change agent can make is to assume that members of the client system do not want to participate actively in various aspects of change programs directed at them and/or that they are unable to provide useful information through participation. An example illustrating that these can be mistaken assumptions is provided by a former teacher in the Peace Corps. This person spent long days for an entire week in a village of approximately 500 adults asking various adults selected on a nonrandom basis how they would go about convincing more parents to send their children to school. Many suggestions were made that were put into operation with good results. Moreover, many adults offered to provide her with assistance both in and out of the classroom, acting as change agents and even what amounted to truant officers. An agricultural change agent in Indonesia reports acquiring, without any overt solicitation, the voluntary services of several farmers as adjunct change agents as a result of his soliciting advice from farmers as to how they would interest and induce other farmers to try new planting techniques.

Failure to Consider Informal System

Another planning deficiency occurs when formal activities are not preceded by adequate "prospecting." The sudden creation of a formal commission, a quick institutionalization of a change team, hiring a consultant, or any other rapid implementation of a vehicle for implementing a formal change program has been associated with problems. The problem may be in the form of resistance, as when cooptation is not used. Also, the formal structure may be imposed with an insufficient appreciation of the way informal social processes function. Knowledge of the informal processes enables social planners to build formal systems that take advantage of favorable processes and try to offset negatively functioning processes. The change agent or team leader is likely to overlook the informal system in his planning if he is overly confident in his own perception of the needs and change strategies or if he has a stereotype picture of the client system. One USAID official complained that one of the biggest headaches he experiences with consultants or newly arrived staff personnel is that they have done too much homework in terms of generating solutions prior to their arrival on site. In eagerness, they tend to start imposing their solutions rather quickly, having knowledge only of the formal patterns of authority, communication, and so forth.

Pressure for rapid change from the change agency and client system may cause the change agent to start without adequate researching of the informal system. There may be self-induced pressure for rapid positive results stemming from a desire to enhance one's reputation as a change agent.

The informal social system can be very large even when the formal system is small in terms of the number of different roles in the client system. For example, in the policy formulation process, it is not sufficient to be sensitive only to informal flows of information and influence within the agency or department responsible for policy formulation in the social context of concern. It is also necessary to consider the roles played by individuals and groups outside the formal social structure. This may involve members of opposing and supporting political parties, citizen action groups, unions, lobbyists, and even spouses. The change agent must consider how the advocated change will affect these groups and what their perceived needs for change are, as well as what their expectations may be regarding the behavior of the change team and the consequences of the intervention.

Failure to Identify Individuals Open to Change

One error that is sometimes made in the planning process is an inaccurate perception of whose viewpoints in the client system must be considered fixed and be taken as given and whose viewpoints might be the object of change efforts. There are instances ranging from industrial settings to rural communities in which the change team believes that key influentials are persons whose ideas and attitudes cannot be changed in either positive or negative directions. Influentials may be readily open to changing their opinions, given that adequate information is provided in an appropriate manner, for example, in a seemingly unbiased way. If an influential person or group is favorably disposed toward an innovation, reinforcing information should be provided on a routine basis, particularly if the advocated change is controversial and resisting groups are trying to alter the influential party's feelings and beliefs. Thus, during the planning period, influentials must not only be identified, but specific action must be programmed to create or maintain a favorable disposition among them.

MOTIVATION OF CHANGE AGENTS

Direct Financial Incentives

There appear to be basically three methods for motivating change agents: (1) direct financial incentives, (2) indirect financial incentives, and (3) nonfinancial incentives. These methods should not and cannot be used to overcome a basic lack of achievement need. Fixed salary is one of the most commonly used incentives for change agents, ranging from industrial salesmen to field workers in rural family planning programs. Of course, salary should not be considered strictly as fixed but subject to change with an improvement in performance. In many instances there are ranges of fixed salaries that depend on the change agent's length of employment in the program. Fixed compensation is usually desirable where the change agency wants to maintain maximum control over the change agent. For example, the field worker who is paid for the number of persons adopting contraceptives will devote most of his or her time to "easy" prospects when in fact hard-core resistors to family planning should be the targets. Similarly, the salesman who is paid on a straight commission basis is less likely to stress items having small profit for him, no matter how important that product is to the company directly, or indirectly by stimulating sales of other products in the company line. A fixed salary scheme also enables the change manager to specify more exactly what the change agent should do.

Commissions, either alone or in connection with a fixed salary, introduce additional flexibility in compensation. Commission plans are varied and are not presented here. Commissions, however they are based, have the advantage of tying rewards (money) directly to results (sales); they tend to attract self-confident agents, and they tend to retain the most successful change agents. One problem, however, is that agents' attentions focus on immediate sales or adoptions and tend to disregard long-term client relations. For example, canvassers, as men seeking other persons for vasectomies are called, who were paid on a commission basis, tended to focus only on persons who were ready for a vasectomy and ignore those who needed to be educated or better informed. Generally, where considerable service or other follow-up is needed in terms of providing information and postadoption assistance or counseling, a major commission plan is not well advised. This situation is characteristic of many contexts in health, education, and welfare. Often, in commercial contexts of this nature, a combination plan involving fixed salary and commission is used.

Bonuses are another financial incentive. Unlike the commission, which is related to each adopter or client obtained, the bonus is tied to a certain quota of adopters and is awarded when that quota is exceeded. Bonuses may also be given to change agents for each new client who has adopted some practice or for the continued use of the practice by a given client beyond a fixed period of time, for example, one year in the case of oral contraceptives.

Indirect Financial Incentives

There are various indirect financial incentives. One is in the form of fringe benefits such as paid vacations, pensions, and various types of medical and life insurance. In commercial contexts, there is some question about the relative importance of these to management and salesmen. One problem is that employees are generally not fully aware of the fringe benefits, particularly in terms of their real cash equivalence. Some innovative commercial firms are experimenting with various bonus schemes such as increasing fringe benefits as the employee becomes older or as his achievements vary. Varying fringe benefits with performance is expected to provide some of the same stimulus that a straight commission yields.

Another indirect financial incentive widely used in commercial contexts but seldom in noncommercial social change-oriented programs is the use of contests. This technique has high potential in motivating change agents in family planning, education, agriculture, social welfare, and other kindred nonprofit contexts. The values of contests are manifold:

1. They may prevent slumps in client recruitment when quotas are used. This is a common problem in family planning programs using target systems for field workers—once the target is achieved, recruitment of new contraceptive adopters fall off sharply. The contest in this instance would take place after the target has been achieved.
2. If tied to introducing innovations, they may result in a shifting of the traditional diffusion curve to the left. This has worked in agricultural development in Southeast Asia.

3. Contests may also be used to stimulate change agents to promote innovations that they personally do not favor or that are meeting with resistance from the client group.

4. Contests also place greater pressure on change agents to secure an adoption of an innovation by an individual or group more quickly than might otherwise be the case. This, then, is a method for encouraging change agents to prevent protracted decision making by client systems.

5. Contests could also be effective in stimulating change agents to contact hard-core resistors to change.

6. Contests could also be tied to the dissemination of information and educational materials relating to an advocated change.

In commercial contexts, contest prizes have included cash, merchandise, travel, and honorific items such as trophies and plaques. There is some feeling that noncash awards tend to be overvalued by recipients and remembered longer. Awards that involve family or friends have proven to be particularly effective. Contests are not without limitations. A major drawback is that they become expected and are another routine form of compensation and thus lose their incentive power. Care must also be exercised in establishing contests: (1) goals must be explicit; (2) the theme of the contest, for example, attracting resistors, must be clear; (3) the timing and period of the contest must be thoughtfully considered; (4) the contest itself must be promoted to establish not only awareness of it but also the fact that everyone has a reasonable chance of winning; (5) the types of awards must be carefully selected in terms of their significance to change agents; and (6) the procedure for distributing awards must be considered, for example, a major ceremony or use of the mail.

DIMENSIONS OF CHOICE IN SPECIFYING THE CHANGE AGENT ROLE

An initial consideration in designing a program for change is the specification of the type of change agent to be utilized. Although there has been some discussion in the literature of the change agent role (see Lippitt, Watson, and Westley, 1958; Rogers and Shoemaker, 1971), there has been little discussion of the kinds of choices the manager of change faces in specifying the change agent role. This section of the chapter discusses (1) the different dimensions of choice the manager may have in specifying the change agent role, (2) the tasks of the change agent and how these various dimensions affect the performance of the task, and (3) the optimally

Table 9.1   Analysis of Change Agent Dimensions of Specification and Change Agent Tasks

Tasks of the Change Agent	Change Agent Dimensions					
	Internal Change Agent vs. External Change Agent		Single Change Agent vs. Change Agent Team		Homophilous Change Agent vs. Heterophilous Change Agent	
I. Establishment of a relationship						
A. Acknowledgement viewed as legitimate	−	+	−	+	−	+
B. Sharing of expectations						
1. Change agent view and methods of operation	−	+	N/D	N/D?	−	+
2. Client system expectations	−	+	N/D	N/D	−	+
C. Sanction power base of change agent	−	+	−	+	N/D	N/D
II. Diagnosis						
A. Understanding what problem change issue is	N/D	N/D	−	+	+	−
B. Independent data collection						
1. Change issue	−	+	−	+	+	−
2. Need for change	−	+	−	+	+	−
3. Openness to change	−	+	−	+	+	−
4. Resources available to change	N/D	N/D	−	+	+	−
5. Commitment to change	−	+	−	+	+	−
C. Methods that can be used in diagnosis						
1. Observation	+	−	−	+	+	−

specified change agent role. The objectives here are twofold: (1) to sensitize the manager to the various choices one might consider in specifying the change agent role and (2) to raise some issues about the structure of the change agent's role that might stimulate future research.

Some of the dimensions the manager in the client system can or should consider in specifying the change agent role are (1) whether the change agent is an actual member of the client system (i.e., internal) or whether he is someone from outside the client system (i.e., external) who has been called into the client system; (2) whether the change agent is a single person or a team of change agents is used; (3) whether the change agent, as Rogers and Bhowmik (1971) have indicated, is similar to members of the

Table 9.1 (Continued)

Tasks of the Change Agent	Change Agent Dimensions					
	Internal Change Agent vs. External Change Agent		Single Change Agent vs. Change Agent Team		Homophilous Change Agent vs. Heterophilous Change Agent	
2. Interview	–	+	–	+	+	–
3. Collect questionnaire data	N/D	N/D	–	+	N/D	N/D
III. Select correct helping role						
A. Feedback of diagnosis	–	+	–	+	+	–
B. Playing different helping roles	–	+	–	+	+	–
1. Expert	–	+	–	+	+	–
2. Catalyst role	N/D	N/D	–	+	+	–
3. Process consultant	–	+	–	+	+	–
IV. Determining change objectives						
A. Nature, scope, involvement	N/D	N/D	–	+	+	–
V. Dealing with resistance						
A. Identify sources	+	–	–	+	+	–
B. Understand clients' perception	+	–	–	+	+	–
C. Anticipate sources of resistance	+	–	–	+	+	–
D. Identify long-run benefits	N/D	N/D	–	+	+	–
VI. Maintenance of change						
A. Institutionalize change	N/D	N/D	–	+	+	–
B. Internal support	+	–	–	+	+	–

client system in attributes (homophilous) or is different with respect to certain traits (heterophilous). Is the change agent perceived by the client system as being representative of them?

The following discussion focuses on six tasks of the change agent (see Lippitt et al., 1958) and discusses for each one the relative advantage that each aspect of these three dimensions has on the particular task. This process is summarized in Table 9.1, which relates the different dimensions to the change agent tasks and identifies by a + or a – the advantage-disadvantage aspect of a dimension. N/D means that, on that particular change agent task, no distinct advantage can be given to one or the other aspect of the dimension in facilitating the performance of that task.

CHANGE AGENT TASKS

Establishing a Relationship with Client System

In accomplishing this primary task of establishing a relationship, there are several things the change agent must do. For example, (1) the client system must come to see the change agent as a competent and legitimate person whose task is to help it change its organization. (2) There must be an exchange of expectations about the change process between the change agent and the client system. By this exchange, the client system can get a better idea about what the change agent feels he can do for the system. By finding out the expectations of the client system, the change agent is better able to correct any unrealistic expectations on the part of the client system that could lead to a feeling of failure and frustration if these expectations are not met. (3) Adequate sanction must be given to the change target.

INTERNAL VS. EXTERNAL CHANGE AGENT. A frequent choice dilemma for the manager is whether to select a change agent from within the target system or to select an external change agent. Certainly there are instances in which an external and internal change agent may be used together, although often the internal ends up playing a coordinator role. Someti nes, too, a gatekeeper or intermediary exists who is a member of the change target as well as a member of the change agency. Just as frequently, or perhaps more so, a change manager is faced with the choice of selecting as an agent of change a person within the target system or someone not recognized as a member of that system. It can be a dilemma in that potential change agents within a target group may lack the technical competence an external agent may have, whereas the external agent may lack the familiarity with the target group the internal agent possesses. The external change agent may have the advantage in both being acknowledged and in sharing expectations. The external change agent is "being called in" and thus comes in as an outside expert with certain skills to be made available to the client system. On the other hand, the internal change agent is a member of the client system who is temporarily occupying the change agent role. It is somewhat more difficult for him to establish himself as the expert, given his other roles in the system (Scurrah et al., 1971). Although the external change agent is seen as an outsider, his involvement may still be desirable. He may establish himself as more objective by virtue of his lack of prior or existing client system ties. He may also have ascribed to him a higher degree of professionalism, which enhances his credibility. In the sharing of expectations with the client system, the external change agent is again given the advantage, although this may be true primarily in formal organizations such as business firms and hospitals and much less true in rural communities in developing societies. He is seen as being more objective and professional, and thus his motives for involvement as a change agent are less open to suspicion by the client system. Government representatives in community development, on one hand, may be viewed with considerable skepticism at the outset of their community work. The internal change agent, on the other hand, is an insider and may be seen as having more of a stake in the change; thus his objectivity and motives might be questioned. Also, because the external change agent is seen as more objective, members of the client system are likely to be more open in sharing their expectations concerning the change program. This implies, too, that where data considered as confidential by the members of a client system are sought, an outsider may be preferred.

With respect to the sanction or power base of the change agent, it would appear that, again, the external change agent has the advantage. There are several bases of power such as reward, coercive, expert, referent, or legitimate, as defined by French and Raven (1959). Although both internal or external change agents may possess some formal power (reward and coercive), the external change agent is somewhat more likely to develop some additional power base by being seen as an expert, and, as pointed out above, more legitimate than the internal change agent. Thus it appears that the external change agent would have a broader base of influence over the client system. Again, this generalization is more tentative with conventional formal organizations in relatively well-developed societies. Whenever a member of a target system is provided, through training, with the skills normally possessed by outside change agents, he will often be the preferred source of assistance in any change venture, even those undertaken by external change agencies. Training members of school systems in group problem solving is another example of providing expertise to insiders who supplement external change agents.

SINGLE VS. TEAM CHANGE AGENTS. The change agent team is likely to have the advantage over the single agent in being seen as legitimate. Because there are several change agents, the client system is more likely to see them as bringing more skills and expertise to bear on the change issue than a single change agent could. In the sharing of expectations, no relative advantage is seen between the single versus the change agent team. The only problem might be for the change agent team. It is important that the members of the change agent team be consistent among themselves concerning the expectations they communicate to the client system. If inconsistencies in expectations are communicated, these could cause some ambiguity among client system members as to just what the change agents were trying to accomplish and could cast doubt on the abilities or intent of the change agents.

Because of the broader skills that they could bring to the client system, it would be possible for teams to develop a broader power base to include expert and legitimate power. This would be in addition to any formal power they might have.

HOMOPHILOUS VS. HETEROPHILOUS CHANGE AGENT. Homophily-heterophily is concerned with the relationship between source and receiver in a communication exchange. The notions of heterophily and homophily are important because the change agent is often perceived to differ from the client system. This can have important implications for the selection, training, and effectiveness of change agents. Rogers and Bhowmik (1971, p. 529) have indicated that, when communicators are homophilous, that is, share common meanings, attitudes, and beliefs, communication is likely to be more effective. In heterophilous relationships, the interaction is likely to cause some message distortion, some restriction of communication channels, etc. (see p. 529).

With respect to acknowledgment of having expertise and being seen as legitimate in the change agent role, it may be that the heterophilous change agent has the advantage. This is likely if he is perceived as different, because he is seen as being an expert in the particular area in which the organization is considering some changes. It may be that the change agent can be most influential here if he is perceived as heterophilous in the sense of being an expert while at the same time he is seen as sharing many of the same attitudes, values, and beliefs of the client system.

The heterophilous change agent might be more successful in communicating his views to the client if he is seen as an expert. Again, he is likely to be seen as more legitimate in this role than just "another member" of the client system. This same argument holds in explaining why the heterophilous change agent might get a more open sharing of the client's expectations regarding the change. With respect to sanction or power base, there does not appear to be a clear advantage for either the homophilous or heterophilous change agent. Given that both of them would have some formal base of power from the client system, there would be some trade-offs on the other bases of power. For example, the heterophilous change agent could establish some power if he was perceived as being an expert. On the other hand, the homophilous change agent could establish some referent power, since client system members would perhaps be better able to identify with him given their perception of their similarity to him.

The principle of status inconsistency raises an interesting implication for the selection and/or training of change agents. Status inconsistency refers to discrepancies among various indicators of an individual's social status. Rogers and Bhowmik suggest the following proposition based in part on this principle: "Heterophilous communication is more effective when source and/or receiver are status inconsistents (p. 533)." The implication of this finding is to recruit change agents who are internally heterophilous if that agent must work with groups of widely ranging statuses. The change agent should be trained to enact the status that is most consistent with the group or individual with whom he is working. Thus, within a relatively short period of time, as the change agent works with different groups, he will display varying statuses corresponding to his client's statuses.

Diagnosis

The objective of diagnosis is that the change agent gain an understanding of the issues around the change the client system is contemplating. It is important that the change agent collect some data beyond the immediate views of one or two people in the system. Data should be collected from a sample of client system members concerning their perception of some of the following factors: (1) the change issues, (2) the need for change in the system, (3) the openness to change in the system, (4) the resources available for change, (5) the commitment of members of the social system to the change process. The sample selected may vary in terms of representativeness. For example, a random sample may be desirable to gauge openness to change in the system, whereas a panel of experts might be used to identify sources of problems.

INTERNAL VS. EXTERNAL CHANGE AGENT. In looking at this dimension under diagnosis in Table 9.1, there appears to be no overall advantage for either the internal or external change agent in understanding the change issue that the client system has identified. The external change agent has more objectivity in looking at the organization and can take a more flexible approach to understanding the change issue. On the other hand, the internal change agent, because he is a member of the client system, is likely to be more involved in the change issue and thus may be more biased in his perceptions. However, being a member of the client system might initially provide the internal change agent with more information concerning the change issue.

In terms of independently collecting data from the client system, again there are trade-offs between the internal and external agent. In collecting data about the change issue, the perception of client members of the need for change in the system, the openness to change, and commitment to change, the external change agent may have the advantage. Because he is not a member of the system, he is likely to be seen as more objective and not having any vested interest in the change. Thus client system members might feel safer in confiding in him. This is particularly true if the problem is an emotionally charged one. With respect to identifying resources available for change, both have some advantages. The internal change agent knows the system better and at least initially may have a more thorough understanding of the resources available. However, the external change agent, because of his expertise, might be able to help the client system develop more resources for change. For example, an external consultant may be able to do some team building sessions with the client system to help them improve their interpersonal relationships, which could then help them interact more effectively (Beckhard, 1969).

In utilizing different data collection methods during the diagnosis, there are again trade-offs between the internal and external change agent. With respect to observation of the day-to-day activities of the client system, the internal change agent would have the advantage. He is a member of the system and would be less conspicuous than an outsider trying to observe. In using interviews of client system members, the external change agent, if perceived as an expert, would have the advantage. He would more likely be seen as more objective than the internal change agent, and respondents would have the tendency to be more open with him. In utilizing questionnaires neither change agent here is seen as having a particular advantage.

SINGLE VS. TEAM CHANGE AGENTS.   In all aspects of the diagnostic function, the change agent team is seen as having a distinct advantage over the single change agent. The change agent team simply is likely to have more skills available and through a division of labor can carry out more comprehensively the various tasks under the diagnostic function.

HOMOPHILOUS VS. HETEROPHILOUS CHANGE AGENT.   This dimension focuses on the relationship between the change agent and the client system concerning how similar (homophilous) the members of the client system perceive the change agent to be on attitudes, beliefs, and so on, with them. As Rogers and Bhowmik (1971) indicate, the more homophilous the relationship, the more effective the communication between the parties. Thus it appears that, in the diagnostic stage in which communication between the client system and the change agent is critical, a more homophilous change agent will have the advantage. He will be able to communicate much more effectively with the client system; thus his diagnosis will potentially be more accurate. Thus, the more similar the change agent can be perceived by the client system in terms of attitudes, values, beliefs, and appearance, the more effective will be his interaction with the client system. Even if the change agent is seen as an expert, which may make him heterophilous in one sense, he should try to empathize with the client system, which will make him at least social-psychologically more homophilous (Rogers and Bhowmik, 1971, p. 535). It is important, however, that empathy not be allowed to overshadow expertise. Empathy or identification may produce a common perception of a problem, but what may be in some way objectively necessary is a perspective significantly different from the client's.

Selecting the Correct Helping Role

There are two main tasks here: (1) Feeding back the diagnosis to the client system and (2) choosing which of several helping roles he might perform. There are several role choices: playing the expert role in which he provides direct solutions to the client system; playing the catalyst role in which he becomes an advocate of the change; and playing the process consultant role (Schein, 1969; Zaltman et al., 1977).

INTERNAL VS. EXTERNAL CHANGE AGENT. In feeding back the diagnosis, the external change agent seems to have the advantage. He is not a member of the client system, so again he is likely to be perceived as objective and as having little motive for biasing the feedback. A bias could be ascribed to an internal change agent, who might be identified with certain factions in the client system. An external change agent may have a bias ascribed to him by virtue of his association with a change agency or the person within an organization who brought him in as a consultant. Under such circumstances, it is necessary to indicate, within the bounds of accuracy, that the agent has at least some degree of independence from his sponsor's position. This is especially important because, although various organizations and individuals within those agencies may be required to take a public position, they may also be tolerant and even actively (but quietly) supportive of a very different position or policy.

In terms of filling any of the different roles of expert, catalyst, or process consultant, neither internal nor external change agent has a clear advantage. The external change agent who is called into the organization will have an easier time playing the role of the expert. He has usually had specialized training and is more clearly perceived as an expert. On the other hand, the internal change agent is, just another member of the system. Thus it will be more difficult for him to establish himself as the expert.

In filling the role of the catalyst or advocate for change, there does not seem to be a clear advantage to either the internal or external agent. The external agent is seen as the expert, and he can use this power base in trying to win support. On the other hand, the internal change agent is an ongoing member of the system; thus he can play the advocate role indefinitely.

In filling the process consultant role, it appears that the external change agent has the advantage. His potential of being seen as objective and neutral will facilitate his ability to help the client to perceive, understand, and act on process events (interpersonal relations, communications, intergroup relations) that occur in the client system (Schein, 1969, p. 9).

SINGLE VS. TEAM CHANGE AGENTS. Once again, as indicated in Table 9-1, the change agent team, because of the greater resources and potential for division of labor, is seen as being more effective for feeding back the diagnosis and in selecting the correct helping role. For example, the clear advantage of the change agent team might rest in its ability to have different members play the expert, catalyst, and process consultant roles. This approach would be more comprehensive than what a single agent could perform in occupying just one of the change agent roles.

HOMOPHILOUS VS. HETEROPHILOUS CHANGE AGENT. In feeding back the diagnosis, the more homophilous change agent might have the advantage, since he can communicate more effectively with the client system. There would be, at least initially, more openness to the more homophilous change agent.

In playing the different helping roles, the advantages again seem to favor the more homophilous change agent. In playing the expert role, catalyst role, or process consultant role, which creates some heterophily, the more the change agent can be similar in other attitudes, values, beliefs, appearance, the better the communication between him and the client system. Playing any of these roles may create some heterophily, but the more the change agent is similar on these other dimensions and empathizes with the client system, the more the change agent will be perceived overall as homophilous.

Determining Change Objectives

The specific issues at this stage of the change process are: What is the nature and scope of the change program going to be? Who are the people that will be affected by the change? What exactly is the change program trying to change—attitudes, values, beliefs, or behavior? What are the leverage points (Lippitt et al., 1958, p. 100; Zaltman et al., 1977); that is, who are the persons or subgroups that the change agent should work with if he is to have some impact on the system?

INTERNAL VS. EXTERNAL CHANGE AGENT. In determining the nature and scope of the change program, we are concerned with such things as: Who will be affected by the change program? Who are the influential people in the client system that the change agent should focus on in his influence attempts? Both internal and external change agents have advantages here. The internal change agent has some clear advantages in that he knows the system well and should be in a position to know the key influential people in the client system who might be the initial change targets. These individuals could then be used as allies by the change agent. On the other hand, the external change agent with special training and prior experience and potentially a more objective view of the client system may be in a better position to determine the nature and scope of the change within the client system.

SINGLE VS. TEAM CHANGE AGENTS. Again, the change agent team with its potentially greater resources and division of labor should be more efficient than the single change agent in determining the scope and objectives. Because of their larger numbers, they are better able to deal with a larger number of leverage points in a social system.

HOMOPHILOUS VS. HETEROPHILOUS CHANGE AGENT. The homophilous change agent is seen as more effective in determining the scope, nature, and involvement in the change program. This is based on the more effective communication flow that this type of change agent is predicted to have with the client system. The homophilous change agent is likely to get more accurate and valid data from the client system.

Dealing with Resistance to Change

There are several things that the change agent may do in attempting to reduce resistance to change: (1) Find out what the sources of resistance are. (2) Try and reduce these sources of resistance rather than putting more pressure on the system to change. Resistance can be reduced by the change agent by (1) involving members of the client system in the determination of the change objects; (2) finding out client members' perception of the anticipated changes in order to see if there are any misunderstandings about the change program; (3) trying to anticipate sources of resistance on the part of the client system and thus address these before the client system can mobilize them; (4) pointing out the long-run benefits of the change to the system and how they outweigh any immediate problems (see Watson, 1966).

INTERNAL VS. EXTERNAL CHANGE AGENT. The internal change agent might have some advantages in dealing with resistance. He is a member of the client system; thus he is in a potentially better position to identify those people or groups who might be resistors. Also, because he at least shares some values and norms in common with system members, he is in a better position to understand how clients may perceive the change. As Watson (1971) has indicated, resistance can often be reduced if the change agent can anticipate sources of resistance or deal with misunderstandings about the change issue when they arise. The internal change agent, by virtue of his membership in the system, is potentially in a better position to do this.

In identifying the long-run benefits of the change to the system, neither the internal or external agent has a clear advantage. The internal change agent has initially more in-depth knowledge about the client and should be able to point out ways that the change can be useful to the organization. On the other hand, the external change agent can bring his experience from other systems to bear and thus point out long-run benefits that client system members may never have thought about.

SINGLE VS. TEAM CHANGE AGENTS. From a resource standpoint, the change agent team is again seen as more effective. Each member of the team can concentrate on fewer aspects of the resistance. Additionally, a change team is more likely to have a particular skill necessary to reduce resistance than will the individual change agent.

HOMOPHILOUS VS. HETEROPHILOUS CHANGE AGENT. The homophilous change agent is potentially more effective in dealing with resistance. This is because the homophilous change agent can more effectively communicate with the client system.

Maintenance of Change

A final task facing the change agent is how to maintain or refreeze a change into the ongoing structure and function of the client system. It is at this point in the cycle that many change attempts falter (Zaltman et al., 1977, especially Ch. 8). A change or innovation is introduced into a system and perhaps has an immediate impact, but then in a short time the change fades and eventually disappears. The task of maintaining change is important from two perspectives: first, from the perspective of the client system, it is important if there is some concern for obtaining long-term change; second, from the perspective of the change agent, the maintenance issue is important because it affects the timing of the termination of the relationship.

In trying to enhance the maintenance of change in a system, there are several strategies open to the change agent. First, and perhaps most appropriate, is to attempt to institutionalize the change process. The institutionalization of change is the development of a set of shared, learned norms among members in the system that define the new change or innovation as a legitimate aspect of carrying out one's role as a manager. This institutionalization process focuses on both the individual and the structure of the client system. It focuses on the individual in that it emphasizes the legitimacy of the new change and creates the expectation among organizational members that this change is part of one's job definition and that one should expect to carry out this activity in one's role. This institutionalization process also focuses on the structure of the organization in that it creates a set of role expectations on the part of organizational members that supports and reinforces this change by other organizational members where role expectations are congruent with the change.

Second, the change agent can remain in the client system and provide support as they work through difficulties they may experience as they implement the change.

In any event, the change agent is always faced with a difficult choice in deciding when to terminate as a change agent with the client system. There is always the potential problem of the change fading out when the change agent is not around to provide support and expertise.

INTERNAL VS. EXTERNAL CHANGE AGENT.   The internal change agent has a differential advantage in attempting to institutionalize change. He is better able to maintain pressure for maximum involvement and to sense, by virtue of his greater familiarity with the system, hidden obstacles to institutionalizing change. The internal change agent is also available as a more or less continuing source of support for the advocated change.

SINGLE VS. TEAM CHANGE AGENTS.   Again, the change agent team, because of its potential resources, is seen as having a clear advantage over the single change agent.

HOMOPHILOUS VS. HETEROPHILOUS CHANGE AGENT.   In maintaining change, the homophilous change agent is seen as having a clear advantage. Because of his similarity to client system members, his communication with them is improved. Also, because he is perceived as similar to client system members and thus communicates more effectively, he is better able to persuade them to change and then maintain that change.

(Rogers and Bhowmik, 1971, p. 534). The perceived similarity of the change agent facilitates identification of client system members with him and his views. This can facilitate not only the change process (Bennis et al., 1968), but also the refreezing process, since the homophilous change agent is likely to be seen as more supportive and as a more credible source of reinforcement for the change itself.

OPTIMALLY SPECIFIED CHANGE AGENT ROLE

Having discussed the various dimensions along which the change agent role can be specified, what conclusions can be drawn? In reviewing Table 9.1, it is clear that no dimension has a clear advantage in performing the six tasks of the change agent. However, from the preceding discussion, it appears that we can specify the ideal change agent as follows:

First, from Table 9.1, it can be recalled that the change agent team had, in general, a clear advantage over the single change agent, because the team had the potential for having more resources and a better division of labor.

Second, the team should consist of an internal and external change agent. The external change agent brings a certain degree of objectivity and broader perspective to the client system. An internal change agent can complement this with his understanding of the system. Also, the internal change agent, by virtue of being in the system, is in a better position to help maintain the changes that take place.

Third, although homophily is particularly important, it is best to think in terms of optimal mixes of homophilous and heterophilous attributes between change agent and client. Homophilly facilitates communication, whereas heterophilly in expertise may facilitate client reliance on the change agent as a source of information and hence a person with whom it may be desirable to interact.

This chapter can be summarized by the following general principles:

1. The change agent(s) should be sensitive to the needs and perspectives of the change target system in designing solutions to change situations.
2. The change agent(s) should build capabilities within the change target system so that a vacuum is not created when the change agent(s) leave(s) the system.
3. Change agents should always seek the simplest solution when working with a change problem.
4. Change agents should have administrative capabilities so that the change process can be managed effectively.

5. The change agent(s) should strive to maintain good interpersonal relations with persons in the change target system during the change

6. The change agent(s) should consider the change problem as stated by the target system as a hypothesis and then should seek additional information regarding the change situation before selecting a particular course of action.

7. The change agent(s) should be sensitive to and tolerant of the constraints that govern the change situation.

8. The change agent(s) should be prepared to operate under stressful conditions without getting defensive when confronted by the client system as it deals with its anxieties and frustrations associated with the change.

9. The change agent(s) should have the self-confidence and positive self-image to accept setbacks with poise and not project anger or frustration on to the change target.

10. The change agent(s) should be able to define the change program in a manner that is attractive to the various constituencies involved in the change program.

11. Change agents should strive to maximize their credibility in the eyes of the change target system in terms of the change agent's motives, competence, and truthfulness.

12. The change agent(s) should work through opinion leaders in the change target system.

13. The change agent(s) should expect an initial period in which change efforts will not be rewarded in terms of quick change in the target system.

14. To maximize the change target system's cooperation with the change agent(s), the change target should perceive that it has some freedom or free choice in entering into a relationship with the change agent(s).

15. The change agent(s) should get the change target system involved in problem definition and need specification to develop target system commitment to and trust in the change agent(s).

16. To maximize the change target system's cooperation with the change agent(s), the target system members should perceive that the change agent(s) is helping them to gain more influence over the change process.

17. The more similar (homophilous) in attitudes, values, and beliefs the change is to the change target system, the more motivated the target system will be in working cooperatively with the change agent(s).

18. There should be an exchange of expectations regarding the change process between the change agent and the change target system so that no misunderstandings occur regarding the change.

19. Change agents should strive to create as broad a power base (i.e., referent, legitimate, expertise) for their position in the change target system as possible.

20. In determining change objectives, the change agent(s) should consider the nature and scope of the change, the target of the change (attitudes, values, beliefs), who will be affected, and the initial key people to work within the change target system.

21. The optimally structured change agent would be a change agent team consisting of an internal and external change agent who are homophilous with the change target system.

Characteristics of Change Targets: Individuals

In this and the next chapter, we focus on the change target. In this chapter, we focus on the change process as it affects the individual and those characteristics of the individual that can affect change. In Chapter 11, we look at the nature of social systems and their characteristics as they can affect the change process. The objective in both this chapter and the next is to present the reader with a better understanding of the factors associated with the change target that must be considered in planning for change.

The process by which an individual adopts some new mode of behavior has been the subject of much interest among social scientists. Several

225

models of innovation adoption processes have been suggested during the past several years (Beal and Bohlen, 1962; Campbell, 1966; Howard and Sheth, 1969; Lavidge and Steiner, 1961; Robertson, 1971; Rogers, 1969, 1971). In most instances, these models have produced fruitful research but often have led to conflicting findings (Bell, 1964; Jacoby, 1971; and Zaltman, 1965). A number of criticisms (Miro and Bogue, no date; Zaltman and Stiff, 1973) of these models have been made and are noted briefly here. The criticisms are not totally applicable to all models, and where exceptions exist they are noted. Many of the models are insufficiently explanatory and predictive, an important attribute of any scientific conception. The models are good descriptions of overt, but not covert, behavior, with the possible exception of the Howard-Sheth model (1969) and the Robertson model (1971). With the exception of Campbell (1966), they all assume a highly rational decision maker, which, although consistent with contemporary social theory, tends to limit the varieties of the behavior studied. Generally, they assume a passive actor; they do not consider explicitly the possibility of planned intervention (Argyris, 1970; Havelock, 1971); with the possible exceptions of Eichholz (1963) and Zaltman and Stiff (1973), they do not take into account rejection or resistance processes; and, finally, apart from explicit learning behavior models, the adoption models often display too little social psychology and sociology theory.

The purpose of the first section of this chapter is to present an alternative model which, at least partially, avoids the chief limitations of other existing models. This model is presented in Figure 10.1. The solid lines represent the normal progression toward adoption or rejection, and the dotted lines represent the feedback mechanism. The particular model here has received some direct research support (Kennedy, 1974) and indirect support from a variety of research reports documenting the existence of one or more of the stages shown. Other research supports the stage sequencing (O'Brian, 1971).

PERCEPTION

The decision process begins with perception. Both the innovation and the need must be perceived by the individual for eventual adoption to occur. There is no clear evidence indicating whether the outcome of a decision is influenced by whether a need is perceived prior to the awareness of an innovation or afterwards. A felt need or the recognition that a given innovation is relevant to a felt need may develop only after considerable familiarity with the innovation is achieved (Zaltman and Pinson, 1974). If the client does not recognize the existence of a need, the change agent's task is to identify for the client exactly what the need is. Differences in change agent and client percep-

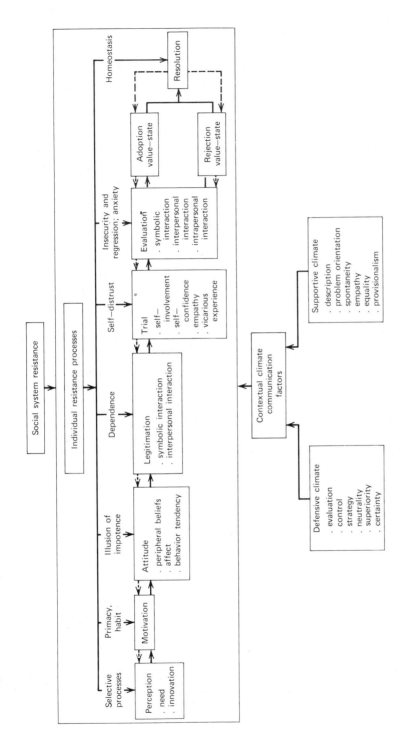

FIGURE 10.1 A resistance/adoption model.

227

tions of a problem will produce different definitions of what "real" needs are and hence differences in opinions about the appropriateness of a given advocated change.

Often investigators impose their perceptions on others in order to achieve "proper" classification, when the important perceptions are actually those of the client (Rogers and Shoemaker, 1971; Zaltman and Lin, 1971). Different research outcomes can be expected depending on whose view is employed. An outstanding example of this problem is given by Bennett (1969) in classifying newcomers in a dynamic society. They may be "precocious" when they adopt a behavior on their early exposure to it in the society, despite the fact that the product had already achieved wide acceptance. The precocious individual who adopts almost immediately because of the newness of the product to him is more likely an innovator, but may be classified as a laggard by the investigator who is unaware of the situation. Thus the investigator's perception of the situation may have an important, but perhaps unfortunate, impact on the research findings. We elaborate on this later in this chapter when discussing the current status of early adopter theory.

It it likely that perceptions of an innovation by the individual adopter change as he or she moves through the various stages of adoption. What is perceived as new or different in the early stages may eventually be perceived as commonplace in the later stages and vice versa. Changes in perception such as these would have important effects on individual behavior. Thus perception should be traced through the entire process for possible changes. Do innovators change their perceptions more rapidly than others? Or do they adopt because of the newness involved, whereas others require more familiarity with the product? Conceivably, too, early adopters may be such because they see the item as familiar (Zaltman and Dubois, 1971).

Many other perceptions are important. Self-perceptions by the individual or organization concerning the capacity to encompass and utilize an advocated change are important (Niehoff, 1966). Also, because an advocated change logically fits a particular need, it does not necessarily imply there will be an active interest in the change (Tanker, 1973). Another factor to consider is the change target's perception of the degree of commitment required by a change. If a major commitment of resources is perceived necessary for adoption, it is less likely that further interest will be expressed about the change without the direct intervention of the change agent at this stage. Yet another very important factor is the degree to which a client perceives himself or itself as having control over the change process. Resistance to incorporating a patient advisory council in a community public health agency was a partial result of a perceived lessening of control by agency officials (Bradshaw and Mapp, 1972).

The obvious factors affecting the individual at this stage are the selective processes, both selective perception and selective retention. More than just a result of currently held attitudes, these processes may originate as a result of the cultural (Rogers and Shoemaker, 1971), social (Flinn, 1970), or communicative climate (Gibb, 1961).

The major goals of the change agent with regard to this stage are to create or stimulate perception of a problem and solution and to shape the perceptions in a desirable way.

MOTIVATION

A necessary step in overcoming natural resistance to change is motivation. Behaviors that are comfortable (habit) are normally resistant to change, as are those which represent the first successful attempt at solving a given problem (primacy). Individual perceptions of the existing need and the innovation must be such as to provide the impetus for further action, thus overcoming these, and other resistances. The greater the sense of deprivation inherent in a felt, unmet need, the greater the propensity to seek a solution (Morrison, 1973). For sustained motivation throughout the decision-making process, there must also exist a belief that suitable solutions exist in the environment or that a particular innovation is a good candidate for solving a problem. The degree to which individuals feel they have control over those things that affect them appears to influence their motivation to search for or be open to ideas for social change. Niehoff (1966, pp. 26–31) and others (e.g., Rokeach, 1968, and Barnett, 1953) have identified many types of needs that serve as motivations, including educational, medical, economic, convenience, and achievement needs.

The major goal of the change agent at this stage is to actively encourage the potential adopter to be concerned about solving a problem.

ATTITUDE

The next stage of the process is attitude. The attitude stage contains three components that have been suggested by numerous sources: cognitive, affective, and behavioral. As the individual moves through this stage, he or she develops beliefs about the innovation based on information he developed in social interactions, in reading advertisements, in reviewing reports on the product published in various consumer magazines, and so on. Beliefs of this type have been classified as peripheral beliefs. They are based on information supplied by authorities who are viewed by the consumer as trustworthy (Rokeach, 1968). Work by Jacoby is enlightening for authority-based beliefs and action tendencies (Jacoby, 1971a, 1971b). Such beliefs are subject to change if the authorities from which they are derived change position, or they may be replaced if more basic beliefs contradict the peripheral beliefs.

The affective component at this point probably is not strong. It may be limited to liking or disliking the product or idea. The strength or intensity of this emotional component supplies the impetus for the behavior that follows. Greater intensity may mean a more rapid progression through the remaining stages of the process. Research on this component could reveal the actions of such factors as acceptance or repression of emotion in the behavior of innovators and laggards, as well as the effects of the strength of this emotional component on rapidity of adoption or rejection.

An important type of resistance at this stage is the individual's illusion of his own impotence. It seems quite likely that this could act to reduce the affect felt by the individual, and consequently reduce the likelihood of his behaving in a way the change agent desires. Such an outcome at this stage might cause the process to remain incomplete (Zaltman and Stiff, 1973).

Kelman and Warwick (1973, p. 25) identify five types of processes that help persons to cope with or adjust to their environment as it affects their achievement of certain goals. Attitudes which emerge from these coping processes tend to be reinforced, making change difficult. The five coping processes are "(1) efforts to manipulate aspects of the external environment to facilitate goal achievement; (2) efforts to come to grips with inner conflicts that cannot be resolved in direct and conscious ways; (3) efforts to find meaning and order in the environment; (4) efforts to enhance one's self-esteem and to actualize one's self-concept; (5) efforts to relate one's self to important others in the environment, particularly groups in which one holds or aspires to hold membership" The task for the change agent is to identify with which coping processes particular attitudes are connected and to adapt the advocated change accordingly. For example, in apparently direct sensitivity to item (3), developers of planned communities, particularly retirement communities that require major changes in life-style among new members, have attempted to present these communities in a manner that provides order and certainty for a period of life that may contain considerable uncertainty for couples nearing retirement. Similarly, commercial advertisers have long utilized the concept of the reference group (item 5) in developing appeals and designing copy for promoting new products. The technique followed is essentially grounded in balance theory (Heider, 1957; see also Homans, 1974). A favorable association is established in the advertising between the new product and an esteemed reference group with the intention that the consumer develop favorable attitudes toward the product. Resistance to innovative family planning techniques such as vasectomy and laparoscopy appears to be expressed by highly rational attitudes that, on in-depth probing, are "covers" for inner conflicts that are not resolvable in conscious thought. Attempts to confront such attitudes threaten the holder, who often becomes more tenacious in holding the attitude (see Kothandapani, 1971).

Attitudes must be approached carefully by the change agent. Attitudes alone, whether they are favorable or unfavorable to an advocated change, cannot be used to predict behavior. Nor are attitudes unalterable. Thus a change agent cannot assume that adoption will occur because favorable attitudes appear to exist, nor should efforts to alter attitudes unfavorable to change be considered impossible or fruitless. A particular attitude toward an innovation may be inconsistent with stronger competing attitudes, competing motives, inabilities to convert attitudes into behavior because of personal inabilities and/or externally imposed constraints, unexpected consequences of other behaviors, diffuseness of the attitude or attitude object, unforeseen extraneous events, and so forth. A very large literature exists on this topic (see Wicker, 1969, 1971; Weinstein, 1972; and Liska, 1975). Liska (1974) further details other problems such as the validity of attitude and behavior measures and the difficulty in treating multiple attitudes (both toward the attitude object and toward the manifest behavior—differences have been noted between attitudes toward vasectomy and attitudes toward the act of obtaining a vasectomy), particularly the assigning of relative weights to attitudes and isolating multicolinearity effects. Social support is very much related to the next stage in the social change decision process and very relevant to the attitude-behavior consistency problem. Congruence between social support and attitude is likely to produce attitude-behavior consistency. However, it is unclear whether, or perhaps when, (1) social support that is consistent with an attitude has an independent effect on behavior and for this reason helps create an attitude-behavior consistency situation, (2) social support acts directly only on attitudes, or (3) social support impacts directly on both behavior and attitude (Liska, 1974). From the change agent's view, the situation in which social support has direct positive impact on both attitudes and behavior is preferable. Thus, where favorable attitudes toward a proposed change exist, the change agent should demonstrate that there is considerable social support for the set of attitudes in question, thereby implying that the advocated behavioral change is also socially supported. Although their ethics are questionable, at least one abortion clinic in an Eastern city is known to have counseled unmarried, pregnant, potential clients during the initial contact, that a "recent study"—which was a total fiction—had shown a substantial minority of parents in the United States in favor of abortion and another substantial minority undecided. The intended result was to lessen the perceived stigma—the perceived lack of social support—by indicating that many more members of an apparently important reference group were, if not approving, at least not disapproving of the apparently favorable attitudes toward abortion that were held by the unmarried clients seeking counsel and medical assistance. The source of this example strongly believed that this markedly reduced the inner conflict experienced by many of the clients she counseled. The conflict was described as being between the client's own attitudes and values and those of significant others in her life.

LEGITIMATION

The legitimation stage occurs when the individual seeks reinforcement for an action being considered. The appropriateness of the action is of prime importance. This may be determined by the individual by observing the performance of the behavior by others in his group, or by seeking affirmation from his peers, or relatives, etc. Social interaction is an important criterion for recognizing occurrence of the legitimation stage. Although their study was based on another model, the work of Pareek and Singh (1969) lends support to this position. Their study showed increasing interactions that peaked just prior to the trial stage. Further research in this area is desirable.

The resistance process most likely to occur here is dependence. The fact that the individual looks to others for support is a reflection of an earlier stage of life in which parents supplied emotional support to the child. When people receive such support from others, they tend to adopt the outlook of those others, thus maintaining a conservative position toward change and perpetuating the status quo.

It is important at this stage for the change agent to demonstrate the social acceptability of an advocated change, as the abortion clinic just cited did. This may be accomplished by bringing the potential adopter into contact with people who have already successfully adopted the change.

TRIAL

In the trial stage, the individual will put the innovation to a personal test prior to complete acceptance. Occasionally, the nature of the innovation (Zaltman and Lin, 1971) or the situation may make personal testing impossible, and the individual may "try" the innovation through vicarious experience, with results similar to those of personal testing. The individual who lacks confidence in his own ability may have unsatisfactory experiences here.

At this stage, the change agent must facilitate experimentation with the change by the change target. As a minimum, the change agent should help his client imagine what it would be like to adopt or accept the change.

EVALUATION

Evaluation is a necessary formal step between the trial and adoption stages. Following trial, the individual will review the pros and cons of continued or increased use. The existence of this stage was uncovered in the Pareek and Singh study; their evidence was a high level of interaction following the trial stage in which interaction was almost nonexistent. Although it is quite likely that informal or very brief evaluation follows each stage in the adoption process, to review the situation to that point, a formal evaluation is probably necessary before a formal commitment is made. Therefore, evaluation should precede adoption in the model.

Symbolic adoption is very important at the evaluation stage and an excursus on this topic is in order here.

THE CONCEPT OF SYMBOLIC ADOPTION

The concept of symbolic adoption refers to a state in which the target person or group would accept a change if they could. The concept of symbolic adoption has been developed by Klonglan and Coward (1970).

Symbolic Adoption as a Point of Lag

Klonglan and Coward appropriately point out that there are cases in which individuals hold a favorable orientation toward an innovation but are not using it. They refer to cases of this nature as incomplete adoption for which they postulate two special forms: (1) constrained adoption, in which the individual is unable to use the innovation because of its nonavailability; and (2) anticipatory adoption, in which adoption has not occurred because of an inappropriate situation.

Ehlich (1969) specifies several somewhat more specific factors that may inhibit the behavioral expression of acquired attitudes or that otherwise create discrepancies between attitudes held and behavior. Among these intervening variables that contribute to incomplete adoption are: the inherent inexpressibility of the attitude, lack of clarity of a channel for behavior, inaccessibility and invisibility of acceptable means for implementing the attitude, the absence of learned ways of implementing the attitude, and physical and social barriers. The nature of the attitude object, that is, the innovation, will partially determine which of these intervening variables will be present. Theoretically, all the variables could operate as inhibitors when the attitude in question involves overt action or practices as well as verbal statements. The change agent must be sensitive to these intervening factors and attempt to reduce their impact.

Behavioral Adoption as a Point of Lag

It has been widely observed that there are many instances in which individuals may not hold a favorable attitude toward some idea, practice, or material object, and yet he or she may behave in a manner favorable to the attitude or object. Thus symbolic adoption, that is, "the adopter unit's (attitudinal) acceptance of the idea component of an innovation," may follow behavioral acceptance or may never occur at all despite behavioral acceptance. This raises the very important question as to what psychological, sociological, and situational conditions are instrumental in determining whether symbolic adoption of an innovation precedes or follows its behavioral use, or for that matter, whether symbolic acceptance ever occurs at all? A second question related to the first is, how might the adoption process be conceptualized to handle either basic case? From the change agent's point of view, it is important to know when and for whom behavioral adoption may precede attitudinal adoption. The change agent would only want to use as opinion leaders or influencers persons who have adopted attitudinally. Persons who have adopted behaviorally but not attitudinally may be sources of negative word-of-mouth communication.

Several aspects of symbolic adoption need more research. The first need concerns the modes of acquiring and changing attitudes. Kelman's (1958) three processes of attitude acquisition involving compliance, identification, and internalization seem particularly relevant. Following his lead, further research should focus on the differential impact of these three processes on symbolic adoption. Such research could proceed along the lines of isolating the differential impact of the initiators of each of the processes. If, as is usually the case, the goal of a change agent is to accomplish symbolic and behavioral adoption of an idea, the manner in which the new idea is presented to potential adopters becomes crucial.

A second aspect of symbolic adoption (or rejection) is represented by the essential elements of attitudes. The elements most commonly discussed (as mentioned earlier) are affect, cognition, and predisposition to action. Here we pose the following question: What set of interrelationships among those components will produce favorable or unfavorable attitudes;

that is, what are the internal dynamics of attitudes in a given situation that will precipitate symbolic adoption or symbolic rejection of an innovation?

The three dimensions of attitude acquisition and change are closely related to attitude components. Consider, for example, the role of empathy. Empathy is the ability to place oneself in the context occupied by another. Empathy functions to link the identification aspect of attitude change with the affective component of the attitude itself. Identification as defined by Kelman occurs when an individual accepts influence in order to establish a satisfying, self-defining relationship with another person. The ability of a potential adopter to empathize with a current or past user of an innovation is then an important factor in this decision-making process (Barnett, 1953; Lerner, 1958; Rogers, 1969). To the extent there is identification with an actual adopter, the potential adopter is, in effect, experiencing adoption through the most readily available symbol of the innovative act, some other user. This might be called vicarious adoption. Identifying with another person is a form of interaction with that person, and consequently, as balance theory would predict, feelings toward that person (who has adopted) will influence how the potential adopter feels toward the innovation in question. Thus the more homophilous the adopter-potential adopter dyad is on relevant characteristics, the more likely the potential adopter is to identify favorably (i.e., accept influence) with the adopter and hence develop positive affect toward the innovation. Positive affect, of course, contributes to symbolic adoption.

A third aspect of symbolic adoption requiring additional research concerns the innovation itself. The nature of the innovation will influence the means by which the attitude is acquired or changed and the manner in which the components for that attitude function (see Zaltman and Lin, 1971). If the innovation has high communicability, identification with others who have adopted is made easier, since adopters can be located and observed readily. This facilitates identification, which initiates the process just elaborated above. Other dimensions of innovations may stimulate means of acquiring attitudes other than identification, which may in turn interact with other attitude components. The nature of the innovation is also an important determinant of whether symbolic adoption precedes or follows behavioral adoption. If the innovation is a public innovation, that is, one that becomes available to everyone if it is adopted by just one or a few individuals or groups, behavioral adoption may come first if the individual is against the innovation. Fluoridation of water supplies is an example of a public innovation. Also, when an innovation has few, if any, intrinsic rewards, behavioral adoption may occur first because the adopter seeks approval from significant others. In this instance, the compliance process of attitude change is probably the most likely to become operative.

Another consideration for research in symbolic adoption concerns the sequence of antecedents and consequences. This can be viewed as the sequence of an initial symbolic adoption → behavioral adoption → subsequent symbolic adoption or modification. There are two aspects of this; both involve the issue of perceived inherent newness of the attitude object. In the first case, an item or idea is perceived as new, an attitude is developed, the innovation is adopted, and, as a consequence of the ensuing experience, the initial attitude is altered in substance and/or strength. In the second case, an item is not perceived as an innovation, thus an existing attitude is involved rather than a new one being acquired. Use occurs, during which the item is indeed found to be innovative in some significant way, and the initial attitude is altered. It might be argued in this second case that, since the item was not initially perceived as an innovation, symbolic adoption only occurs after use. This argument restricts the use of the term "adoption" to refer only to perceived innovations. A very important point is raised here. If the subsequent symbolic adoption state differs from the initial symbolic adoption state, in addition to inquiring about the time sequence of symbolic versus behavior adoption, we must also ask whether the nature of the symbolic adoption state differs depending on whether it precedes or follows behavioral adoption. It seems highly probable that it does make a difference, and we must direct future research toward determining what the possible differences are and the circumstances in which they are likely to occur.

ADOPTION/REJECTION

The adoption stage represents a level of commitment by the individual, with repeated or continued usage. This is also a stage with cognitive, affective, and behavioral components. However, there is a difference between this stage and the attitude stage. The cognitive component at the adoption stage contains beliefs based on personal experience. These beliefs are more basic and strongly held than the beliefs of the attitude stage (Rokeach, 1968). They may supplement or supplant beliefs that are less basic.

The emotional component is also likely to be stronger than in the attitude stage. It is only logical that it takes a greater incentive to commit oneself to a continuous behavior pattern of use than merely to continue consideration of such use.

The depth of commitment and the belief change move this stage toward what might be termed a value-state. This approaches greater centrality in psychological position than does an attitude, and becomes a basis against which other possible states are measured. The stronger the adoption commitment that is made, the more central will be the value-state achieved. An innovation that would enable the individual to achieve some deeply held goal would be difficult to replace.

The alternative to adoption, at this point, is rejection. Unsatisfactory outcomes in the process prior to this stage may result in the achieved value state being negative. The components of rejection are the same as those of adoption, and achieve similar intensity.

RESOLUTION

The final stage is that of resolution (Campbell, 1966). This concept includes dissonance reduction, but is more inclusive. Dissonance is not the inevitable result of adoption. Some innovations are adopted without regret, even enthusiastically. Dissonance may result when one is forced to choose between two or more attractive alternative; but some innovations may be far superior to anything previously known, or may be the only known alternative available to solve a problem. Thus the concept of dissonance may be somewhat overly restrictive or misleading. The broader perspective suggested by resolution could lead to fruitful investigative results. The idea of perpetuating the achieved state is represented here by the resistance process of homeostasis.

The change agent should consider the resolution stage as a very important stage that he or she may influence. Making certain that the change is implemented and evaluated properly (formally or informally, evaluation does occur) is important not only for continued use by the same change target but for positive word-of-mouth communication by users.

Future Research on Adoption Processes

There are a number of directions future work on adoption processes should take in addition to addressing the problems cited at the beginning of this chapter. First, there is a serious need for empirical testing of the different models. This has been almost totally absent to date. It is necessary not only to test the existence of certain stages but to test the sequence of stages as well. One likely finding is that different situations require different models, although, at present, we have little information concerning this.

A second research theme should focus on parallel decision processes. For example, although a person may proceed from initial awareness to active interest to active evaluation and so forth, he may simultaneously be forming attitudes (however weak), engaging in symbolic trial and so forth. This is important to study because the nature of the process whereby awareness is established, interest stimulated and satisfied, and so on, may affect the formation of attitudes and the nature of symbolic and actual trial.

A third theme or direction should concern the notion of critical threshold. We know very little about conditions that affect movement from one stage to another. How much (and what kind of) awareness must exist to trigger movement to an interest stage? What degree of favorable attitude must exist for precipitating a legitimation process stage? What produces a critical threshold? Answers to such questions as these should enhance the effectiveness of change agent intervention. Related to this theme is the phenomenon of feedback. Not everyone passes through all stages in an inexorable way. What happens, for example, when unfavorable attitudes are developed? Does the individual remain at that stage or is there a "feeding back" to, say, a perception or awareness stage? Does activity at an earlier stage, entered into again for the same innovation, differ from the activity associated with that stage the first time it was evoked?

Another research question is: Do different adopter categories, such as innovators, early majority, laggards, and so forth, have different decision-making models? Present models do not differentiate among different adopter or potential adopter groups except with regard to the time required to reach decisions and the relative degree of reliance on different sources of information. One might expect differences with regard to the role of problem perception, for example, where early adopters might require the presence of a problem, wheras late adopters may not require a problem perception stage as a prerequisite to adoption.

Similarly, the different models do not really differentiate among innovations. Research is required before we can answer the important questions: Do different innovations evoke different decision-making processes? For what kinds of innovations can behavioral change precede attitudinal change? When is legitimation or trial not present or relevant? A related consideration involves the impact of situational factors as mediating forces influencing which decision-making process the particular innovation evokes.

Yet another topic in need of explanation concerns decision-making processes for innovations versus noninnovations. Are the various adoption models significantly different from models one would construct that focus on goods, services, and ideas that are not perceived as innovations? If the processes do differ, an interesting question is: What does the decision process look like when the change is initially perceived to be an innovation? For example, as familiarity with the change increases, it may be judged not to be an innovation and vice versa.

A major decision that must be made by a change agent is whether to follow (1) a strategy that focuses primarily (initially at least) on targets who are most likely to adopt a change, (2) a strategy that focuses primarily (initially at least) on those targets who are most resistant to change, or (3) a strategy that gives more or less equal treatment to both early and late adopters. The usual choice of strategy is to focus primarily on the early adopter. This is done for a variety of reasons. First, potential early adopters are important sources of information for later adopters. The earlier potential early adopters become aware of and accept a change, the earlier the late adopters will be to accept the change. Thus the overall diffusion process is hastened. Second, since early adopters are important sources of information, it is important to provide them with favorable information and to correct inaccurrate perceptions. Third, change agents are able to show early successes connected with their efforts.

Extensive research has been conducted on the traits and behaviors of early and late adopters (e.g., Lin and Burt, 1975). Rogers and Shoemaker (1971) present an excellent summary of this work. However, there are several flaws with early/late adopter research, two of which require special mention. One flaw arises from the way the early adopters are most frequently identified. The method consists of introducing the new idea in a test market or region and then obtaining the names of the first persons to try or adopt the change. A sample of these people are interviewed, and their common characteristics are noted. It might be found, for example, that they have a higher-than-average education. Then the change agent assumes that early adopter types have a higher-than-average education, and the change agency undertakes to aim more of its communication at the higher-than-average educated group.

This information collection design contains a major fallacy. It rates people on their time of adoption from the time of the product introduction, not their time of adoption from the time of their exposure. The advocate of change really must identify the group of people who responded early after exposure, not the group that responded in the early period! Virtually all research that has been done has been based on early adopters relative to the time the product or innovation is introduced rather than the time of first awareness. This distinction is of crucial importance in media planning. There should be more homogeneity between those who adopted early after exposure, and it is their common traits that should be identified. This means that greater attention must also be given to learning which media delivers messages more quickly to people. Media may differ in terms of who they reach first. Also overlooked is why people rely on particular media and why this differs, if it does, between early and late adopters.

The second major problem concerns the magnitude and persistence of adoption. Almost totally absent in existing research is any consideration of the degree of commitment involved in adoption (as opposed to trial and evaluation) (Kiesler, 1972). Change targets, when trying to satisfy some need or desire, may vary from infrequently using a change, to using it often, to using it all the time. Media factors may be important here (Maloney and Schonfeld, 1973).

Other problems can be noted briefly: (1) There is a lack of attention given to interpersonal relations between early adopters and others. The call for more "relational analysis" in which relationships, not individuals, are the unit of analysis, has been issued by many persons recently (e.g., Rogers, 1969, 1973). (2) There is an underemphasis on resistance or nonadoption and discontinuance relative to full adoption, and on the role of various channels of communicatin in creating resistance. (3) There has been only limited research about how the perception of different adopter categories differs (a) for perceived innovation versus noninnovation and (b) across different types or categories of innovations. (4) The communications-mix implications of feedback loops in adopter and resistor decision making has not yet received serious attention. (5) Another topic needing research is whether particular media have differential effectiveness for innovations as compared to noninnovations and does this vary (a) by stage of the adoption process, (b) by adopter category, (c) by type of innovation, or (d) according to salient attributes of the product or service. (6) More work is needed to study the interaction effects, if any, between personal and mass media channels of communication.

One problem with the conventional treatment of target or adopter categories is the assumption that early adopters are identical to the best early prospects for a social change. Early adopter types may not be the best candidates or prospects on which to focus change efforts (Kotler and Zaltman, 1976). Assume that a state agricultural extension agency is interested in introducing an agricultural innovation and that this agency could ask its extension workers to collect certain information. What should this information be? First, it would be desirable to ascertain the early adoption propensity of potential adopters. Early adoption propensity is the probability that the farmer would be an early user of the innovation after exposure to it. Second, it would be useful to learn the volume propensity of the farmer. This refers to how much the farmer will be likely to buy per year if he adopts the innovation. A third desirable piece of information concerns the farmer's influence propensity, which is the additional purchasing per year this farmer would be likely to stimulate in others through interpersonal influence. A fourth information item is the cost of an effective communication exposure. Thus the value of a particular farmer might be expressed as:

Value of farmer = adoption propensity (volume of own use +
volume of use stimulated in others) –
cost of effective exposure

Thus adoption propensity, although important, cannot be considered as the only determinant of the value of a possible adopter. In fact, some early adopters may well be low loyals, that is, easily switched to another competing innovation or product.

It may be useful to examine some of the components of the elements that determine the value of any given potential adopter. Early adoption propensity would seem to be a function of (1) the extent to which the advocated change is perceived as highly likely to satisfy a key need (need fulfillment potential), (2) the degree to which a person is oriented toward innovation (innovative disposition), and (3) the extent to which the potential adopter can readily adopt the innovation or implement the change (decision enactment). Heavy volume propensity is influenced by (1) the likelihood that satisfaction at the trial stage will be high enough to warrant repeated use (trial satisfaction), (2) the frequency of use, and (3) the average amount used per use occasion. Influence propensity depends on (1) the degree of social interaction based on number of acquaintances and frequency of interaction with acquaintances, (2) the influence over people who would not otherwise try the innovation, and (3) the average volume an influenced person buys within a specified period. Prospect communication cost is a function of the probability of message exposure, perception, comprehension, evaluation, retention, and actual delivery cost.

Factors Associated with Change

A wide array of factors have been found to be associated with social and individual change. One of the best discussions of these factors can be found in Rogers (1969). Although Rogers speaks primarily of change in peasant societies, there is strong evidence that the same factors may be found relevant in industrialized societies. A chief difference, however, between the two types of societies is in the way the particular factors are operationalized in the different contexts. Table 10.1 lists some of these factors taken from Rogers (1969).

It is also necessary to consider various factors involved in the general climate for change. Several factors can be identified (See Table 10.2). One,

*Table 10.1   Individual Factors Associated with Societal Change: Special Reference to the Peasant Society**

Literacy	The degree to which an individual possesses mastery over symbols in their written form (p. 51).
Mass media exposure	Exposure to such communication channels as newspapers, radio, films, and so forth (p. 52).
Cosmopolitaness	The degree to which an individual is oriented outside his social system (p. 52).
Empathy	The individual's ability to identify with others' roles, especially with those who are different from himself (p. 53).
Achievement motivation	A social value that emphasizes a desire for excellence in order for an individual to attain a sense of personal accomplishment (p. 54).
Aspiration	Desired future state of being (p. 54).
Fatalism	The degree to which an individual perceives a lack of ability to control his future (p. 55).
Innovativeness	The degree to which an individual is earlier than others in his social system to adopt new ideas (p. 56).
Political knowledge	The degree to which an individual comprehends facts essential to his functioning as an active and effective citizen (p. 56).
Opinion leadership	The ability to influence informally other individual's attitudes in a desired way and with relatively high frequency (p. 88).

* Adapted from Rogers, 1969.

perhaps the most widely discussed of all, is the perceived need for change. This refers to the extent to which an individual or group experiences a problem. Openness to change concerns the readiness to accept change. This disposition is not one in which *any* change would be adopted, but rather one in which a person would adopt a proposed solution to a problem if that solution appeared to satisfy the need reasonably well. An individual or group that rigidly adheres to a strict agenda of requirements for a solution, no matter how severe the felt need, can hardly be said to be open to change in a conventional meaning of the term. Potential for change refers to the capacity to accept and implement change. Capacity can consist of financial resources, material resources, human resources such as skills and value, and belief systems compatible with the advocated change. Perceived control over the change process is another important factor. This factor concerns the extent to which the change target feels it can influence the selection and implementation of change. We stress the importance of perception here. It seems important for individuals or groups to believe that they are active participants in all stages of the change process, even when it is difficult for them to accurately assess the extensiveness of their impact. There are a few instances on record in the community development area in which major decisions were made by certain officials and community groups were coopted afterwards, and their involvement structured in such a way that the groups came to approximately the same conclusion as that reached earlier by the key decision makers. The decisions were presented publically as being the product of various community groups, even though nonvisible steps were already underway to implement the earlier decision. Commitment to change is the recognition of the importance of undertaking remedial action in response to a problem.

These five factors are obviously related. Without a perceived need for change it is unlikely, although not impossible, that an openness or readiness to accept change will exist. The experience of a problem, however severe, does not necessarily create a readiness to accept change. Religious beliefs, for instance, may prevent the use of drugs to help cure physical or mental illness among members of the religious denomination. Even when a readiness to accept change exists, the lack of resources for acquiring and implementing change may prevent change from occurring. The perceived need for change and the readiness and capacity to accept change may be present in a change target, and yet change will not occur because the target group feels that it has little control over the change process. Members of one school district in which the authors have worked resisted a problem-solving technique that stressed their own control over change precisely because they felt that in their school district effective control over important issues lay with the School Board. In this case, despite the nature of the organizational

Table 10.2   Factors Involved in the Climate for Change

Perceived need for change	The extent to which an individual or group experiences a problem
Openness to change	The readiness to accept change
Potential for change	The capacity to accept and implement change
Perceived control over the change process	The extent to which the individual or group feels it can influence the selection and implementation of the change
Commitment to change	The recognition of the importance of undertaking remedial action (social change) in response to a problem

development technique that stressed teacher control, teachers simply did not believe that in important matters the School Board, or in some cases, school principals, would relinquish their own control. On the other hand, perceived control over the change process can lead to an openness to change.

Each of the factors associated with individual change is related to one or more factors associated with the general climate of change. These factors are listed as follows. Here we discuss only the most prominent association between the two sets of factors. Literacy appears to have its strongest association and causal impact on openness to change and on potential for change (see cross marks in Table 10.3). Becoming literate appears to unlock creative abilities that enable an individual to consider creative activities such as the coupling of himself with an advocated change, that is, the association of two previously unrelated things. Literacy enhances the potential for change by satisfying the first criterion—being able to read—for reaching people through the printed media. The addition of this channel of communication increases the ability of the individual to comprehend change.

Mass media exposure is especially important for establishing a sense of perceived need for change. To the extent that the mass media make individuals aware of more desirable alternatives for satisfying current needs, they can create or enhance the perceived need for change. The cause of this enhanced need is dissonance with current practice or solutions that fall well short of newly identified alternatives. The greater the degree of dissonance, the greater the recognition of the importance of undertaking remedial action. Thus mass media exposure can enhance commitment to change.

Table 10.3   Climate for Change

Indiv. Change Factors Associated with	Perceived Need for Change	Openness to Change	Potential for Change	Commitment to Change	Perceived Control over the Change Process
Literacy		X	X		
Mass media	X			X	
Cosmopolitanism	X			X	
Empathy			X		
Achievement motivation	X				
Aspirations	X				
Fatalism				X	X
Innovativeness		X			
Political knowledge					X
Opinion leadership					X

Cosmopolitanism appears to function in a manner similar to mass media exposure. Both factors broaden experience and exposure to new solutions or alternatives relevant to current problems. To the extent that the alternatives are preferable to current practice, there may be an increased commitment to actions supportive of the new alternative.

Empathy is probably most closely associated with potential for change. The greater the degree to which a client can identify with a change agent and view his own situation from the vantage point of the change agent, the better able he is to accept and implement change. Empathy also affects potential for change in another related way. It permits the client with empathetic abilities to experience vicariously with an existing user of the change the consequences of adopting the advocated change.

Achievement motivation enhances openness to change to the extent that it produces a higher sensitivity to or search for opportunities that would be instrumental in achieving desired goals. When aspirations are high relative to one's current situation, openness to change may also be high.

Fatalism and the highly related concept of locus of control (the extent to which people feel they have control over those factors which affect them) are most relevant to perceived control over the change process and commitment to change. Strong fatalistic attitudes are reflected in action apathy: "Since-there-is-nothing-I-can-do-about-it-why-bother?" Fatalistic attitudes deny the possibility of remedial action by the individual, hence there may be little commitment to change. An external locus-of-control mental set denies that the individual or group can influence the selection of the advocated change. There is an element of the self-fulfilling prophesy involved. If individuals feel that they have no influence in the selection and implementation of change, they may not make the effort to become involved in the selection and implementation processes, and by default all responsibility shifts to the change agent. Thus the change target in fact does not have effective involvement and influence in the change process.

Innovativeness is probably most related to openness to change. The greater the readiness to accept change, the earlier the person is likely to adopt change. The prime causal direction in this case would be readiness to accept change causes earlier adoption. An argument can be made, however, that innovativeness is a composite variable that represents factors such as novelty needs or needs for stimulus variation, which in turn produces a general readiness to adopt change that promises to satisfy that need. Clearly, the nature of the relationship between a disposition to adopt early and a readiness to change requires further research.

Political knowledge is the degree to which an individual comprehends facts essential to his functioning as an active and effective citizen. Apart from change involving public affairs, political knowledge may be a good indicator of other less readily measured factors such as social integration or anomie. In this regard, political knowledge may be particularly related to perceived control over the change process. The greater the degree of participation in and felt integration with society in general, the greater a person's sense of influence or control over the change processes affecting him.

Opinion leadership is defined as the ability to influence informally other individuals' attitudes in a desired way and with relatively high frequency. The greater a person's awareness of this influence, the greater his perceived control over the change process will be.

This chapter can be summarized by the following principles:

1. The change agent must help establish awareness and recognition of a need to change and connect the advocated change to this awareness.

2. The change agent should not confuse or equate his perceptions with those of the target group.

3. The change agent should monitor the target group's perception about the change throughout the decision and postdecision process, since perception may frequently change.

4. Where appropriate, the change agent should stress the availability of a change or the ease with which a suitable change can be developed to address a need. This can stimulate the target system's motivation to change. The change agent can stimulate this further by demonstrating how the target system can control the change effort (if indeed they can).

5. The change agent should be wary of using attitudes to predict behavior and vice versa.

6. Every effort should be made to reinforce the target system's plans. Stressing the legitimacy of the change is one way of doing this.

7. The change agent should make every effort to have the target system try the advocated change on a restricted basis. Trial may be symbolic or behavioral.

8. The change agent should be particularly concerned with the target system's evaluation processes after the trial stage. The change agent may provide assistance to the target system while it evaluates the trial experience and information from other sources.

9. After change has been adopted, it is very important to provide support for the decision or action to reinforce enthusiasm or to counteract dissonance.

10. The change agent should provide assistance to help the members of the

target system to cope with their environment as it affects goal attainment.

11. The change agent must consider explicitly the strategic implication of each stage in the overall adoption/rejection decision process and adapt his actions according to the requirements and special processes of each stage.

12. The change agent should target initial efforts toward those persons who adopt innovations early upon awareness, and try to foster early awareness for such persons who are otherwise late knowers.

13. The change agent should also distinguish between light users of a change or innovation versus heavy users, with relatively more emphasis on the latter.

14. It is also desirable for the change agent to focus on interpersonal relations rather than individuals alone.

15. When analyzing the target group, the change agent should attempt to identify individuals or types of individuals who have high influence propensity, that is, who are active opinion- or fact-giving people or who are role models. The members of this group must be among the very first to be reached.

16. A useful activity for a change agent, perhaps working with a group of other change agents or persons knowledgeable about given target groups, is to construct an index or a rating chart to predict the likelihood of change in a target group. This requires joint consideration (Table 10.3) of individual factors associated with change (Table 10.1) factors affecting the climate for change (Table 10.2).

CHAPTER ELEVEN

Characteristics of Change Targets: Organizations

Chapter 10 discusses the adoption of change from the standpoint of the individual. Seldom if ever do individual decisions occur in a social vacuum. The various stages describing individual adoption processes are influenced

248

by other people and phenomena, such as values and norms emerging from social interaction. As individuals proceed through an adoption process as discussed in the preceding chapter, but now in conjunction with other individuals facing common tasks and goals, another decision process emerges. This process is the group or organization decision-making process. It does not necessarily replace the individual's decision-making process, although it does alter how that process is engaged or enacted.

In this chapter, emphasis is placed on the decision-making process as it occurs or emerges from a formally organized group of individuals. The objective of this chapter is to provide insight into the processes involved in getting organizations to change. The change agent involved may be someone who is a member of that organization or an outsider to the organization.

CHARACTERISTICS OF ENVIRONMENT

Organizational Environments Defined

Organizations or subunits of organizations do not exist in isolation from the larger environment. Rather, they exist in an environmental setting. Organizations or subunits are open systems that must exchange inputs and outputs with the environment to survive over time. Figure 11.1 provides an example of an automobile company and the input–output exchange process with the environment.

Figure 11.1 indicates that, for this organization to survive over time, it must produce cars that the public is willing to buy. Unless it does this, it becomes difficult for the organization to exchange these outputs with the environment for inputs such as capital, new orders, etc. The recent oil

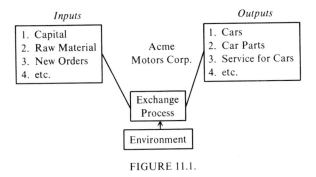

FIGURE 11.1.

shortage has drastically changed the environment's overall acceptance of the outputs of the major car producers in the United States during 1974 and 1975. The organization's environment has radically changed and requires some change on the part of the organization.

Before we look more closely at the role of the environment as it affects the change process, we must first define more clearly the environment and its major components and characteristics.

Various organizational theorists have dealt with the concept of the environment, but few have attempted to define it (Lawrence and Lorsch, 1967; Emery and Trist, 1965; Terreberry, 1968). A notable exception is Dills (1958) concept of the task environment, which focused on only those parts of the organization's external environment (customers, suppliers of labor, capital, materials, competitors for both resources and markets, and regulatory groups) that are relevant or potentially relevant to the organization's goal setting and goal attainment activities.

We define organizational environment as the totality of physical and social factors that are taken directly into consideration in the decision-making behavior of individuals in the organization. If the environment is defined in this way, there clearly are factors within the boundaries of the organizational activities that must be considered as part of the environment. Therefore, a differentiation is made between the system's internal and external environment (Duncan, 1972a).

The internal environment consists of those relevant physical and social factors within the boundaries of the organization or specific decision unit that are taken directly into consideration in the decision-making behavior of individuals in that system. For example, the internal environment for an industrial organization would consist of the various departments and personnel making up that organization.

The external environment consists of those relevant physical and social factors outside the boundaries of the organization or specific decision unit that are taken directly into consideration in the decision-making behavior of individuals in that system. For example, the external environment of an industrial organization would consist of suppliers, customers, regulatory groups, and so forth.

Components of the Environment As They Affect Change and Innovation

Once the environment has been defined in this broad sense, the next question is, what are the components of the environment—what makes up the environment? Awareness of these components will help the change agent or manager of the change process identify various pressures for or against change. Table 11.1 presents a list of components of the organization's internal or external environment. The list presented here may be particularly

Table 11.1   Factors and Components Comprising the Organization's Internal and External Environment*

Internal Environment	External Environment
Organizational Personnel Component	Customer Component
A. Educational and technological background and skills	A. Distributors of product or service
B. Previous technological and managerial skill	B. Actual users of product or service
C. Individual member's involvement and commitment to attaining system's goals	Suppliers Component
D. Interpersonal behavior styles	A. New materials suppliers
E. Availability of manpower for utilization within the system	B. Equipment suppliers
	C. Product parts suppliers
	D. Labor supply
Organizational Functional and Staff Units Component	Competitor Component
A. Technological characteristics of organizational units	A. Competitors for suppliers
B. Interdependence of organizational units in carrying out their objectives	B. Competitors for customers
C. Intraunit conflict among organizational functional and staff units	Socio-political Component
D. Interunit conflict among organizational functional and staff units	A. Government regulatory control over the industry
	B. Public political attitude toward industry and its particular product
Organizational Level Component	C. Relationship with trade unions with jurisdiction in the organization
A. Organizational objectives and goals	Technological Component
B. Integrative process integrating individuals and groups into contributing maximally to attaining organizational goals	A. Meeting new technological requirements of own industry and related industries in production of product or service
C. Nature of the organization's product service	B. Improving and developing new products by implementing new technological advances in the industry

*Source: Duncan, 1972a, p. 315.

relevant to industrial organizations and could vary for other types of organizations.

It should be emphasized that no decision maker would be expected to identify all these components as part of the organization's internal or external environment for a specific decision situation. Rather, the makeup of the environment is expected to change over time. For example, a programming and planning decision unit's environment in developing one type of product might focus on customer demand and the marketing and production departments. In planning and developing programs for a different type of product, the relevant environment to be considered in decision making may have changed to include government regulatory agencies with jurisdiction over this type of product. This specification of what part of the environment on which to focus in decision making is the domain problem.*

The domain problem is most crucial. Defining the domain indicates what part of the environment the organization must consider relevant in its decision making. In studying how organizations scan their environment in decision making, Aguilar (1967) has investigated the relative importance of different types of information from the environment. Aguilar's findings are presented in Table 11.2. They indicate that the most important source of information for the organization is the market. In fact, the market accounted for 58% of all responses, which is more than three times the amount of information coming from the technical area. This dominance of market information leads Aguilar to conclude that ". . . companies tend to react to current conditions rather than to innovate" (p. 54). These results thus indicate that the most important domain for the organization is the market.

The importance of this external information is also supported by the research on idea generation by Baker et al. (1967). This research indicates that there are two kinds of information required at the idea generation stage in innovation or change. First, there must be some knowledge of a need that is relevant to the organization. Second, there must be some knowledge of a means or technique for satisfying the need (p. 156). The results of their research on 300 ideas created in a divisional laboratory of a large company indicated several things. Need events stimulated 75% of the ideas, whereas 25% of the ideas were stimulated by a knowledge of a means, which was then followed by knowledge of some need (p. 160).

The Marquis and Myers (1969) study of innovation in United States firms and the Carter and Williams (1957) study of British firms also support these findings. Table 11.3 summarizes both studies. It can be seen that the major sources of ideas for innovation came from marketing factors, as opposed to technological factors. Other evidence indicates that the extent to which a new product is developed on the basis of an identified need rather than on the basis of an available technology to produce it is the major factor distinguishing commercially successful from commercially unsuccessful new products (Schewing, 1974).

These studies thus indicate that, in innovation and change, the organiza-

* The following section draws from pp. 116–120 in Zaltman et al., 1973.

*Table 11.2   Relative Importance of Areas of External Information: Overall Data,
All Managers (Percent of Responses)**

Area of External Information	Percent of Total Responses	Information Category	Percent of Total Responses
Market tidings	58	Market potential	30
		Structural change	10
		Competitors and industry	4
		Pricing	4
		Sales negotiations	6
		Customers	4
Technical tidings	18[a]	New products, etc.	14
		Product problems	2
		Costs	1
		Licensing and patents	3
Broad issues	8	General conditions	4
		Government actions, etc.	4
Acquisition leads	7	Leads for mergers, etc.	7
Other tidings	9	Suppliers, etc.	3
		Resources available	2
		Miscellaneous	2
Total percent	100		102[a]
Total number of responses	190		190

[a] Error due to rounding.

* *Source: Aguilar 1967, p. 48.*

tion is reacting to a need rather than being innovative in the sense of taking existing knowledge or means (i.e., technology) and trying to create a new need (i.e., output demand) whereby it can use its technology in a creative way.

It is also apparent from existing research that the sources of ideas for innovation come from outside the organization itself. In studying the 25 major process and product innovations in DuPont, Mueller (1962) indicated that 56% (14 cases) originated outside the organization. Marquis and Meyers' (1969) analysis of 157 case studies of innovation has shown that 61% (96 cases) of the ideas for new innovations came from outside the organization. Utterback's (1971b) study of innovation in the instrument industry found that of "... 59 pieces of information incorporated in the ideas for 32

Table 11.3   Sources of Successful Commercial Innovations*

| Source of Innovation | Percentage of Innovations | |
	Carter and Williams	Myers and Marquis
Adopted (not original)	33	23
Technical factors; the desire to use the work of research and development departments	18	17
Marketing factors	32	35
Demands by customers for new types or qualities of product	12	
Direct pressure of competition (copying or forestalling rival/rms)	10	
Desire to meet excess demand	10	
Production factors	17	23
Desire to overcome labor shortage	5	
Desire to overcome materials shortage	12	
Administrative factors	n.a.	2
Total	100	100

* Source: Utterback, 1971b, 148.

new scientific and measuring instruments, 66% (39 pieces) came from out-side the firm that developed the idea" (Utterback, 1971, p. 145).

It can also be seen from Table 11.3 that both the Carter and Williams (1957) study (33%) and the Myers and Marquis (1969) study (23%) indicate that a small percentage of innovations are adopted from outside the firm. Rather, the majority of innovations are developed, tested, marketed, or incorporated in existing operations by the firm itself. These findings do not contradict the preceding discussion, in which we discussed how the sources of ideas and information concerning innovation come from outside the organization. Rather, it appears that innovative organizations must be open to outside sources of information to get ideas for innovation before they then develop the innovations themselves. For example, Miller's (1971) study of 16 United States and Western European steel firms has also indicated that organizations do not innovate by the introduction of technology developed within their own organizations but rather by technology developed outside the firm. "Innovation behavior . . . follow(s) from a capacity to adapt and utilize technical possibilities offered by the environ-ment, rather than from internal creativity" (Miller, 1971, p. 107).

Thus from the foregoing discussion it is evident that the interaction between the organization and its environment is crucial to the change and innovation process. The organization continually must obtain several kinds of information from the environment. First, it must determine the kinds of outputs the environment seeks that may require change or innovation to be more readily received by the environment. Second, it must discover the kinds of technology or means that may be required to produce the change or innovation—what are other organizations doing; are there existing changes or innovations that the organization might adopt to facilitate its response to these needs? Third, once the organization does in fact implement the change or innovation, is the change or innovation effective in meeting the demands of the environment? Here it is necessary for the organization to get feedback from the external environment.

The environment can have an impact on the change-innovation process in at least two ways. It can be the initiator of change, or it can be a factor that mainly reacts to some change. Some examples should clarify this distinction. Certain components of the organization's external environment can put pressure on the organization to change. For example, members of the community may put pressure on a police department to be more sensitive to the community's needs, particularly those of minority groups. This pressure from outside the police department then serves as a stimulus to change. The department may react in several ways. This external pressure may be seen as legitimate by the department, and a real commitment to change may occur. On the other hand, if the department does not perceive the pressure as legitimate, strong forces of resistance may be generated against the change. The main factor here is that the external environment can be the key initiator of the change process. This is compatible with the earlier discussion of idea generation, which showed how the external environment plays a central role in generating the stimulus for innovation.

The external environment may not be the source of change, but it still plays a key role, since it will react to any change or innovation by either supporting it or resisting it. Thus even in those cases in which the environment does not initiate the change, it still must be considered in terms of how it might react to the change. In one organizational development project in a school district made up of some 22 schools the impetus for the organizational development program came from within the district. However, one of the problems the change agents are facing in the first year of the program is that some teachers and principals are concerned that parents might not support the organizational development program because it might be seen as something that "takes teachers away from their most immediate classroom activities." Thus a source of potential resistance is the external environment.

The joint programs an organization has with other organizations can also be an environmental factor that stimulates change and innovation. Organizations may often engage in joint programs as a method of achieving some program change at a reduced cost. Hage and Aiken (1970) point out several consequences of this interorganizational involvement that can stimulate change or innovation: (1) Increased interaction with other organizations can lead to increased awareness of opportunities for change and innovation. (2) Internal communication within the organization is increased to cope with increased coordination problems that occur as a result of this interorganizational activity. The result is that there is a greater potential for information exchange within the organization. (3) New specialties may be introduced in the organization to help with the joint programs. Again, the result is that more and varied information is available to the organization (Hage and Aiken 1970, pp. 80–81).

The major point in the preceding discussion is that the environment plays a key role in the change and innovation process; it can either initiate change or support what change does take place. There are some important implications in terms of resistance to change. Where the stimulus for change comes totally from the external environment personnel within the organization can feel pressured or coerced into changing. The result may be resistance to change. For example, community groups might put considerable pressure on a police department to revise its patrol practices. If there is no initial internal support from police personnel for this change, they may react defensively and strongly resist this "interference from the outside."

Environmental State and the Change and Innovation Process

So far we have talked about the components of the environment. However, we have not discussed the state of an organization's external environment that might affect the change and innovation process. For example, how does the complexity and dynamics of an organization's environment affect it, and what are the implications for change?

One might argue that the more complex the environment, the more likely that an organization might be under some pressure for change. A complex environment is one in which there are a large number of factors in the environment that make demands on the organization or some subunit of it to respond in some certain way. Given a wide divergence of perspectives and interest among these environmental components to which the organization must respond, there are likely to be situations in which the organization will have to moderate its behavior in order to respond to some particular group. The organization may have to change its response to react to whatever particular component in the environment is exerting the most pressure. Thus the more components there are in the environment, the more likely that the organization will have to change. For example, the environment of an automobile company is very complex in that there are many diverse groups (customers, ecologists, governmental regulatory agencies, etc.) the organization must respond to. Often these components of the environment will have conflicting demands for the organizations. Ecologists want clean and more economical cars, whereas customers may want bigger and faster cars. The diversity of the environment requires that the organization be flexible in how it responds.

The complexity or degree of heterogeneity of the organization's clientele can also be a pressure for change. The studies of Israeli service organizations' reactions to new imigrants indicate how the impact of this new clientele created a structural change resulting in a modification of role prescriptions at lower levels (Eisenstadt and Katz, 1960; Bar-Yosef and Shild, 1966).

The dynamics of the environment also put some pressure on the organization to change and innovate. The static-dynamic dimension of the environment indicates the degree to which the factors of an organization's environment remain basically the same over time or are in a continual process of change (Duncan, 1972a). For example, the rapidly changed environment of the recession of 1974 and 1975 and resulting uncertainty placed considerable pressure on planning groups in organizations to develop new planning techniques that could deal with this high level of uncertainty.

CHARACTERISTICS OF ORGANIZATIONS AFFECTING CHANGE AND INNOVATION: AN ORGANIZATIONAL DESIGN APPROACH

The key question now is how to structure the organization or its subunits so that it can effectively deal with the environment during the change-innovation process. What must the change agent know about the structure and climate of the organization so that he or she can better facilitate change and innovation? We are not saying that every change agent should be prepared to change the organization's structure or climate. We are simply saying that there are important characteristics of organizations that can both facilitate and impede change. Hopefully, by being more aware of these characteristics, the change agent will be in a better position to effect change.

The Stages of Change

Before discussing the organizational characteristics affecting the change process, it is important to define briefly the stages of the change process. There are two general stages: initiation and implementation. Each of these stages has substages.

The initiation stage has three substages. The knowledge-awareness substage focuses on how the organization becomes aware of an innovation that may produce change. Performance gaps occur when a discrepancy exists between what the organization is doing and what its decision makers believe it ought to be doing (Downs, 1966, p. 191). The performance gap then increases the search for alternative courses of action—the gap is a stimulus to change. Knowledge-awareness of an innovation can occur in two ways, depending on how the performance gap is identified. The organization might have a need to change that increases the search for alternative ways of dealing with problems. This increased search then brings the organization into contact with a new practice. For example, an organization might realize that its information-gathering capabilities are not what they should be (i.e., a performance gap exists) and thus searches for alternative ways of dealing with a problem. A second way in which knowledge-awareness can occur is if the organization initially becomes aware of an innovation. Once aware of this innovation, the organization may decide that it could be performing better—a performance gap leading to a need is produced after awareness. For example, an organization may initially have no particular need to change. However, as its managers travel to conventions, they might become aware of a new, more sophisticated or more efficient information system. Being aware of this new system, managers realize that their organization could be much more effective in gathering and processing information. This awareness has changed performance expectations upward so that a need now exists to change by implementing the new information system. Thus knowledge-awareness can occur in two ways: (1) some need to change exists that creates search which then leads to awareness; or (2) awareness of a better alternative leads to dissatisfaction with existing practice which then creates a need to change.

The attitude formation substage focuses on organizational members attitudes toward change. These attitudes, which constitute the climate for change, play an important role in whether the organization decides to implement or reject the change. As pointed out elsewhere, there are five dimensions of this climate for change: (1) the perceived need for change on the part of the organization, (2) perceived openness to change—the readiness to accept change, (3) the perceived potential for change—the perception on the part of organizational members that they have the capacity to accept and implement change, (4) perceived control over the change process—the extent to which organizational members feel they can influence the change process, and (5) perceived commitment to change—a feeling on the part of organizational members that they can deal with the problems that might occur during change.

These attitudes toward change are important in that they create the setting in which the change takes place. For example, Hage and Dewar (1974), in studying innovation in health systems, have found that values supporting change among executive directors and others who participate in strategic decisions are a more important predictor of the implementation of change than are such structural variables as complexity, decentralization, or formalization. Mohr's (1969) study of health departments in the United States and Canada also indicates the importance of top level support for change and innovation. Mohr's study found that an activist ideology characterizing executive's attitude toward change was important in the actual implementation of change. It may be that executives and key elites in an organization have more impact in creating change, since they are not seen as "deviants" to traditional norms when they do support change and innovation. They have more formal power to implement these supportive attitudes to change and thus may be seen by the rest of the organization as communicating a new direction or ideology for the organization (Rothman, 1974). Apparently these attitudes communicate commitment to a change process that motivates others in the organization to accept change.

The implication here for the change agent is clear. The more the change agent can create a need to change, openness to change, a perceived potential for change, perceived control over the change process, and commitment to change, the more successful the change attempt will be. Our chapters on strategies for change focus on how to create the optimal change climate.

The decision substage focuses on the choice to implement or not to implement the change. During the decision substage, the information-gathering and processing needs are great as organizational decision makers decide whether or not to implement a change.

Implementation is the second major stage of the change process and consists of two subphases. The initial implementation substage is that in which the organization implements the change on a trial basis to determine if it is practical before a long-term commitment is made to utilize the change or innovation. The continued-sustained implementation substage occurs when, after a successful trial, the organization decides to formalize the implementation of the innovation into the ongoing processes of the organization on a long-term basis. For example, an organization might try a new management information system in one of its divisions. After a six-month trial, the organization may consider this innovation to be successful and therefore decide to implement it in the rest of the organization.

In summary, then, the process of change and innovation has two distinct stages. Initiation is concerned with how the organization becomes aware of change, forms attitudes toward it, and then makes a decision about implementation. Implementation is concerned with the process by which the organization integrates the change into its ongoing processes.

The argument now to be developed is that, given the very different tasks that must be performed in the initiation and implementation stages, different types of organizational structures are most effective in these two stages. In fact, the existing research seems to indicate that the very structural factors that facilitate initiation may impede the implementation of innovation. This is the change design dilemma. Even if the change agent is not in a position to alter the structure of the organization to facilitate the change process, it is still important that he be aware of the constraints that organizational structure can create during the change process.

The two stages of change and innovation can be related to Ansoff and Brandenburg's (1971) operating and strategic design problems, which may conflict with each other. The organization must be strategically responsive in making major changes, while at the same time it must be concerned with carrying out its activities in the most efficient manner. The initiation stage of the change-innovation process really deals with the strategic responsiveness of the organization. What is the organization going to do, how is it going to make major changes in its fundamental mission in response to new demands? The implementation stage then begins to deal with the operating problems. How çan these changes be integrated into the ongoing activity of the organization so that they work in the most efficient manner? Different structural configurations are presented as facilitating the accomplishment of these two processes.

Various organizational theorists (Burns and Stalker, 1961; Lawrence and Lorsch, 1967; Duncan, 1973 and 1974) have indicated that there is no one best way of structuring the organization for all tasks. Duncan's (1973) study of decision-making groups further elaborates on the notion of differentiation in organizational structure in finding that the same organizational unit utilized different structures for making routine and nonroutine decisions. For routine situations for which the information needs were low, a more bureaucratic structure was most effective in decision making. On the other hand, for nonroutine decisions, in which information needs are greater, the more nonbureaucratic structure, with its more open channel flow of information, was more effective in making decisions. This finding is similar to Zand's notion of the collateral organization, which is a more loosely structured organization to parallel the more formal organization (Zand, 1974). The more loosely structured collateral organization is viewed as being most appropriate for ill-structured problems. Thus the above research supports the view that the structure of an organization is comprised of a variety of dimensions that may vary from one situation to the next.

Several organizational characteristics are now discussed in terms of how they affect the initiation and implementation stages of change. The characteristics are: complexity, formalization, and centralization. In this section, the role of these various organizational characteristics is discussed vis-a-vis change and innovation, and the implications for the change agent are identified. The issue raised here is that the change agent is faced with a dilemma. The very organizational characteristics that facilitate the initiation of change impede the implementation of change and vice versa. Thus the change agent ideally will want to create different organizational structures for the two stages of change and innovation. If he cannot change the structure, the change agent must still be aware of these structural factors, since they may create problems as he goes through the change process.

Complexity

The degree of complexity of the organization can have various effects on the change process. Complexity is defined here in terms of the number of occupational specialties in the organization and their professionalism (Hage and Aiken, 1970, p. 33), and a very differentiated task structure (Wilson, 1966, p. 200). In a highly complex organization, either its output or its technology (the process by which the output is produced) may not be highly specified. Thus as Wilson (1966, p. 201) indicates, each individual will have some latitude to tailor his task to fit his needs. Given the diverse nature in which individual tasks may be defined, it becomes very difficult for close supervision to occur. The supervisor simply may not be aware of the specific way each member has tailored his job. The result is that the individual has greater opportunity to discover areas that require change.

A high number of occupational specialties with professionalism will result in organizational personnel placing a high value on specific knowledge and information to utilize on their jobs. This diversity in occupational backgrounds can then bring a variety of sources of information to bear that can facilitate awareness or knowledge of change at the initiation stage. These varied sources of information simply increase the opportunity for different types of information to be made available to the organization as it considers changing. For example, Dearden (1972) emphasizes the complexities of designing management information systems by indicating that a variety of individuals with understanding of the different information needs of the organization are required. These diverse individuals are required to make sure that the information system reflects the complexity of the information needs of the organization. The external organizational linkages of professionals are also important in stimulating change. These external professional linkages provide an opportunity for personnel to become aware of new developments in their fields (Aiken and Hage 1968). For example, the study by Morris and Randall (1965) of community organizers found that organizers who worked from an external position and were linked up to a variety of other social welfare organizations in the community were more effective in stimulating new programs for the aged. Feller and Menzel (1975) suggest that the larger the number of professional associations in a context such as police or fire protection, the more change oriented agencies in that context will be.

However, this diversity or complexity creates a problem for the organization. Wilson (1966, pp. 200–204) has indicated that high diversity in the organization will lead to organizational members conceiving and proposing more innovations but not adopting these innovations. Wilson's argument is that the high diversity (complexity) makes it difficult for any one source of authority to try and force some consensus toward agreement as to which of the many proposals should be implemented. Thus there appears to be a basic conflict between the process for the initiation of change and its implementation. This organizational dilemma is clearly exhibited in Sapolsky's (1967) study of innovation in department stores. The diversity in the reward structure in terms of increased professionalization on the part of retail controllers led to proposals to separate buying and selling functions, the use of computers in merchandising, and the use of sophisticated decision-making techniques such as PERT and operations research in merchandise problems. However, the diversity in department store structural arrangements, decentralized decision-making authority, and the existence of a large number of equally situated subunits frustrated the implementation of these proposed innovations.

It appears that the complexity of the organization can have both positive and potentially negative effects on various stages of the change process. At the initiation stage, highly diverse organizations are apparently able to bring a variety of bases of information and knowledge to bear that can increase the awareness and knowledge of change and the need to implement it in an organization. However, at the implementation stage, high complexity, because of potential conflicts, makes it more difficult for the organization to actually implement the change.

The strategy implications for the change agent concerning the complexity dilemma are several. First, the diversity of an organization or subgroup of an organization might be increased in order to increase the number of changes proposed during the initiation stage. For example, a project group might be composed of several engineers, production managers, and marketing managers to develop a new management information system. The heterogenous background of this group, with its diverse sources of information, could generate many new ideas for the system and its components. To facilitate the actual implementation of change, team-building activities could be provided to the highly complex group. Team building would help them develop their interpersonal skills and trust so that they could deal openly with conflicts and disagreements they would experience in trying to agree on the specific aspects of the innovations to be implemented (see French and Bell, 1973, pp. 112–121). This team-building activity could help individuals "deemphasize" the complexity of their work group, by having individuals focus on their similarities and emphasize their common perspectives. This could help them to see things in a somewhat similar manner; thus there would be more opportunities for agreement on implementation. Another alternative would be for the initiation and design of the management information system change to take place in a highly complex-diverse project group. Once the proposals for the new system were developed, the less diverse, more homogenous functional groups, that is, marketing, production, and so on, would work on the design for implementing the system into their respective areas.

Formalization

Formalization is concerned with the emphasis that is placed within the organization on following specific rules and procedures in performing one's job. The assumption is that strict emphasis on rigid rules and procedures may prohibit organizational decision makers from seeking new sources of information. Thus there is simply less opportunity for them to become more aware of the need for change or the technology required for it.

In considering the relationship between formalization and change, it is again necessary to consider the particular stage of the change process. For example, Shepard (1967, p. 474) had indicated that low formalization might be most appropriate during the implementation phase. During the initiation stage, the organization must be as flexible and as open as possible to new sources of information and alternative courses of action in considering change. Many rules and procedures might become constraints under which the organization might have to operate. This seems to be the case in Hage and Aiken's (1967, p. 511) study of welfare organizations. They found a negative relationship (r = −0.47) between job codification and program change.

During the implementation stage, however, Shepard (1967, p. 474) indicates that a singleness of purpose is required. Radnor and Neal's (1973) study of the successful implementation of operation research-management science activities in large industrial organizations indicated that, in order to bring the change into practice, specific formalized procedures had been developed to facilitate implementation. These procedures covered such factors as formal project selection, long-range planning, scheduling, and regular progress reports (Radnor and Neal, 1973). The formalization procedures seemed to have reduced the problems of implementing OR/MS activities. Apparently, these formalized procedures provide both information and specific techniques that facilitate organizational personnel's ability to utilize the change. The lack of these more formalized procedures at the implementation stage is likely to lead to both role conflict and role ambiguity. Role ambiguity could result since, without more formal procedures, the individual is likely to be unclear concerning how the change is to be implemented and how it will affect job activity. Role conflict could occur also, because the lack of specific procedures concerning how the change is to be implemented may lead to conflict with existing rules and procedures.

For example, a new, more formalized management information system might be introduced into an organization. Previously, information channels may have been decentralized and informal, and the information system was fairly unstructured. The new information system is thus a significant change from the old one. To be successfully implemented, it would be necessary that quite specific procedures be communicated to personnel, so that they know that they are now expected to effectively utilize the more centralized communication system. They are thus made aware that the role expectations of their superiors concerning the information system have changed, and that the older informal system is no longer sanctioned.

Once again, the change agent is faced with a dilemma—low formalization seems to facilitate the initiation stage, whereas high formalization facilitates the implementation stage. In attempting to reduce this dilemma, the change agent might propose using different degrees of formalization at the initiation and implementation stages of the change and innovation. For example, a broad set of operating guidelines might be established for a group that reduces formalization during the initiation phase which, as indicated above, would potentially stimulate change and innovation proposals. However, once the proposals were generated, the group could then focus on specifying operating rules and procedures as to how the change-innovation would actually be implemented. This greater formalization of the change process at the implementation stage would tend to reduce the ambiguity and potential conflict that individuals would experience as they implemented the change. They would have potentially a better understanding of how to use the change or innovation and how its implementation fit into existing procedures in the organization. In studying the attempted implementation of a new teacher model, Gross et al. (1971) found that some of the barriers to implementation were that either teachers were not clear about the kinds of role performance required to carry out the change or that there were incompatible demands made on them that did not support the new mode.

Centralization

The centralization dimension is conceptualized here in terms of the locus of the authority and decision making in the organization. The greater the hierarchy of authority (higher in the organization in which the decision takes place), and less the participation in decision making that exists in the organization, the greater the centralization. During the initiation stage, there is likely to be a good deal of uncertainty experienced by organizational members (Knight, 1967; Duncan, 1972a). The result is that the information needs for the organization are likely to be high.

A strict emphasis on hierarchy of authority is likely to cause decision unit members to adhere to specified channels of communication and to selectively feed back only positive information regarding their jobs. They could be neglecting any negative feedback that might actually help the organization better identify both the need and technology for change. Also, Shepard (1967, p. 471) has pointed out that the ideas for change are often generated at some distance from the power center in the organization, and must be communicated up the hierarchy. The more bureaucratized the authority structure, the more channels of communication through which the idea must travel. This then increases the probability that the proposal may get screened out, because it violates the status quo in the organization.

Greater participation in decision making may bring new insights and new sources of information to the change process. For example, Dearden (1972), Zani (1970), and Zannetos (1968) indicate that, because of the complexity of designing an overall management information system for the entire organization, a variety of individuals from the user functional areas must make inputs. These inputs are needed to determine what decisions are to be made, what are the important factors in decision making, how and when the decisions should be made, and what information is useful (Zani, 1970, p. 98).

Although the available research is less clear on this, we might suggest once again that the effects of centralization might vary depending on the stage of the change process of the organization. It is clear that less hierarchy of authority and more participation in decision making can increase the information available to the organization and thus increase knowledge-awareness at the initiation stage. However, as Shepard (1967) has indicated, when the organization gets to the implementation stage, a more specific line of authority and responsibility is required to reduce the role conflict or ambiguity that might accompany implementation of the change. Also, Sapolsky's (1967) study of innovation in department stores found that decentralized authority and decision-making structures frustrated attempts at implementation because it was difficult for the organization to gather enough influence over participants. Hage and Dewar (1974) indicate that having more people involved may reduce the possibilities of implementation, since people may have different priorities.

Once again, it appears that the design strategy for reducing this dilemma is to utilize different degrees of centralization during the different stages of the change process. During initiation, more autonomy could be given to participants in order to facilitate the awareness of the need and technology for the development of a change. Then, during the implementation stage, the decision process could become more coordinated and centralized as the change is implemented. Another procedure that the change agent might consider would be to use different task groups for the initiation stages. A less centralized structure would facilitate an R & D group's ability to scan its environment and become more aware of innovations as well as the need to change. Then, when implementation was to take place, the adopted units could be more centralized so that clear cut instructions and procedures could be communicated to the users such that their ambiguity regarding how to use the innovation would be low.

We now have specified that there are different configurations of organizational structure that facilitate the change processes in its two stages. Specifically, it was emphasized that, with regard to stimulating the initiation of change, a higher degree of complexity, lower formalization, and lower centralization facilitate the gathering and processing of information that is crucial to the initiation stage. It was also emphasized that, with respect to the implementation phase, a higher level of formalization, centralization, and lower level of complexity was likely to reduce role conflict and ambiguity that could impair implementation. The structure of the organization is thus contingent on the particular stage of the change process. This conclusion implies that the organization must shift its structure as it moves through the various stages of change.

FACTORS FACILITATING DIFFERENTIATION OF STRUCTURE FOR CHANGE AND INNOVATION STAGES

Given the preceding argument that the organization should differentiate its structure during the two stages of the change process, the key argument now becomes, how does the change agent get the organization to adopt this differentiated structure? There are two aspects involved here in attempting to get the organization to implement this structure. First, the organization and its key decision makers must be presented with the rationale for this dual structure as it is outlined above. This will provide some basic understanding about the reasoning behind the use of different structures. Second, the change agent must be aware of and be able to manipulate certain key variables to facilitate this differentiation in structure.

We now discuss several capabilities that are required of the organization if it is to try to use different structures for dealing with the two stages of the change process. It is the task of the change agent to see that these capabilities exist or are developed to facilitate this differentiation in structure. There are four such abilities: (1) the ability to deal with conflict, (2) the ability to maintain effective interpersonal relations, (3) the ability to create adequate switching rules for utilizing different structures, and (4) the ability to institutionalize the dual organizational structure changes. The entire model is presented in Table 11.4.

Dealing with Conflict

Utilizing different structures for the two stages of change is likely to have some potential for conflict. For example, if the initiation stage takes place in one group—say an R and D group—and then is implemented in a second group—manufacturing—conflict might occur along several dimensions. As a result of their tasks and operating structures, these two groups might be highly differentiated from one another. The R and D group might have a much longer time horizon as it works on developing ideas, whereas the manufacturing group is much more sensitive to getting things done quickly to meet output demands. The goals of the group concerned with the initiation stage might be different from the goals of the group concerned with developing product improvements that are really new and different. However, the group concerned with implementation is likely to be more concerned with the ease and speed with which the change can be integrated into existing procedures. In this example, the manufacturing group is also likely to be concerned about change and innovation. However, it is also likely to be concerned that the changes or innovations not be too radical in altering how it operates.

Table 11.4   Contingency Model for Designing Organizations for Innovation.

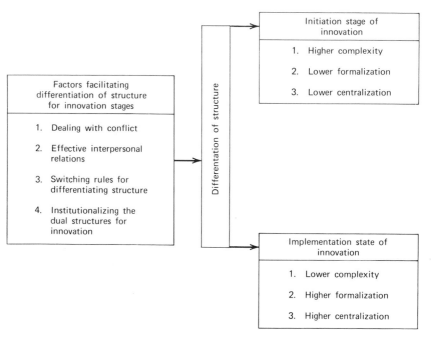

Being able to deal with conflict is also important if the same organizational unit is involved in the initiation and implementation stages. For example, a production unit in an organization might be concerned with identifying new products that they could produce. During this initiation stage, the unit might be decentralized into subgroups to work on new product ideas. Then, at the implementation stage, at which certain new products or product improvements were going to be adopted, the unit would operate under a more centralized structure so that the operating procedures for implementation could be clearly developed. As the unit shifts to the more centralized structure during the implementation stage, there are likely to be conflicts and disagreements about how the selected product innovations are to be implemented. Also, there are likely to be conflicts and disagreements as the unit becomes more centralized in its structure during implementation. Organization members who had been more involved in the initiation stage are likely initially to resist more centralization in rules and procedures, and decision making. Thus it becomes important for adequate conflict-resolution mechanisms to exist.

Various strategies have been discussed in the literature for dealing with conflict (Burke, 1970; Lawrence and Lorsch, 1967; Robbins, 1974). These strategies range from avoidance, smoothing–playing down differences, confronting the conflict, to forcing resolution through some coercive means.

Confrontation has been found to be a most effective way for dealing with conflict. With confrontation, conflict is seen as a legitimate organizational phenomenon that should be acknowledged and dealt with. The process of confrontation involves placing the relevant facts before the parties involved in the conflict and then discussing the disputed issues until some agreement is found (Lawrence and Lorsch, 1967). When individuals are able to confront conflict effectively, there is a better opportunity for the organization to resolve disputes such that they will not occur again. For example, when a conflict occurs between individuals regarding how to implement a new products innovation, confrontation requires that both parties present their viewpoints. The objective is to discuss these viewpoints until a common ground for agreement develops.

It may also be necessary for the change agent to create structurally certain mechanisms to confront conflict. For example, liaison positions may be established to coordinate activities between units that initiate and units that then have to implement change. These liaison persons might act as third party negotiators to help the units resolve disagreements (Walton, 1969; Galbraith, 1973). It is also possible that the change agent could be the liaison person. For example, an R and D group might be quite rigid in its development of an innovation that a production department might implement. The liaison person could bring the two groups together and work with them in trying to get a better understanding of the position of each group.

The implication here for the change agent is very important. Because the confrontation of conflict is a learned skill, it is likely that the change agent will have to provide some opportunity for organization members to learn this process. Therefore, it may be necessary to conduct conflict resolution exercises for organization members whereby they are able to gain some experiential learning of these processes.

Effective Interpersonal Relations

Whether the initiation and implementation takes place between organizational units or within the same organizational unit, dealing with interpersonal issues is important in facilitating this differentiation in structure between the two stages of the change process. For example, where initiation takes place between an R and D group and then implementation takes place in a second group—say, a manufacturing unit—it becomes very important for interpersonal issues to be dealt with (Bean and Romagosa, 1975). Members of the R and D group who have initiated the changes may have a strong identification with certain innovations when they are turned over to the manufacturing unit to implement. Members in the R and D unit might be motivated not to cooperate with manufacturing unit personnel, or they might react quite defensively when manufacturing personnel question the practicality of some of the changes. Thus it becomes important for each party to develop a capacity to express to one another how they feel about problems in their relationship in such a manner as to help those with whom they are communicating to express themselves in a similar manner (Argyris, 1965). Interpersonal issues are also likely to occur when the initiation and implementation is taking place within the same organizational unit. Again, organizational respondents, having been more involved in the initiation stage of the innovation'process, are likely initially to resist more centralized decision making in the implementation stage of innovation. Dealing with these feelings then becomes important so that they do not develop into full-scale resistance.

Once again, it may be necessary for the change agent to provide training for organizational personnel whereby they can learn how to deal with interpersonal issues. Team-building activities that focus on understanding group process and dealing with interpersonal issues as they affect task behavior become an important component of the design process.

Switching Rules for Differentiating Structure

In developing this contingency model of designing organizations for change, we have indicated that organizational structure should be differentiated for the initiation and implementation stages. Perhaps the most critical factor for the change agent is to develop switching rules indicating when the organization should differentiate its structure between the two stages of the innovation process. In this discussion, we suggest three switching rules for differentiating the organization structure: (1) the need for change and innovation, (2) the uncertainty associated with the situation, and (3) the complexity of the change or innovation.

SWITCHING RULE 1. The greater the need for change, the more the organization should differentiate its structure for initiation and implementation. A greater need for change implies the need for a quick response. For example, oil shortages have created a very high need on the part of many petro-chemical-dependent companies to develop, if possible, petroleum substitutes for their products, and to do this rather quickly. Using a more decentralized and informal structure during the initiation stage can facilitate the gathering the processing of information required to identify change needs. Once identified, a more centralized and formalized structure could be utilized to implement these petroleum substitutes or modifications in the production process. This more centralized implementation stage is likely to reduce the ambiguity and uncertainty individuals may have about how to utilize the change.

SWITCHING RULE 2.   The greater the uncertainty associated with the change situation, the more the organization should differentiate its structure for initiation and implementation. When uncertainty regarding a change is high, information needs during initiation will also be high. It is necessary for the organization to have a more flexible (low centralization and formalization) organizational structure as well as more information inputs, that is, more complexity to generate the information necessary to identify new areas for change. Very rigid rules and procedures might simply impede scanning the environment for new sources of information.

Scanning the external environment is very important in the change process. Various studies have indicated that the sources of ideas for innovation come from outside the organization itself (Zaltman, 1976). These studies have also found that the major sources of ideas for innovation come from market factors (customer demand, pressure from competitors) as opposed to technological factors (the desire to use the work of research and development departments). Thus the organization should monitor the environment to identify opportunities and intiate the related change process. During the implementation stage, it is necessary that specific rules and procedures be followed. This reduced ambiguity surrounding implementation is important.

SWITCHING RULE 3.   The more radical the change, the more the organization should differentiate its structure for initiation and implementation. Radical changes are characterized by their novelty and uniqueness. The more a change differs from existing alternatives, the more radical it is likely to be perceived (Lin and Zaltman, 1973). When an organizational unit is required to initiate a radical innovation, the information gathering and processing needs will be very high. Once again, the more flexible organizational structure of low centralization, low formalization, and high complexity is going to facilitate the gathering and processing of information. Here, task forces or project groups might be established to brainstorm and develop these new solutions. The organizational structure here would be very flexible. However, when shifting to the implementation stage, it again becomes necessary to develop a somewhat more centralized, formalized, and less complex structure. This is needed to provide the necessary rules and procedures whereby organizational members can fully understand how the innovation affects their role performance. This greater formalization and centralization again will have the potential for reducing ambiguity regarding how the innovation is to be implemented.

Obviously, the above switching rules are not exhaustive. However, they are the key rules that the designer might use in deciding to differentiate between structures. When the need to change is high, or uncertainty is high, or the change is radical, differentiation in structure between the two stages is appropriate. Obviously, when all three of these characteristics occur, the case for differentiation is even stronger.

Institutionalizing the Dual Structures for Innovation

Given the "radical" nature of this strategy of utilizing different structures for the two stages of the change process, it becomes important to integrate or institutionalize this process. By institutionalization we mean that this dual structure for change be integrated into the ongoing activities of the organization so that it is seen as legitimate. It is important for organizational members to see this differentiation of structure as the best way to deal with change. This differentiation of structure is then part of their role expectations. To fully institutionalize this, it is necessary for top level managers to support this process through their own behavior and organizational reward systems. Hage and Dewar (1974) have found in health and welfare organizations that top level values supporting innovation in the executive director and those who always participated in strategic decisions were more important in predicting innovation than simply structural factors such as decentralization or formalization. From this study, we can see that these supporting values are important factors in facilitating innovation.

In order to utilize this dual structure for dealing with the two stages of the change and innovation process, it is necessary to develop a climate in the organization that supports this process. To create this supporting climate, it may be necessary for organizational members to undergo training to learn how to implement the dual structures (Duncan, 1974).

In summary then, the factors facilitating the differentiation of organizational structure between the initiation and implementation stage of change-innovation are: (1) the ability of the organizational unit to deal with conflict, (2) effective interpersonal relations, (3) the development of switching rules, (4) institutionalizing the dual organizational structures for innovation. These are the four factors with which the change agent must work to facilitate the differentiation of organizational structure.

The Environment Viewed from a Public Sector Perspective

Public sector organizations may often differ from private sector organizations and warrant special treatment. Some of the differences might be mentioned here. (1) Public sector organizations may be more responsive to changes in federal funding than the private sector organizations. (2) Adoption of a new product is more a function of the physical depreciation of the product being replaced in the public sector organization than in the private sector organization. (3) The impact of a purchase on organization goals is less easy to evaluate in a public sector organization than in a private sector organization. (4) Finally, there may be differences in the kinds of people working in the two different areas.

Two researchers, Feller and Menzel (1975), have reviewed the literature on innovation in public sector organizations and have collected empirical data concerning the impact of the diffusion milieux on innovation in the public sector. Many of their ideas are presented in the discussion to follow.

Feller and Menzel conceptualize the diffusion milieu to be those external sources of pressure on an organization that cause it to change existing practice, that set limits on what set of alternative practices may be adopted, and that influence the kind of information an organization may receive about an alternative practice. The external forces are filtered through the decision-making process of an organization and help influence the outcome of that process. Examining the diffusion milieu is important with regard to public organizations because in most mission agencies changes or decisions to adopt a new product or service are the result of external pressures rather than internal initiative. It is important to study mission agencies (fire department, environmental control agency, welfare department, police department, etc) because in state and local government the decision to adopt new equipment, the choice of management consulting firms, and so on, lies primarily in the agencies rather than in central administration.

The various components of the diffusion milieu are presented in Figure 11.2. These components are also relevant to profit making organizations. We discuss each of these in turn. First, within an agency a performance gap may be felt (Zaltman, Duncan, and Holbek, 1973). A discrepancy exists between actual and desired performance. Good measures of performance are generally less available for public sector agencies than for private sector agencies, hence it is more difficult to determine what must be changed as well as when and how effective the change has been.

As indicated earlier, technological change may result more from the initiative of external groups such as manufacturers of fire fighting equipment than by any internal initiative. The manufacturer's salesmen make the public sector official aware of new and better alternatives and thus create a

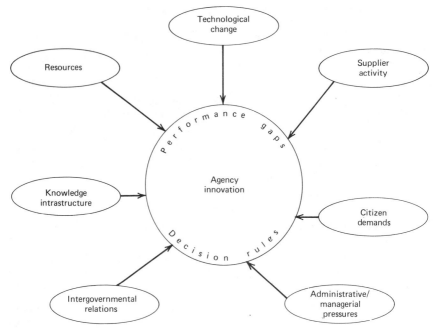

FIGURE 11.2   Conceptualization of a diffusion milieu and agency innovation. Source: I. Feller and D. C. Menzel, "Diffusion Milieux as a Focus on Innovation in the Public Sector," p. 11.

performance gap—the discrepancy between what that agency is doing now and what it could do with the new equipment. Some change agent implications of this are noted below.

A given product may be sold sequentially as the solution to different problems as the manufacturer searches for a satisfactory market response. Thus, a device for regulating the intensity of light from traffic signals was marketed sequentially as a means of obtaining optimal lighting conditions under different ambient light conditions (and thus, reducing accidents), a means for conserving energy, and as a method for reducing bulb fatigue and thus maintenance costs. Elevated platform fire trucks can be sold as providing a more reliable rescue approach or as providing a better deluge system. Along the same lines, a given technique may be marketed on different grounds to different levels of the organization—rapid water might be sold to firemen on the grounds that it reduces the weight of the hose and makes it possible to lay hose more rapidly and with less strain on the fire fighter; it might be sold to the fire chief not only on these grounds but on the basis that it permits a reduction in crew size, and thus makes possible a redeployment of manpower in the face of rising demands and fixed budgets; it might be sold to budget officers on the additional grounds that by reducing the weight of hose it reduces accidents and thus budget outlays for contingency crews and disability pensions.

Administrative/managerial pressures are another source of pressure on mission agencies. Often within the context of budget discussions, the mayor, city manager, town council, and so forth may pressure the fire department to find labor-saving techniques rather than hire more firemen. Thus public agencies may be particularly concerned with labor saving and more reliable machinery. This is especially true when federal funds are available to help subsidize the purchase. It is important for the marketer or other change agent to monitor legislation that may make funds available to an agency to help purchase equipment or special services that the marketer can provide. There is a trend toward increasing involvement of executive branches (mayor, city manager, governor, etc.) in the decision-making process of mission agencies. This makes personnel such as a mayor and his aides and advisors important as gatekeepers, legitimators, and influentials who should be informed about new products or services. A piece of equipment might be shown to reduce maintenance costs because it is of higher quality or involves a technology less susceptible to breakdowns. Certainly private organizations have similar budget-related concerns; however, these concerns may tend to be still greater in public agencies.

Feller and Menzel speculate on the influence of suppliers. Apparently very little attention has been given to understanding the role of suppliers in affecting the adoption and diffusion of innovations by public sector organizations. The following paragraphs dealing with supplier influence are from Feller and Menzel (1975).

The activities of suppliers can exert an independent influence both on the extent of adoption of specific techniques and on the identification of actual adopters in the following ways:

a. Marketing strategy. Firms differ in the extent to which they promote new techniques. Some firms actively promote new products by advertising, field demonstrations, sales representations, and technical courses for potential adopters. Others apparently develop a new product, advertise it, and then rely upon the market to respond. Interestingly, both strategies can rest upon the existence of a demonstration effect among potential customers to spread the word of the value of the new product. In the one case, however, a manufacturer may direct its initial sales efforts at the "1 percent innovators" among the potential clientele, that is, that group of potential customers which by past contact or by professional reputation are regarded by the manufacturer as most likely to be receptive to a "better product." Given acceptance by this group, the manufacturer seeks to convert the presumed favorable reaction of this group into a stimulus to the rest of the group. The other approach is more passive. It is almost as much a matter of taking orders as of active promotion.

b. How a firm allocates its sales effort by city size or region may also affect the observed pattern of adoption. That is, one might expect adoption of a given technique to be higher among groups that have been contacted by a sales representative than in cities which have not been contacted. There are numerous channels through which firms market their techniques so that it is not possible to say with any degree of certainty that one form of distribution is more effective in promoting adoption than another. However, [there are] "marketing" backwaters, where, either as a function of city size or distance from major distribution points, agencies have not been approached by manufacturers. Moreover, marketing efforts do appear to be correlated with . . . receptivity to new techniques . . .

There also appears to be a pronounced geographical influence on adoption which is traceable to the number and location of firms in the industry. Some multi-product films sell through national networks. For other technologies, firms appear to have discernible geographic areas-of-market con·rol. In still other cases, particularly for firms entering the public sector mar·et, the technique appears to spread outward in a form of contagious diffusion from the manufacturer's locale, with allowances made for spatially discontinuous jumps to distant major markets, (e.g., for example, one-man garbage collection vehicles appear to be diffusing eastward from California).

c. The number of firms which are potential suppliers of a technique may also exercise an independent influence on the receptivity of state or city governments. The general purchasing procedures in these jurisdictions are for the functional agency to determine its technological requirements, to set out the purchase specifications, and for a separate purchasing department to issue requests for bids on these specifications. Competitive bidding is typically required.

When there is only one supplier, as is the case in a number of these technologies in the public sector, an agency can secure a waiver of the competitive bid requirement. There is, however, somewhat of a hesitancy on the part of some agencies to request waivers, at least as frequently as might be required given the possible number of sole source suppliers of new techniques. Two factors appear to shape this hesitancy: (1) the view that the existence of a monopolistic supplier is prime facie indication that the price is too high, and that the best agency strategy is to wait until additional suppliers of equivalent techniques appear and drive down the product price; and (2) concern that too frequent use of sole source contracts will make the agency vulnerable to the change of favoritism.

On the other hand, too large a number of suppliers creates a situation in which agencies have considerable difficulty in keeping pace with the characteristics of the technology, with the reliability of the techniques offered by specific manufacturers, and with the reliability of the promised services of the manufacturer. Such a situation may tend to make agencies more cautious in their adoption of technologies, and thus generate a slower observed rate of adoption.

Still another consideration is resource availability. We have already mentioned that the federal government, by subsidizing purchases, can influence the adoption of innovations by public organizations. There is considerable variation among different public sectors in the resources made available to them by federal, state, and local governments. In general, more is spent on education than highways, and more is spent on highways than police and fire protection. Police agencies receive substantially more funds than do fire departments. Hence if a firm were marketing a communication device, it would do well to offer it first to highway construction departments, followed by police, which in turn would be followed by fire departments.

Knowledge infrastructure is a sixth important element in the diffusion milieux. The greater the number of public and private professional associations in a given context, the more readily a new product or service will be diffused in that context. Professional associations serve as connective tissue among consultants, researchers, practitioners, etc. and relay information that tends to reduce uncertainty and risk about new products. Another important aspect of knowledge infrastructure is the general state of knowledge. In some areas, information about solving problems may be scant. Moreover, information or knowledge about the nature of a problem may be scant.

The primary differences between the functional fields appears to be that where there is a highly developed knowledge infrastructure, performance norms and technical certification will occur at the national level. Manufacturers in approaching individual buyers will relate their product to these norms, augmented perhaps by references to the experiences of their other purchasers. Contact between potential purchasers and past purchasers may occur, but the need for such communication on the part of the perspective customer is lessened by the national norm. Where such norms do not exist, manufacturers will be required to rely more heavily on selling to individual customers, and the role of opinion leaders and the degree of interaction among adopters and potential adopters as a factor in the diffusion process will be increased (Feller and Menzel, p. 25).

Legislation may create a kind of performance gap: the discrepancy that exists between actual and current performance on one hand and what performance must be on the other hand. Air quality legislation has produced considerable research and development among commercial firms seeking to tap the newly created market for air quality control devices. In another area, federal hearings on highway safety increased attention among state and local officials to this problem. This increased attention increases the receptivity of an agency or its officials to purchasing highway safety equipment. Thus intergovernmental relations is a very important aspect of the diffusion milieu. States may influence how local governments can spend federal funds; state highway officials may act as advisors to municipal agencies. It is necessary then for marketers to determine who the influentials are when agencies from different levels of government interact. In some cases the flow of influence may go from a large city to a state official or agency and from there to other municipalities.

Citizen demands are another element of the diffusion milieu. Citizen demands may act primarily as a veto rather than stimulant of change. This component of the milieu may be one of the least important elements and also the component whose main effect is most indirect and least susceptible to analysis.

Finally, we must consider agency decision rules. Feller and Menzel, in summarizing the literature in this area, suggest the following implicit decision rules influencing a public organization's purchase behavior. The marketer should be aware of them. First, agencies typically have budgets that increase only slowly from year to year, and only a small portion of the budget may be allocated for equipment purchases. Although the latitude to buy may be small, it is considerable with respect to what equipment will be purchased. The most likely purchase will be a close functional equivalent of a present piece of equipment and be similarly priced. An agency is likely to purchase less of a product whose price has risen sharply (if it is purchased at all). The level of expenditure will approximate the original cost of the equipment being replaced. Large-scale innovation is unlikely to be considered unless external funds are available to subsidize a significant portion of the expenses.

Feller and Menzel put forth a few propositions, although many propositions are implied in the discussion thus far. *Proposition 1.* New products will be accepted more rapidly in fields in which the need for change has been clearly articulated than in those fields in which manufacturers must work alone on a one-to-one basis with potential customers to develop an awareness of the need to change. *Proposition 2.* New products will be accepted more rapidly in fields in which performance standards are clearly specified and policed by government agencies than in fields in which the manufacturers must work alone on a one-to-one basis with potential customers to develop an awareness of the need to change. Additionally, "the greater the 'legitimacy' within an agency of the external sources [of new products] (government/professional associations/manufacturers), the greater the likelihood of adoption of innovations" (p. 33). *Proposition 3.* Public sector organizations are more likely to wait until a product seriously deteriorates before replacement than are private firms.

This chapter can be summarized by the following principles:

1. It is important for the organization to scan its environment, since this is a critical source of ideas for change and innovation.

2. The organization must scan its environment, since this is an important source of information regarding the technology or means that may be required for change.

3. The organization must be sensitive to the pressures existing in its environment with respect to how these environmental pressures might support or resist some change the organization is contemplating.

4. Organizations can enter into joint programs with other organizations as a way of stimulating change and reducing its financial costs.

5. The organization must be aware that there are two phases to the change process, each with its own set of problems. During the initiation stage, the organization must generate much information regarding the change. During the implementation stage, the problem becomes one of integrating the change into the organization.

6. Organizations with complex environments that are changing in terms of their composition and the kinds of demands being made on them should be prepared to monitor their environment closely and be ready to change.

7. The organization should shift its structure for the initiation and implementation stages of the change-innovation process. During the initiation stage, a higher degree of complexity, lower formalization, and lower centralization should be used, since these increase the information gathering and processing capabilities of the system. During implementation, the organization should shift its structure to lower complexity, higher formalization, and higher centralization, since this ensures that those using the change will have a clearer understanding about how to use the change.

8. To facilitate the shifting of its structure from the initiation to implementation stages, three conditions must exist in the organization: (1) the ability to deal with conflict, (2) effective interpersonal relations, (3) the institutionalization of the dual structures.

9. The greater the need for change, the more the organization should differentiate its structure for the initiation and implementation of change.

10. The greater the uncertainty associated with the change situation, the more the organization should differentiate its structure for the initiation and implementation of change.

11. The more radical the change, the more the organization should differentiate its structure for the initiation and implementation of change.

12. Utilizing the dual structures for the initiation and implementation of change must be institutionalized in the organization. This process must be seen as legitimate and supported by the top level people in the organization.

PART FOUR

Neglected Topics

There are two topics above all that we feel are too frequently slighted in the social change literature. These topics involve research use and value and ethical issues in social change. Chapter 12 focuses on research use and some of the mechanics involved in the translation of research and theory into practice. Several principles and guidelines for effective knowledge use are also presented. If change agents are to make use of new knowledge about human behavior, they must have this knowledge translated into actions. Although it is reasonable that change agents should carry some of the responsibility for this, we also feel that producers of knowledge must share in this responsibility. Hopefully, some aspects of this chapter will be of value in facilitating the mutual sharing of this responsibility. Chapter 13 is concerned with value and ethical issues in planned change. We feel that these issues, in practice, are resolved one way or another. What is not clear is how alternative ways of resolving these issues impact on the change process: Are "good" ethics compatible with "effective" planned change?

CHAPTER TWELVE

Research Use and Translation

A major social science practice has been recurring throughout this book. This practice has been performed either by people whose works have been cited or by the authors. The practice involves one of two things and sometimes both: (1) the translation of theory and research into practice, and (2) the generalization from one applied context to another. Although increasing attention is being given to knowledge utilization practice, relatively little is known about many aspects of this phenomenon (Stankey et al., 1975; and Lyons et al., 1975). In fact, knowledge utilization may be termed the Achilles heel of the social sciences. The intent of this chapter is to provide a variety of perspectives concerning knowledge utilization in the social

283

sciences. In so doing, we hope the reader will not only become more sensitive to and aware of the issues involved, but will also improve his own knowledge utilization capacity in the area of applied social change.

An important first observation that must be made is that a very considerable amount of social science research is oriented toward maintaining the status quo rather than stimulating social change.

> Since social research is often sponsored, conducted, and used by the established segment of society, it tends to be addressed and applied to the problems perceived by these groups. It is, thus, more likely to support the status quo than to support social change, not primarily because of any deliberate attempt to uphold the status quo, but because status quo assumptions tend to be built into the definition of the research question and the frame of reference within which the research is carried out (H. C. Kelman, p. 85).

Neither we nor Kelman mean to imply that deliberate research is not undertaken explicitly for purposes of gaining social change. Considerable change-oriented research is carried out by the major institutions in our society, and bold suggestions are being made to foster such research (Campbell, 1969; Pressman, 1975). However, the full potential of change-related research is far from being used to its maximum degree. Two important causes of the limited use of change-related research are inadequate dissemination of research results and the difficulty both researchers and users have in translating research into practice. For example, laboratory experiments concerning cooptation are relevant to interpersonal problems between stepfathers and their stepchildren; for example, coopting would-be stepchildren during the premarital courtship process considerably lessens the number and magnitude of postmarital parent-child interpersonal relationships. The two key questions are, how does the family counselor learn about cooptation research (the information dissemination problem), and how does the counselor and/or the researcher translate that information into principles for family counseling (the translation problem)?

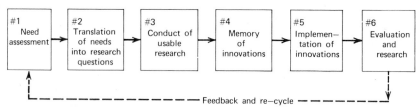

FIGURE 12.1 Source: Rogers, Lin, and Zaltman, 1974, p. 13.

In the first part of this chapter, we focus on factors affecting the dissemination and usability of information. These factors have implications for researchers and practitioners involved in social change. The latter part of this chapter focuses on the translation process.

RESEARCH USE

What is Information Use? *

Information or knowledge production and use is the process whereby information is formulated and then used to some advantage. It can be thought of as a process in which knowledge produced from research, theory building, strategy testing, conceptualizing, observation, professional practice, etc. is translated into useful forms.

Rogers, Lin, and Zaltman (1974) define information use as a "process by which users' needs are determined and communicated to researchers, leading to research designed to meet the needs, and eventually to new knowledge based on research which is communicated to answer their needs." This definition implies the need for direct and open communication between the researcher and user. It does not preclude the case in which the researcher and the user can be the same person. There are several components involved in research use. These components are shown in Figure 12.1 and are defined briefly below.

1. Need assessment involves identifying (1) who the users are, (2) their readily expressed or felt needs, (3) latent needs, and (4) needs that may develop in the relatively near future.

2. The translation of needs into research questions involves redefining problems or difficulties stated by users into a format meaningful to researchers.

3. Conduct of usable research involves doing research related to user needs in a manner amenable to users' capacity to act on research.

4. Creating an innovation memory bank (or inventory) involves gathering the results of research, synthesizing them, storing them, retrieving them as necessary, and formatting the new knowledge in a way that facilitates user interpretation.

5. Implementation of innovations or new knowledge involves helping users put the new knowledge or research results into practice.

* The reader will find *Evaluation* magazine a good source of information about research use problems and strategies.

6. Evaluation and research involves the constant improving of all earlier components.

Rogers (1973) also presents a somewhat different conceptualization. This is shown in Figure 12.2. This figure is presented because it portrays a very real and sometimes very unfortunate situation involving separate systems for research, for linking, and for clients. The key features of Figure 12.2 are the direct or indirect interaction among: "(1) the *research system* which

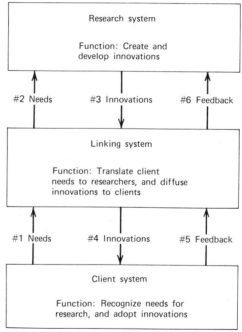

FIGURE 12.2   Paradigm of the research utilization process. Source: Rogers, 1973 (E-W Center Conference).

The communication flows numbered in this paradigm may be identified as follows:

1. Flow of user needs (for information) to linkers.

2. After interpretation and clification, these needs are transferred to the research system.

3. Researchers attempt to provide needed information for users' needs, either from accumulated knowledge or via newly originated research.

4. Linkers distill and interpret this new information (innovations) for users.

5. Feedback from users to linkers on the new information in meeting their needs.

6. Linkers convey users' feedback to researchers, perhaps leading to further user needs and recycling of the entire process.

creates and develops research results into innovations, (2) the *linking system* which performs the function of translating client needs to researchers, and of diffusing innovations to clients, and (3) the *client system* which recognizes needs for research and thus leads to its initiation, and which later adopts the innovations that may result."

There are several other information use perspectives developed by Sieber et al. (1972), Havelock (1973), Clark and Guba (1974), Argyris and Schon (1974), Lazarsfeld and Reitz (1975), and Zaltman, Florio, and Sikorski (1977). These perspectives, like that shown in Figure 12.2, also clearly separate research, linking, and client systems and describe (like Figure 12.2) ideal interactions among them. However, there are real barriers between the three systems, particularly between the research and client systems. Some of these are status barriers, others are language or jargon barriers, and still others derive from self-imposed territorial domains by those called researchers or theory builders and those called practitioners or doers. As a practical matter, most researchers and practitioners interact very little with one another either directly or indirectly through linkers. Change agents as linkers tend to be more practitioner oriented than researcher oriented. Thus change agents tend to diffuse to practitioners knowledge about what other practitioners are doing rather than diffusing new (or preexisting) research information that may be relevant to a problem. We refer to only formal research. Practical experience constitutes an important kind of data for change agents and practitioners and represents an important type of informal research that should not be ignored.

Guidelines for Using Information to Achieve Change

A number of guidelines or principles for effective use of information to achieve social change can be culled from the formal literature, the authors' experiences, and the informally reported experiences of other change agents.* These guidelines are far from exhaustive. Appendix 1 of this chapter lists a still larger set of principles. The guidelines presented here are intended for both the change agency as well as the change agent carrying out the policies of the change agency. The change agency may be the target group itself that is attempting self-renewal, or the agency may be external to the target group. Many of these guidelines may appear too obvious to warrant mention. Sadly, it is surprising how frequently these guidelines are forgotten or simply not considered very seriously.

Guideline One: The application of new information, whether as an idea or as a physical product, should be done in such a manner that minimal dis-

---

* These guidelines are also presented in somewhat different form in Zaltman et al., 1977 (*Dynamic Educational Change*).

ruption occurs in the overall formal and informal social system being changed. For example, one business firm maintained a large secretarial pool, with most secretaries being in close physical proximity to one another; this facilitated the formation of a strong but informal social system in the pool. A major reorganization of the firm resulted in the dispersal of secretaries to different areas of the building where they often sat alone and without sight of more than one or two other secretaries. Morale declined considerably, and a relatively large number of secretaries left the firm in the three months following the reorganization. The reasons given by most secretaries for leaving clearly indicated that the disruption of the social system and a lack of a good alternative one under the new arrangements was an important causal factor.

Guideline Two: Information about a change or innovation should be cast in terms of the vocabulary used by the target or client group rather than that used by the change agent. Researchers may forget that many practitioners do not know the difference between a "chi square" and a "t score." When speaking with persons for whom, say, English is a second language, care must be exercised to use words whose sound could not cause confusion with other words. A very interesting and perhaps apocryphal example of this was related to one of the authors on a visit to Costa Rica. A livestock specialist from the United States was visiting an experimental cattle breeding station in Costa Rica, and, through an unofficial interpreter (a young lady university student), conversed with one of the workers at the station. The visitor mentioned that frozen semen was to be shipped to the station for use in impregnating some livestock in an attempt to improve the quality of cattle being produced. The astonished worker replied, through an embarrassed interpreter, that he had great respect for American science but could not understand at all how anything cold sailors could do would improve the livestock.

Guideline Three: Every effort should be made to lessen the perceived threat of information from an outside change agent and to stress the commonality of the change agent's goals and activity with those of the client The fear that new information or ideas may contain hidden threats to survival is especially prevalent in the field of education. For example, local anti "Program Planning Budgeting Systems" groups were formed to prevent this managerial technique from being implemented in local schools. These groups feared PPBS as a federal move to take over local school operations.

Guideline Four: Responsibility for major change should not be assigned to a new member of an organization unless that person has extensive authority or is working in collaboration with members who do have such authority and/or acceptance by the more established organizational members.

Guideline Five: New information requiring or implying change may best be introduced at a time when new leaders are being installed and when those leaders are expected to bring new ideas. There are many instances in which school boards have adopted procedures recommended by a new superintendent that were also recommended by his predecessor but refused by the school board (Zaltman et al., 1977). It appears, too, that the amount of change in an organization is greater under new leadership when that leadership comes from outside the organization.

Guideline Six: The infusion of new knowledge and associated changes should be attempted first in an organization or part of an organization open to change. The more innovative entity can then serve as a role model for others that are less open to change. This guideline is simply a restatement at the organizational level of the marketing approach to using early adopters as change stimulators.

Guideline Seven: The introduction of new knowledge is most acceptable when a real or potential crisis is evident and the new knowledge is believed to represent a way of solving the crisis or aiding the development of the organization and its members. The adoption of new technologies in industrial settings is correlated with the incidence of economic and competitive pressures (Kelley et al., 1975). Use of consultants is disproportionately high when a new division or department of a firm is in the process of being established compared to when it has been functioning for two or more years.

Guideline Eight: New information can often be gained very quickly by having clients observe an application of the information in a setting similar to their own. Thus change agents should consider bringing potential adopters to sites at which information is being applied effectively. This practice is not uncommon in agriculture and the heavy construction equipment industry.

Guideline Nine: External change agents as well as new permanent or temporary members of the system should be used as a source of new knowledge. The external change agent has the potential for bringing into an organization a variety of informational input; however, he or she may run into many barriers if not familiar with the organization values, self-perception, and the environmental forces it faces. The degree of success of change agents and new personnel in making the organization members open to information will be significantly affected by what Rogers and Shoemaker call "homophily." Homophily is the ". . . degree to which pairs of individuals who interact are similar in certain attributes, such as beliefs, values, education, social status, and the like" (p. 14).

Guideline Ten: It is very desirable to have a special group whose members are carefully selected to serve as a new information research team

Table 12.1   Principles for Effective Information Use in Education*

Principle 1.	The educational system should provide, at various levels of the system, for a knowledge use function.
Principle 2.	The initiator of change should show a general willingness to alter his own operation or behavior.
Principle 3.	Resources (broadly defined) should be available to implement change based on knowledge.
Principle 4.	Resources should be available to develop staff capabilities for change.
Principle 5.	There should be a general sensitivity to the various constituents who could be affected by a change program.
Principle 6.	It is necessary to establish a basic belief among members of the school system that research and new knowledge can improve existing and prospective change efforts.
Principle 7.	The practice of continually translating program needs into research-able questions and projects should be encouraged.
Principle 8.	Experiences with any change should be evaluated by persons with different perspectives, and the evaluations should be translated into recommendations for continuing and/or altering the change program. However, one person or group should be responsible for the evaluation and recommendations while being careful to communicate with persons affected by the evaluation.
Principle 9.	Research staff in an educational system should be trained to be sensitive to the need for developing and expressing the practical implications of their work. Care should be exercised to select research topics that are relevant to specific practitioner problems rather than topics that are of special theoretical or methodological interest.

on a permanent basis, or at least on a temporary basis in response to a problem. The practice of having new information search teams is quite limited. It appears, however, that these teams are most effective when they report directly to senior management.

Zaltman et al. (1977) also suggest a number of principles for effective information use in education. These principles seem applicable to most other contexts and are presented in Table 12.1.

Two Cases of Effective Knowledge Use

Two areas or fields have been regarded as "successful," at least relative to most other areas, in their knowledge use activities. These areas are agriculture and marketing. The agricultural approach has been used as a model in several other contexts, including education, social and rehabilitation

Table 12.1   (Continued)

Principle 10.	Research, teaching, and other staff should be rewarded for breaking out of the bounds of current practice and daring to be different.
Principle 11.	Mistakes in the research, translation, or implementation processes should be identified and the lesson(s) learned from them made clear. However, a general understanding should exist to the effect that mistakes are quite likely, perhaps inevitable, and should be viewed in a constructive spirit rather than in a spirit of criticism and blame.
Principle 12.	Research project priorities should be based on the potential projects have for effective use as well as on the importance of the problem.
Principle 13.	Continuous two-way communication between researchers and users should occur at all stages of the knowledge use or action research project and program implementation.
Principle 14.	Special incentives should be provided to motivate teachers, administrators, and other staff to use such internal resources of change knowledge as (1) teacher/researcher discussions and idea presentations, (2) in-service training, (3) curriculum supervisors and organizational development specialists, (4) library facilities, (5) research-evaluation staff, (6) media specialists, and (7) student discussions and idea presentations.
Principle 15.	Special incentives should be provided to motivate teachers, administrators, and other staff to use such external resources as (1) state education agencies, (2) federal agencies, (3) universities and colleges, (4) professional associations, (5) ERIC, (6) regional educational laboratories, (7) foundations and other private sources, (8) consultants, and (9) publishers.

* *Source: Zaltman, Florio, and Sikorski, 1977.*

services, and family planning (Rogers, Eveland, and Bean, 1974). The marketing approach is lesser known but has received increased attention from nonmarketers in the last few years. Both approaches are discussed here briefly.

The agricultural approach is best viewed as an organizational structure as shown in Figure 12.3. This structure is a blend of federal, state, and local organizations and individuals. There are approximately 17,000 professionals distributed across the various levels shown in Figure 12.3. A key group at the state level is the extension specialist. This specialist is concerned with translating research findings in his area of specialization into meaningful applications for the county extension agents and farmers whenever possible. Extension agents exist for several specialties such as farm management, home economics, poultry, husbandry, and so forth, which reflects the growing specialization of farming in the United States.

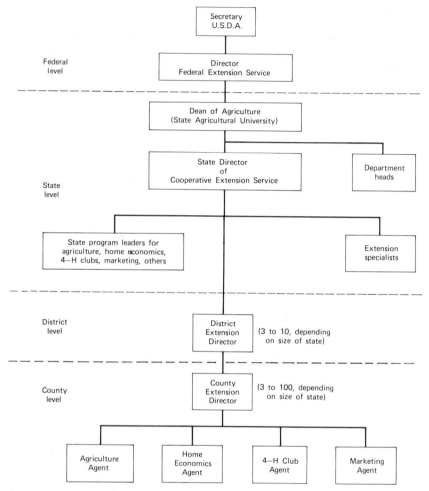

FIGURE 12.3   Organization of the Agricultural Extension Service in the United States.
Source: Axinn and Thorat (1972).

At the county level, county extension directors and extension agents work to disseminate research findings brought to their attention by state extension specialists and others. It is ultimately in their laps that responsibility falls for facilitating among farmers the adoption and diffusion of new ideas about farming and relative areas. County level personnel represent the main link between the farmer and the research producing community as represented by state extension specialists.

Importantly, the county extension agent does not only listen upward and speak downward. The county extension agent usually has an influential advisory group. This group helps determine the county agent's activities by identifying particular farmer needs and helping to structure programs or efforts to meet these needs. For example, if local farmers are having trouble with pests, they will see that the topic of pest control is featured in the agent's agenda for the coming year. The advisory council also has influence on the budget and even the tenure of the agent's job. The advisory group can also initiate a request for information that may or may not require new research at the state and/or federal levels. This does not mean that such requests are honored.

In the late 1960s, extension aides in home economics were added at the county level. The aides are selected from among the client group, that is, low-income homemakers. Aides receive limited training and close supervision. They are expected to visit homes and address a wide range of problems. The territory of the aides includes rural areas and inner city areas.

The relative success of the agricultural extension model seems to be due in part to the following factors: (1) A formal procedure for knowledge utilization existed, that is, specific roles were created to translate research into knowledge. (2) Linkages between researchers and practitioners were established through the office of the state extension specialist, county extension agents, and more recently extension aides. (3) Provision was made through advisory councils to identify client needs and to communicate these needs upward to researchers and to reflect them in county level programs. (4) Researchers have been sensitized to practical needs. (5) The social and psychological aspects of innovation diffusion have been recognized.

The area of marketing is represented as an area in which there is effective knowledge use. Like agriculture, it is far from perfect, although relative to many other areas it is successful in using knowledge. Unlike agriculture there is no massive formal organizational structure ranging from the federal to state or county levels of government. However, within business organizations, departments or divisions that are effective users of knowledge do exhibit a number of common characteristics and practices that seem to account for their effectiveness. These characteristics and practices can be described briefly.

First, virtually all marketing groups have a marketing research expert who is often in charge of a substantial research program. Alternatively, these groups have contractual agreements with advertising agencies and other research organizations to supply market data on a continuing basis. Thus there is a constant flow of research information into the organization.

Second, the information received is in response to very explicit issues or information needs formulated by the marketing decision makers. Often marketing decision makers demand that the actual questions to be asked in an interview or survey be presented to them first for approval. The best researchers in marketing spend considerable time with decision makers finding out what kind of information is desired and in what kind of format and to what degree of precision it is desired. Researchers' jobs are largely dependent on the success they have in reading the decision maker's mind. This does not mean that market researchers will not dispute particular points with decision makers or will not take initiative in providing unrequested data.

Third, the time lag between the request for information and its provision to decision makers is relatively short. The gathering of new data may require less than two months, particularly when subject or respondent pools are maintained on a continuous basis as in the large scale consumer panel. The actual feedback of data is accomplished in multiple ways. Extensive written reports are prepared as well as executive summaries. Additionally, verbal reports using slides, charts, etc. are also made to decision makers by one or a team of researchers. These verbal presentations enable decision makers to obtain clarification of particular points and request specific additional tabulations. They also provide opportunities for decision makers to ask the researchers whether they think the data support particular marketing actions.

Fourth, researchers and decision makers typically share common values about the use of research. More specifically, researchers in marketing generally agree that the data they provide is most often used to the consumers' advantage by further clarifying for decision makers consumers' needs and preferences. Many social science researchers would not agree with this. Additionally market researchers generally share the decision makers' concern with the questions, "So what? What do these data suggest we do?" Thus decision makers and researchers often share a common action orientation. Related to this is the fact that researchers are typically not condescending to decision makers.

Fifth, marketing decision makers in many organizations with substantial marketing personnel often are competent in research methodology or at least conversant with the terminology and basic ideas behind various analytical techniques. Intensive short courses in multivariate techniques, sampling, etc. are quite popular among nonresearch marketing managers.

A sixth feature of marketing settings that is not widely shared by other contexts is that the market responds fairly quickly and in readily measured ways (sales, market share) to particular actions. This in turn allows marketers and researchers to determine whether the right questions were asked, whether the data obtained were sound, and whether the correct interpretations were made. This does not mean that a market failure can always be or will always be linked to research or its use. There are many other sources of difficulty that produce market failures even in the presence of good research used well.

Seventh, when the responsibility for the implementation of research findings lies with a senior level executive, research tends to be used more effectively than when this responsibility is held at lower management levels.

Eighth, the most effective executives at any level are those who interact with researchers directly on a face-to-face basis at all stages of the research. This is most readily done in marketing contexts in which considerable research is done in-house and in which independent research agencies are accustomed to close and frequent contact with their clients.

A few firms have started using debriefing days quite successfully. For example, one of the authors conducts debriefing sessions with managers in which three to four hours a month are devoted to presenting new data from the published literature that are relevant to one or more problems or tasks the managers face. Each finding or set of data is discussed solely in terms of its implications for the managers, assuming the finding is correct. This approach is also being tried by a nonprofit social action agency. The effects of these debriefing sessions in the different organizations has not been measured in a rigorous way. However, there is a strong feeling among managers in the organizations involved that they have in fact undertaken actions they probably would not have tried, that they are much more open to research not conducted by them, and that they tend to think more imaginatively in translating research into action. This last observation brings us to the next section of this chapter.

TRANSLATION

The practitioner of social change approaches social problems by obtaining data believed relevant to a problem, thinking about the data, and initiating action on the basis of his or her thinking. The thinking process or the performance of cognitive processing is of major concern here. How do change agents or clients bridge the gap between theory and/or data on one hand and action strategies on the other hand? The "data" may consist of a theory, a statistical tabulation, a personal experience, or a report of a change strategy in another context. The change agent then translates the data into meaningful action Tripodi and Miller (1966) suggest factors affecting the translation process: (1) the nature of the judgment task, (2) the form of the data, (3) characteristics of the judge affecting cognitive operations, and (4) the context in which translation takes place.

It is neither within the scope of this book nor this chapter to deal in depth with the translation process. There is considerable concern with this process in clinical psychology, although it is evident that even there much ambiguity exists about the process. It will be useful to devote a few pages to the thinking process that appears to be used in translating research and theory into action. Following this, we focus on various techniques for stimulating the translation process.

Two classical views of the thinking process involve deduction and inference. These are illustrated below.

DEDUCTION: 1. Poor and uneducated persons are more likely to respond to incentives for family planning than are wealthy and educated persons.
2. Residents of community X are poor and uneducated relative to community Y.
3. Residents of community X are more likely to respond to family planning incentives than residents of community Y.

INFERENCE: 1. Residents of community X are poorer and less well educated than residents of community Y.
2. Low levels of income and education are characteristic of persons most apt to respond to family planning incentives.
3. Therefore, residents in community X are more likely to respond to family planning incentives than are residents of community Y.

DEDUCTION: 1. The greater the complexity of an organization, the more difficult it is to implement change in that organization.
2. Organization X is highly complex.
3. Therefore, the implementation of change in organization X is likely to be difficult.

INFERENCE: 1. Organization X is highly complex.
2. Highly complex organizations have difficulty implementing internal change.
3. Therefore, the implementation of change in organization X is likely to be difficult.

Change agents and other practitioners of change tend to emphasize the inference rather than deductive process. We examine three perspectives on inference: (1) purposive behavior inference, (2) inference from norms, and (3) a clinical inference perspective.

Purposive Behavior Inference

One type of inference is referred to as purposive behavior inference. Consider a description of an outpatient clinic in a metropolitan city provided to one of the authors by a person who had recently obtained a vasectomy from a private clinic specializing in this operation after being "fed up" with his treatment by the hospital: "The benches were hard and I waited two and a half hours beyond my appointment time before leaving. There was nothing to read and kids were crying all over the place. Five times a nurse told me to wait my turn and not to bother her anymore. . ." The prospective patient made a number of inferences about the hospital. Two of these inferences were that nonmedical patient treatment did not seem important to the hospital, since their actions with respect to this were indifferent at best; and the hospital placed greater value on other activities, since few resources relative to demand were allocated to the outpatient department. Similarly, it can be inferred from the prospective patient's actions that he was strongly motivated by virtue of waiting an inordinately long time and then seeking a more expensive source of treatment, and his choice of vasectomy over other methods indicates something about his values and attitude about husband responsibility in family planning (the person involved had a wife and four children). It is quite possible, of course, that pressure from his spouse was responsible for his seeking a vasectomy, and his persistence reflects not his own values and attitudes, but more those of his wife. Consequently, multiple indicators of attitudes, goals, etc., should be used whenever possible to allow for convergent validity. In any event, change agents do make inferences about client's goals, motivations, values, and action by observing direction and intensity of actions and alternatives foregone by clients. This information is then used in the presentation of an advocated change.

Inference from Norms or Dimensions of Status to a Collective Goal

In this type of inference, the reasoner may make inferences about the character or nature of group norms on the basis of incentives and constraints the group confers on its membership. In turn, inferences are made from these norms to some more basic goal that suggests the norm. Thus the sequence is:

incentives and constraints $\rightarrow$ inference $\rightarrow$ group norms $\rightarrow$ inference $\rightarrow$ group goals

Many effective change agents engage in this type of reasoning quite automatically. We have observed this many times in discussion with change agents in contexts as varied as district medical officers and field workers in family planning in Bangladesh to directors of large industrial corporations. One example of a strategy based on this reasoning process was provided in the political arena, when a change agency inferred from public statements of a state official that he favored somewhat a particular piece of legislation (an ethics bill) with strong populist support on a national level. It was inferred from this and other behaviors that his position and attraction for the issue was motivated in part by national political ambitions. The change agency used this information to help obtain formal endorsement of the official for another item of legislation that the agency was advocating.

Six Stages of Clinical Inference

Sarbin, Taft, and Bailey (1960, p. 47) describe six stages of clinical inference. The extent to which these six stages are applicable to nonclinical settings is an unanswered question at this point. Also, although clinical counseling involves change—often deep and profound change—it is not yet clear how clinical inference applies to social change. However, in the following few pages some aspects of clinical inference are considered that might have value to change agents in other settings. An attempt is made in the presentation to extend the aspects of clinical inference discussed beyond their traditional domain.

The six stages of inference identified by Sarbin et al. are:

STAGE 1.   The *postulate-system* of the inferrer is invoked. Judgments are influenced by a system of postulates that furnish the cognitive background. For example, the inferrer may hold the belief that literacy is directly and causally connected with modernization. This and other aspects of the inferrer's postulate will affect both his acquisition and processing of information.

STAGE 2.   *Construction of a major premise.* Societies with high literacy levels will be modern.

STAGE 3.   *Observation of occurrences* are made that will provide inputs for the minor premise. An observation could be that society X has a literacy level of 75%.

STAGE 4.   *Instantiation* or the conversion of the occurrence into an instance of a general class is the next stage. (The terms categorization and classification have also been used in lieu of instantiation.) Society X has a literacy level of 75%. A literacy level of 70% or more is "high."

STAGE 5.   *The inferential product or conclusion.* Society X has a literacy level associated with modernization.

STAGE 6.   *Prediction.* Society X is modern.

Conclusions are only formed when major and minor premises are placed together. (The minor premise may occur prior to the major premise.) Thus instantiation, which is the process in which the major and minor premises come together, is crucial and is the main concern of the remainder of this section of the chapter.

Instantiation Through Taxonomic Sorting

Instantiation takes place through taxonomic sorting in which an event or occurrence is sorted as belonging to a particular class. Sarbin et al. state it this way (p. 61):

1. A has characteristic X (the occurrence).
2. X is the defining characteristic of species M (the mediation).
3. A is a member of species M (the instantiation).

There are several ways taxonomic sorting can take place:

1. Through the mediation of specific exemplars
2. Through analogical mediation
3. Through metaphoric mediation
4. Through trait substitution
5. Through reductive mediation

We give an example of each of these from the perspective of a person or group interacting with a change agent.

1. *Mediation through specific exemplars.* In this instance the inferrer "uses a specific person as an exemplar of a species for arriving at an assessment of another person" (p. 61). In the example immediately below, A represents a change agent, X is a characteristic, and M is a particular change agent stereotype. Alternatively, M could represent a particular change agency.

1. Change agent A is aggressive.
2. Aggressiveness is a characteristic of change agent B, who is the exemplar of change agent type M.
3. Therefore, change agent A is a member of a particular set of stereotype change agents (type M).
4. Action conclusion (avoid agent A).

Using the homophillic principle:

1. Change agent A is like me.
2. I am a cooperative person.
3. Therefore, change agent A is cooperative.
4. Action conclusion (interact with A).

2. Analogical mediation.

1. Change agent A is aggressive.
2. Aggressiveness and persistence are defining characteristics of change agent type M.
3. Therefore, change agent A is also persistent and is a member of the class of agents of type M.
4. Action conclusion.

3. Metaphoric mediation.   Inherent in certain words or meanings are surplus references that are generalized. Thus:

1. Change agent A is sincere.
2. Sincerity is denoted by honesty.
3. Agent type M is denoted by honesty.
4. Therefore, agent A is a member of the class of change agent type M.
5. Action conclusion.

4. Mediation through trait substitution.   This process involves the inference of one often unobservable trait from more readily observed behavior and traits that are believed to be concomitant or correlated. The formal statement of this follows:

1. Change agent A is sincere.
2. Sincerity is a concomitant or correlate of honesty.
3. Honesty is a defining characteristic of change agent type M.
4. Therefore, agent A must belong to the change agent type M category.
5. Action conclusion.

Another viewpoint is that correlation or concomitant of independent variables is not necessary but *similarity* is sufficient. Thus if a change agent has a trait perceived similar to another trait, he will have that other trait ascribed to him.

5. Reduction Mediation.   In this approach possible alternative conditions are successively eliminated, leaving only the most plausible condition or trait. Thus:

1. Change agent A is sincere.
2. Sincerity is a defining characteristic of change agent types M and N.
3. Credibility is a defining characteristic of change agent type N.
4. Agent A is not credible.
5. Agent A does not belong to the type N category.
6. Therefore, agent A belongs only to change agent type M.
7. Action conclusion.

These considerations are based on person cognition. It is unclear how effective each of these mediation processes can be in stimulating the generation of different social change strategies. However, a few attempts will be helpful to clarify the problem. The particular concern is the use of power strategies.

1. *Mediation via specific exemplars.*

1. Situation X is in need of immediate change or intervention (the characteristic).
2. The need for immediate change is a characteristic of some other situation that calls for the use of a power strategy.
3. Therefore, situation X is an instance of the group of phenomena warranting the application of a power strategy.

2. *Analogical mediation.*

1. Situation X is in need of immediate change or intervention.
2. Need for immediate change and the inability of local resources to solve problems are characteristics warranting the application of a power situation.
3. Therefore, situation X is also one in which the application of local resources is ineffective or not possible and thus warrants the application of a power strategy.

3. *Metaphoric mediation.*

1. Situation X is in need of immediate change.
2. The need for immediate change is denoted by imminent or present disaster.
3. Power strategies are denoted by disaster conditions.
4. Therefore, situation X is within the set of circumstances warranting the application of a power strategy.

4. *Mediation through trait substitution.*

1. Situation X is in need of immediate change.
2. The need for immediate change is a correlate or concomitant of the inability of local resources to take remedial action.
3. The inability of local resources to take remedial action is a condition specifying the use of a power strategy.
4. Therefore, situation X is a situation warranting the use of a power strategy.

5. *Reduction mediation.*

1. Situation X is in need of immediate change.
2. The need for immediate change is a condition warranting persuasive as well as power strategies.
3. The existence of adequate communication channels is a requirement for the use of persuasive strategies.
4. Situation X does not have adequate communication channels.
5. Situation X does not qualify for the use of persuasive strategies.
6. Therefore, situation X only qualifies for power strategies.

Another illustration will be helpful. Consider the proposition that mass media channels are relatively more important at the knowledge stage of the innovation decision process (Rogers and Shoemaker, 1971, p. 255). We will assume that this has been validated in agricultural settings.

1. *Mediation by specific exemplars.*

1. In the innovation decision process, mass media are uniquely related to the knowledge stage.
2.* The unique relationship is characteristic of agriculture, which is believed to be the exemplar of education.
3. Therefore, the unique relationship probably also applies to education.

It is important to note here that a particular finding in one context is believed appropriate to another context by virtue of an assumed or defined similarity between the two contexts.

2. *Analogical mediation.*

1. In the innovation decision process in agriculture, the mass media are uniquely related to the knowledge stage.
2. The knowledge/mass media relationship and the persuasion function/ interpersonal channels relationship are defining characteristics of innovation decision making in education.
3. Therefore, innovation decision making in agriculture involves a relatively more important role of interpersonal communication at the persuasion function and the overall decision process applies to education.

It is important to note that here a new relationship was taken into account, the persuasion function/interpersonal channels relationship. This relationship, along with the knowledge/mass media relationship, was identified as

---

* The limitations as well as foundation of this observation can be found in Rogers, Eveland, and Bean, 1974.

applicable to education. The new relationship was also made part of the innovation decision process in agriculture. Thus what is known about the innovation decision process has been broadened and made applicable to at least two areas.

3. Metaphoric mediation.

1. In the innovation decision process the mass media are uniquely related to the knowledge stage.
2. The knowledge/mass media relationship is denoted by communication behavior set Y.
3. Communication behavior in education is denoted by communication behavior set Y.
4. Therefore, the innovation decision process in agriculture is applicable in education.

What has occurred in the above process is the establishment of a set of behaviors, Y, as denoting both the knowledge/mass media relationship in agricultural innovation decision making and in the context of education, thereby establishing equivalence of the two situations.

4. Mediation through trait substitution.

1. In the innovation decision process (in agriculture) the mass media are uniquely related to the knowledge stage.
2. The knowledge stage/mass media relationship is a concomitant of the awareness creation and mass media relationship.
3. The mass media-awareness relationship is characteristic of the innovation decision-making process in education.
4. Therefore, the unique knowledge/function mass media relationship is present in education.

In this exercise, one aspect of the innovation decision process in agriculture was observed to be concomitant with another characteristic of innovation decision making in education and through substitution was also assumed to be characteristic of the decision process in education.

5. Reductive mediation.

1. In the innovation decision process (in agriculture) the mass media are uniquely related to the knowledge stage.
2. The knowledge stage/mass media relationship is characteristic of innovation decision making among professionals in education and medicine.
3. Awareness creation (as opposed to knowledge building) and mass

media relationship are not uniquely related (i.e., mass media are not differentially more important at the awareness stage) for medical personnel.

4. The innovation decision process in agriculture does involve a unique awareness function/mass media relationship.

5. The innovation decision process in agriculture does not apply to medicine.

6. Therefore, the innovation decision involving the unique knowledge/mass media relationship applies to education.

In this exercise, the unique knowledge/mass media relationship (in agriculture) was found present in education and medicine, but education did not share all traits with medicine; the innovation decision process in agriculture was still applicable to education. Medicine, however, relative to agriculture, has been demonstrated to be a less useful area for implication for education.

Translation Stimulation Techniques

What are some techniques used to stimulate the translation process? Figure 12.4 presents a number of techniques that are used by change agents. These techniques are discussed in the remainder of this chapter.

STRATEGIC ANALYSIS. Strategic analysis is a method often used in marketing contexts. This method involves raising specific strategic questions concerning responses to some problem or opportunity. The basic reasoning process is shown below.

Basic Strategy: Improve the company's position with its present products in its current markets (market penetration strategy).

Fundamental Marketing Question: What should be done to gain further acceptance of existing products among the target audience?

Question Asked of Data: Do the data indicate an increase in advertising or some other specific form of sales promotion?

1. Strategic analysis
2. Decision trees
3. Reasoning from models of customer behavior
4. Technological forecasting
5. Schools of thought role playing
6. Reasoning by analogy
7. Morphological analysis
8. Translating problems into theory
9. Simulation
10. Linkers

FIGURE 12.4   Methods for Generating Strategies from Data

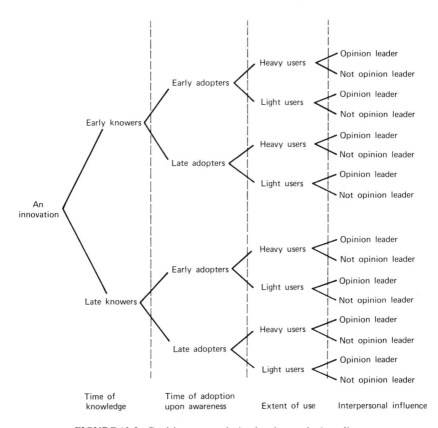

FIGURE 12.5   Decision tree analysis of an innovation's audience.

Table 12.2 presents an application of these strategies in a family planning communication context and in the context of a nutrition improvement program. Many social marketing questions could be raised with regard to any particular basic marketing strategy. The main point here is that it is desirable to use each basic strategy to stimulate questions. It is then necessary to ask whether the available data contain answers to the questions raised. If the question is important and cannot be answered, additional research of a rather specific nature becomes indicated.

DECISION TREES.   The decision tree approach involves a mapping of all realistic alternative strategies available. An example is shown in Figure 12.5. When constructing a decision tree, the research user should describe each event or decision fork in the tree so that they are mutually exclusive.

Table 12.2   Marketing Strategies and Communication-Related and Nutrition-Related Social Marketing Questions

I. *Market Penetration*
  1. Basic Strategy: Improve the company's position with its present products in its current markets.
  2. Marketing Question: What should be done to gain further acceptance of existing products?
  3. Social Marketing Question:
      1. Do the data indicate what appeals might stimulate more mothers in rural areas to practice family planning?
      2. How can we get more mothers in rural areas to use protein-enriched flour?

II. *Market Development*
  1. Basic Strategy: Find new classes of customers that can use the company's present products.
  2. Marketing Question: What new users are there for the present product line?
  3. Social Marketing Question:
      1. Do the data indicate to whom (what other market segments) present advertising should be directed without a change in copy?
      2. Should we direct our promotional efforts concerning protein-enriched flour to low-income families in urban areas in addition to rural areas?

III. *Reformulation Strategies*
  1. Basic Strategy: Improve present products to increase sales to current customers.
  2. Marketing Question: What product changes are suggested?
  3. Social Marketing Question:
      1. Do the data identify promotional themes that should be altered to attract more attention from the present target audience?
      2. Can the taste and texture of the protein-enriched flour be improved to make it more appealing to mothers in rural areas?

IV. *Market Extension Strategies*
  1. Basic Strategy: Reach new classes of customers by modifying present products.
  2. Marketing Question: What changes in goods and services are necessary to reach nonusers?
  3. Social Marketing Question:
      1. What additional mass media channels do the data indicate as necessary to reach discontinuers or resisters?
      2. Can taste and texture be changed to appeal to mothers in low-income urban areas?

V. *Replacement Strategy*
  1. Basic Strategy: Replace current products with new improved products.
  2. Marketing Question: What should replace present products?
  3. Social Marketing Question:
      1. Do the data suggest alternative substitute appeals or themes?
      2. Should vitamins be added to the product?

Table 12.2 (Continued)

VI. *Market Segmentation/Product Differentiation Strategy*
1. Basic Strategy: Attract new customers by expanding the assortments of existing product lines.
2. Marketing Question: Should products be modified to appeal to different groups of buyers and/or users?
3. Social Marketing Question:
    1. Do the data indicate that different types of appeals should be used simultaneously to attract different groups of family planning adopters?
    2. Should different types of enriched flour be used to appeal to different groups, e.g., vitamin-enriched flour to some groups and protein-enriched flour to other groups?

VII. *Product Line Extension Strategy*
1. Basic Strategy: Add more products based on the same technology.
2. Marketing Question: Should more goods and services be made available to current consumers?
3. Social Marketing Question:
    1. Do the data indicate that the advertising or communications mix should refer to ideas other than family planning?
    2. Should enriched soft drinks be distributed *in addition* to enriched flour to rural mothers?

VIII. *Concentric Diversification Strategy*
1. Basic Strategy: Attract new customers by adding products that have synergistic effects with the present line.
2. Marketing Question: Are there new products suggested by the research that will attract consumers who will also start buying existing items in the product line?
3. Social Marketing Question:
    1. Do the data suggest or identify appeals or themes associated with nonfamily planning services that would enhance the impact of family planning appeals (i.e., have a positive interaction effect when placed together in the same basic communication effort)?
    2. Can the introduction of enriched soft drinks in rural areas lead to greater use of enriched flour in these areas?

IX. *Forward/Backward Integration Strategy*
1. Basic Strategy: Improve efficiency by directing efforts to activities prior to, or following, current efforts.
2. Marketing Question: What opportunities exist for vertical expansion of the company's activities?
3. Social Marketing Question:
    1. Are there appeals in messages to ultimate users of contraceptives that might also be effective in motivating field workers, informal community leaders, and so forth?
    2. Should the manufacturer of enriched flour expand activities to include the distribution of the product?

This means combinations of alternatives must be presented as a separate alternative. Also, the research user must enumerate all possible choices or events at each fork in the tree. The sequence of events or decisions does not have to be in chronological order. It is the time at which the decision maker receives information about events that counts and not the time at which they occur. For example, if research relating to push versus pull promotion strategies is received first, this decision set may constitute the starting point for a decision tree for product line extension.

It should be clear now that the main advantage of a decision tree is to increase the number of questions asked of research data and, where answers are not readily forthcoming, to indicate areas in which additional research is necessary to make informed decisions. The large number of questions a decision tree stimulates ensures better coverage of the problem being investigated and helps extract the maximum information a given social research effort can provide. Thus with reference to Figure 12.5, the change agent with limited resources must decide whether to direct his efforts at early knowers of an innovation or late knowers. The change agent must then decide whether to focus on early adopters or late adopters or both.

USE OF BEHAVIORAL MODELS. One very useful and increasingly popular technique for generating marketing strategies from existing data is the use of one or more of the various behavior models that have appeared in the past few years. The basic procedure involves three steps:

STEP   1. Identify variables or concepts in the model that are pertinent to the social problem at hand and determine the relationships between or among these variables.

STEP   2. Identify variables in the study that correspond to the concepts in the model and determine the nature of the empirical relationship between the variables.

STEP   3. Derive the change strategy implied by the empirical relationships.

Thus one asks what relationships are there in the model that may correspond to and therefore suggest relationships not already considered in the available data? Then comes the question, "So what?" What do the relationships in the data mean in terms of a change strategy?

LINKERS. It is often desirable for the practitioner utilizing marketing research to seek the assistance of a "linker." A linker is an intermediary between the researcher and/or research results and the practitioner. For many reasons, only one of which is mentioned here, there are gaps in effective communication between the research-producing data and the practitioner who must utilize the data. This gap is generally bridged, if at all, by additional persons who may be specialists in linking researchers and their data to practitioners (Havelock et al., 1971).

A linker is necessary when mutual understanding between researcher and user does not exist. There are four possible relationships between researchers and users. Following Churchman and Schainblatt (1965, pp. B69–87), we can depict these relationships in Table 12.3.

Doktor and Hamilton's study (1973) suggests that a mutual understanding occurs when both the researcher and marketing manager have identical cognitive styles, that is, when they are both analytically (A) oriented or both are heuristically (H) oriented. A heuristic orientation is one in which a person tends to reason by broad rules of thumb. Mutual understanding can also occur when the manager and researcher have different basic styles but also have an orientation or openness to the other's basic style. When the manager has an (H) cognitive style and the researcher an (A) style but with a small (H) approach, a persuader relationship is likely in which the researcher tries to persuade the manager. When the manager is (H) oriented but with a slight (A) orientation and the researcher is (A) oriented, a communicator relationship is likely to occur. Finally, when the researcher is characterized as a pure (A) style of thinking and the manager characterized by a pure (H) style, a separate functionalist situation is likely.

Several different linking roles have been identified that can be helpful in shifting managers and researchers to a state of mutual understanding. These are viewed here briefly. Change agents or program directors should seek and choose the type of linker whose function most closely satisfies their research utilization needs.

The conveyor. The most elementary linker role is the conveyor who simply passes information from expert to nonexpert users. This passing or transfer function is often performed by salesmen, for example, from various research organizations who sell data such as credit ratings. *The consultant.* The consultant's function is to help define problems, identify existing related information and assist in adapting existing information to the practitioner's needs. Management consulting organizations generally satisfy these needs. Academicians also are used frequently as consultants by social action agencies. *The trainer.* The function of the trainer is to transfer to a

Table 12.3   Researcher Understands Manager's Needs

Manager Understands Researcher and Data	Yes	Mutual understanding	Communication
	No	Persuasion	Separate function

user an understanding of an area of knowledge or practice. *The leader.* The leader role functions by directing utilization efforts or by setting examples of how to utilize data. The leader exerts influence within his own group. Evidence indicates that managers do set for their subordinates examples of how to utilize research effectively. *The innovator.* The innovator stimulates utilization by establishing precedent and example in the use of social research. He is much like the leader when the leader demonstrates how to use research.

Other Data Utilization Techniques

TECHNOLOGICAL FORECASTING. Technological forecasting is a series of human/computer techniques to forecast changes in technology that are likely in certain time frames. Usually a series of experts develop scenarios about changes that they feel are necessary, feasible, and desirable. Typically, an "envelope" or range of technical specifications is produced and the probability of achieving the limits of the envelope is created by the experts. A computer then combines the envelope specifications with the probability to forecast an aggregate-time to occurrence for technology. An example of this would be the forecast of the occurrence of the development of ground transportation that does not use wheels, but rather uses magnetic fields to support the vehicle. The experts would assess what new electromagnetic developments would be needed, together with the time of development with a probability. The results of a number of experts would be combined to get an aggregate forecast of time to development.

SCHOOL OF THOUGHT ROLE PLAYING. This technique consists of the user asking how people with different perspectives, including different theoretical orientations, would interpret the data. How, for example, would an exchange theory advocate interpret the data about change agent-client interaction? How would a motivational researcher interpret the same data?

REASONING BY ANALOGY. This approach essentially involves finding parallels between a process or phenomenon in different contexts. The most familiar form of this is the use of behavioral laboratories involving humans and/or rats to gain insight into real world activity.

MORPHOLOGICAL ANALYSIS. This technique involves identifying the important dimensions of a problem and examining their interrelationships. If there are three dimensions to a problem, a three-dimensional table would be constructed to facilitate analyzing the various interactions.

TRANSLATING PROBLEMS INTO THEORY.   This involves translating a problem into a theoretical question. A problem involving the introduction of a new product could be cast into a diffusion theory or perhaps social change theory question. The problem then becomes imbedded in a theoretical system in which there may be a wealth of concepts and findings pertinent to the basic problem. These concepts and findings may have been overlooked had the theoretical question not been posed.

SIMULATION.   Simulation involves establishing an artificial reality, that is, a reasonable representation of the real world. Simulation may be in the form of kits (see Rogers and Havelock's simulation of change agents) involving very·simple computations, or in the form of highly complex computer programs. Various strategies are undertaken in the simulation and the probable result of the strategy is fed back to the user.

SUMMARY

Throughout this book we have discussed the state of the art regarding the change process. This chapter focused on two critically important areas in this knowledge utilization process. The discussion of research utilization identified the effective guidelines for the change agent to consider in using information to achieve change. The discussion of the translation process focused on the problem that change agents face in translating theory and/or data about a given change problem into some immediate action to deal with the change problem.

APPENDIX 1
Principles Related to Knowledge Acquisition and Use*

1. *Principles That Should Guide Need Sensing, Activating and Communicating Activities*
   a. Multiple inputs: no one user group, no one set of experts, and no one data collection strategy can provide a complete and satisfactory picture of user needs.
   b. Continuing input: needs change over time, and the pattern of most important and persistent needs only emerges from sensing and assessment activities taken at regular intervals.
   c. Present and future: the system should be responsive to existing needs but should also be preparing users for situations they may be moving

* Source: Havelock, 1968

into. Moreover, the R and D cycle in some areas is so long that the system must be able to predict significant needs several years in advance if it is to be in a position to provide relevant assistance when those times arrive.

d. Representative of the total system: an assessment must bring to the foreground those needs which are truly most important for the most people in the system, or which relate to functions (such as effective leadership or management) that are central to the health of the system as a whole.

e. Qualitative and quantitative: needs assessments should not consist merely of statistics or of random anecdotes. Wherever possible, impressions should be supported by appropriate enumerations that render them credible, and enumerations should be supported by descriptive analyses that render them meaningful.

f. Annual report and review: there should be a regular public report on the current status of user needs that summarizes all NSA and C activities of the preceding year and revises forecasts and priorities based on those inputs.

g. Promote policy dialogue: needs assessment and analysis procedures should be stimulative and provocative of discussions at the highest policy levels within SRS.

h. Promote dialogue between SRS and the R and D community: These procedures should also present a meaningful challenge to researchers and developers in relevant disciplines, and should guide them to propose germane and creative solutions.

2. *Principles That Should Guide Knowledge Production Activities*

a. Knowledge that enters the system should as much as possible meet high standards of quality (validity and reliability), relevance (pertinence to the real needs and problems of identified SRS users), and utility (pointing the way to practical action steps to relieve user needs and solve user problems).

b. Some amount of knowledge of high quality, relevance, and utility should be available in each of the need or problem areas designated as highest priority for each of the major SRS audience groups (as listed in Chapter 1). In other words, R and D production should aim at a certain degree of comprehensiveness in its coverage. There should be no large gaps in priority areas.

c. R and D producers should be well linked to priority users both directly and via the R and D sponsoring divisions of SRS.

d. Linkages as specified in (c) should ensure two-way dialogue and stimulation in which users can communicate their needs and R and D producers can communicate accurate and positive images of solution possibilities derived from their output.

e. There should also be continuing dialogue between the RU unit and the R and D sponsoring units within SRS.

f. Special research knowledge should be generated regarding the process of research utilization, itself, to assist in the future guidance of the RU system as a whole. Such knowledge should meet the same standards of quality, relevance, and utility as R and D in other areas.

3. *Principles That Should Guide Knowledge Storage and Scanning Activities*

a. Comprehensiveness: all materials that have potential relevance or utility to any of the potential users of SRS should either be included or at least referenced by the archive. This includes government and nongovernment, copyrighted and noncopyrighted, U.S. and non-U.S., English language and non-English language materials.

b. Central access: would-be users should not have to search many different locations to find the information they need. It should all be available or at least referenced at one source. Even though collections of hard copy may be scattered, they should be centrally catalogued and available (such as in the British University Lending Library system).

c. Rapid input: within weeks of the appearance of a new publication or product, it should be entered in the system, and new entries should be reported at frequent intervals.

d. Rapid access: would-be users should be able to conduct bibliographic searches and review personalized bibliographic listings with a maximum turn-around time of a week, and eventually (assuming remote terminals) on the spot. Microfiche or hard-copy of documents should also be accessible within a week's time. With current NTIS capability, these are not unrealistic objectives, although NTIS is limited to federal R and D reports only.

e. Problem oriented: thesaurus and search terms must be similar or identical to the terms and phrases typically employed by users in their everyday work situations. In some cases, therefore, policy makers, administrators, practitioners, and clients will all have languages different from each other and most certainly different from the R and D community that made the original input. Thus, in the long run, we are faced with a cataloging-indexing job of major proportions.

f. Accurate contemporary labeling: key descriptive terms employed by the researchers, themselves, or by cataloguers for old entries, may be incomplete or inappropriate for the SRS users of today. In order to allow for rapid access to all material, especially old material, relevant to a given topic, another recoding operation will have to be undertaken at some point, possibly aided and guided by the mini-encyclopedia project discussed earlier.

4. *Principles That Should Guide Knowledge Processing Activities*

   a. Maximize the relevance of the material to the intended user. There is a tendency in many quarters to cling to the old nostrum that if you build a better mousetrap the world will beat a path to your door. This may work fine for mousetraps, but in the more complicated world of R and D in which SRS is involved, it may not even be clear to the uninitiated that the item is a mousetrap. Careful thought must be invested by someone in spelling out the multitude of potential connections between a given piece of research and applications by various users.*

   b. A corollary of the above is to minimize the amount of irrelevant material that accompanies the essential message. Thus, if the intended audience is policy makers, technical details that help the researcher or practical details that might help the practitioner should probably be eliminated or kept to a minimum.

   c. Maximize the utility of the knowledge to the particular intended user. Particularly for practitioners-users and clients, the fact of relevance is not enough. Implications also may have to be spelled out in terms of "do's," "don'ts," and "how to's."

   d. Minimize the effort required of the intended user to dig into the material. This principle will be most appreciated by newspaper editors and headline writers. A good headline is a summary of a story: it gives an accurate set to the reader and is enough of a stimulus to tell him whether he should read the rest of the article or go on to something else.

   e. Minimize the potential dangers of misapplication or premature application. When new knowledge is deemed to have important action implications or might tempt such applications, serious thought should be given to the potential consequences, positive and negative. A good RU system does not indiscriminately strive for maximum use at maximum speed for all knowledge but only that which, on balance, improves the condition of people and of the society as a whole.

5. *Principles That Should Guide Dissemination Activities*

   a. Always consider who you are trying to reach. Any dissemination activity should have an audience in mind and should therefore be writ-

---

* This type of analysis has been carried furthest by the NASA Technology Utilization Program in their "Technical Briefs" aimed at a multitude of private sector spin-offs from space-related, government-supported development work.

ten for that audience and transmitted on channels to which the audience is habitually tuned.
   b. Make dissemination purposeful. Not only have a specific audience in mind, but also have a good reason for reaching them.
   c. Wherever possible, link dissemination to other activities, especially activities that will lead the way to utilization by those most relevant.
   d. Use dissemination activities to increase the awareness of all members of the SRS system of the importance of all RU system functions, and of the availability of whatever services are currently available.
6. *Principles That Should Guide Utilization Activities*
   a. Optimally, the user should be a wise judge of R and D knowledge, not merely a passive adopter. It is important that users be sophisticated about R and D knowledge, to know what is relevant, useful, and of high quality to them: it should not be the purpose of an RU system to maximize utilization of R and D, but to optimize it. This is an important distinction.
   b. We should not over-help the user. Providing direct utilization services to individual users on a one-to-one basis can be tremendously costly even though it is sometimes necessary. Therefore, we should not attempt to go further in helping a user than we know is necessary. We should be leading horses to water, but that is all. Another way to say this is that there must be limits on the extent to which an RU system can service its users. Some users will need more help than others, and new users will need special help, but the system should be designed to move people toward independent self-initiated knowledge retrieval as an ultimate goal.
   c. In line with the above principle is the oft-cited and still valid maxim that we should "help people to help themselves." In the context of RU this may mean (1) helping them to understand and diagnose their needs and problems, (2) creating or increasing awareness of the resource universe, or (3) encouraging users to risk trying something new if it promises improvement in their situation.
   d. The utilization component should, in various ways, strive to improve the process of problem solving among individual users at all levels within user organizations, and in the system as a whole.

APPENDIX 2

A Guide for Selecting Variables Affecting the Change Situation

An important aspect of translation is selecting variables and hypotheses to be used when deriving specific tactics for the purpose of stimulating the adoption and diffusion of change and innovation. The object in this section is to outline six steps in this process while pointing out to the reader that this presentation suffers the usual drawbacks associated with work in an area having few guidelines to follow or models on which to build. The six steps are:

Step One   *Identification of plausible variables affecting innovation and change adoption and diffusion for the situation under study.*

Step Two   *Systematic evaluation of each variable according to a comprehensive variable evaluation scheme.*

Step Three   *Establishing relationship among variables.*

Step Four   *Developing a theoretical paradigm.*

Step Five   *Systematic evaluation of each hypothesis according to a comprehensive evaluation scheme.*

Step Six   *Derivation of tactic statements from hypotheses.*

It is assumed that there is an explicit statement of the situation under study before the six steps are initiated. For example, the situation might be the general phenomenon of the acceptance or rejection of changes and innovations originating with noncommercial R and D organizations. Moreover, only those innovations may be considered which are intended for use at the elementary school level where the the unit of analysis may be allowed to vary from the individual teacher to school district. The innovations may be further divided into special categories:

STEP 1.   Identification of plausible variables affecting market behavior for the situation under study.

These variables are to be preliminarily screened at this stage on the basis of the investigators' experience and a literature review.

STEP 2.   Systematic evaluation of each variable according to a comprehensive variable evaluation scheme.

An evaluation scheme is presented in Table 12.4. Table 12.5 defines and illustrates the evaluative criteria. This scheme requires constructing a variable value index. The higher the score, the more important the variable is for inclusion in the final theory. Two caveats are necessary. First, variables that would be listed in Table 12.4 would have received a preliminary screening as per Step 1. Second, a numerical threshold would be established, and all variables whose numerical index falls below that threshold would be eliminated from further analysis. The caveat implicit in this is that some exceptions should be expected. Scoring is, at best, only a rough indicator of probable value; expert judgment should not be suspended when examining total scores. The main reason for not adhering unwaveringly to the cutoff point established by the threshold is that the variable evaluation criteria

Table 12.4   Variable Evaluation[a]

| | Evaluation Criteria | | | | | | |
Variable Name	Observ-ability	Content Validity	Predic-tivity	Dis-crimi-nant Power	Systemic Value	Substan-tiality	Total Score
Degree of central-ization in school district	3	2	3	1	1	3	13
Risk-taking propensity	1	1	3	2	1	2	10
Etc.							

[a] 3 = good, 2 = fair, and 1 = poor.

presented in Table 12.4 are unweighted. In fact, they may be of unequal importance depending on the particular circumstances of the investigation and, ideally, the criteria should be weighted accordingly.

The example in Table 12.4 assumes for convenience that all criteria have equal weights. The variable labeled "degree of centralization," for instance, is readily observed or measured; it has moderate content validity, that is, the total meaning of centrality or all its facets are reasonably but not fully represented by the operational measures; it is a good predictor of innovations or other variables directly related to innovativeness; it has poor discriminant power, that is, it is somewhat close to the variables of complexity and formalization; it is not good at linking what appear to be otherwise unrelated variables; and it does not distinguish well between various market segments. It will be noted that there is a certain relatedness among the evaluative criteria. A variable rated highly on observability, for example, is unlikely to receive a low score on content validity.

STEP 3.   Establishing relationship among variables.

This step involves the construction of a matrix stating relationships among the variables remaining after Step 2. The matrix below (Table 12.6) will serve as an example. Variables a, b, c, and d will represent, respectively, degree of centralization of decision making in the school district, degree of exposure to innovation related information, the extent to which teachers perceive innovative behavior as being rewarded, and the innovativeness of the school district. The relationships to be discussed here are logical possibilities; they are presented for illustrative purposes, and may have no substantive base. By reading the matrix, certain causal hypotheses become

Table 12.5

Criteria	Definition	Examples in the Context of Innovation
Observability	The degree to which a variable is reducible to observations.	To what extent are the traits cosmopoliteness, empathy, and dogmatism observable, that is, how readily can we "see" these concepts with the best instruments currently available? Are valid concepts only those which are observable?
Content Validity	The degree to which an operationalization represents the variable about which generalizations are to be made.	To what degree does a self-rating scale focusing on perceived expertise in an area actually represent the abstract concept of opinion leadership? Does time of purchase relative to other members of a social system represent the degree of innovativeness rather than venturesomeness displayed by an individual?
Predictivity	A subtype of criterion-related validity in which the criterion measure is separated in time from the predictor concept.	To what extent does time of awareness of an innovation influence time of first purchase? To what extent do initial attitudes predict time of first purchase? How adequately do the scores obtained in measuring propensity for innovative behavior predict actual purchase of innovations?
Discriminant Power	The extent to which a variable differs from other variables.	How different is innovativeness from venturesomeness? Is opinion stating the same as advice giving, that is, does the use of the term opinion leadership adequately separate itself from leadership in giving solicited advice?
Systemic Value	The degree to which a variable enables the integration of previously unconnected variables and/or the generation of a new conceptual system.	How many plausible new propositions can be established and existing propositions connected by introducing the concept of reciprocity in interpersonal selling situations involving innovations?
Substantiality	The extent to which the variable contributes to the identification of a discernible substantial market segment	Does risk-taking propensity clearly distinguish between adopters and nonadopters of an innovation? To what

Table 12.6   Relationships among Variables.

Casual direction ──────────────────→

Key: lower case letters are variable codes
+ = positive linear or curvilinear relationship;
− = negative linear or curvilinear relationship,
  = negative relationship evolving into positive relationship,
  = positive relationship evolving into negative relationship

evident. For example, the greater the degree of centralization (a), the lower the degree of exposure to innovation related information (b), the more teachers perceive innovative behavior as being rewarded (c), the more innovative the school district will be (d). There is reciprocal causation as well, as indicated by codes in cells following below the diagonal; the more innovative a school district is (d), the more teachers will perceive innovative behavior to be rewarding (c).

STEP 4.   Developing a theoretical paradigm.

A theoretical paradigm can be derived from the matrix in Step 3. Variable d, school district innovativeness, is the dependent variable, although it is also suggested as having a direct causal effect on variable c, teacher perceptions about innovative behavior. Standard statistical techniques, for example, path analysis, are available for studying the assumptions about causality in this theoretical paradigm or model (Figure 12.6). Obviously, the addition of even one more variable could produce a much more complex model. It is suggested that, to the extent possible without seriously violating the explanatory power of the model, only variables that have a maximum of a two-step connection with the dependent variable such as a $\xrightarrow{1}$ b $\xrightarrow{2}$ d (ignoring in this case the direct a → d connection) be included in the model.

STEP 5.   Systematic evaluation of each hypothesis is according to a comprehensive hypothesis evaluation scheme.

Since the number of variables and hence the number of hypotheses will be large, it will be necessary to screen out the least promising hypotheses. This

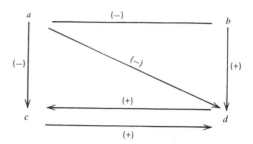

FIGURE 12.6   Theoretical paradigm.

involves another set of criteria, this time for evaluating hypotheses. These criteria should also be reemployed for evaluating the state of the final theory after testing. Each connecting line in Figure 12.6 implies an hypothesis stating a relationship concerning at least two variables. Thus there are a minimum of six hypotheses. This can also be derived from the matrix shown in Table 12.6. A list of evaluative criteria are presented in Table 12.5. Table 12.7 provides a basis for an overall evaluation of each hypothesis in terms of the criteria in Table 12.5.

Not all criteria will be applicable to all hypotheses. The score for a particular hypothesis should be the average rating (with 3 = good, 2 = fair, and 1 = poor) using as "n" the number of applicable cells. As before, it may be desirable to weight each applicable criterion differently in terms of importance, and the numerical threshold established as a cutoff point for discarding propositions should be viewed as a guide to be tempered with experience.

STEP 6.   Derivation of tactic statements from hypotheses.

Each hypothesis should imply a tactic or strategy. The focus of the tactic may be to manipulate a phenomenon peculiar to the target group, such as the level and kind of information possessed. On the other hand, the tactic may focus on a change agent variable, such as the frequency with which contacts with the target group are attempted. Broadly interpreted, change agent variables would include such factors as selecting promotional appeals to be consistent with known personality traits of the target group or market segment or adapting the nature of the innovation to be compatible with their risk-taking propensities.

Operationalizing hypotheses in the form of actionable statements is a specific instance of the so-called action level epistemic gap. Just as there is no perfect correspondence between theoretical statements and their operationalization (the "operational research epistemic gap"), there is no perfect one-to-one correspondence between the operational hypothesis and the derived tactic. Only intuitive intellectual judgments may be made, although guides for improving those judgments can be identified. A given hypothesis may imply more than one tactic for a given change agent or for different change agents. It is altogether possible that conflicting tactics may be derived. The criterion to be employed in making a selection in such cases concerns the compatibility of the alternative tactics with those other tactics which have been derived from other hypotheses in the same theory. The most compatible tactic is the one to be selected.

Perhaps the simplest procedure for analyzing and selecting tactics is to enumerate them on the basis of intuitive yet informed judgment and then apply a set of tactic selection criteria. Consider the hypothesis derived from Table 12.7 in Step 3: the greater the degree of centralization (of authority) in a school system (say the school district), the lower the degree of exposure among decision makers to innovation-related information. This hypothesis has received reasonably good support in both educational and other contexts. One relatively obvious actionable implication of this would seem to be the following: in a highly centralized district do not maximize the number of points in the system (e.g., teacher, principal, department chairman, curriculum consultant, etc.) into which information could be fed, but concentrate initial communication efforts at the district supervisor level. This may involve a second tactic: the use of change agents rather than mass media as vehicles for disseminating information.

Thus the one hypothesis suggests two tactics, each of which introduces a new consideration. The first tactic introduces the notion of points of entry into a system, whereas the second tactic involves the interpersonal versus mass media notion. This suggests a more expedient means of formulating tactics. A number of considerations, such as number of points of entry, interpersonal versus mass media, etc., could be enumerated as tactic

Table 12.7   Hypothesis Evaluation

	Evaluative Criteria					
Hypothesis	*Internal Consistency*	*Strength*	*Falsifiability*	*Scope*	*Etc.*	*Total Score*
A						
B						
C						
D						
E						
F						

Table 12.8   Tactic Stimulator Grid

	Hypothesis			
Tactic Stimulator	*1*	*2*	*3*	*Etc.*
Decision stage				
Timetable				
Points of entry				
Mode of communication		a		
Resource availability		b		
Nature of innovation				
Incentive system				
Etc.				

formulation stimulators. Table 12.8 is an example. The columns 1, 2, 3, etc. refer to various hypotheses, and the small letters within the cell refer to tactics derived from the various hypotheses on the basis of the tactic factors. Assume, for example, that hypothesis 2 represents the relationship between degree of centralization in school districts with exposure to innovation information. The small letters a and b are the two tactics just mentioned. The basic procedure for using Table 12.8 is to ask, for each cell, "What action, if any, does this hypothesis suggest in terms of this tactic factor?" Does, for example, our hypothesis suggest any special action depending on the particular decision-making stage the central authority is in? In the example used here, this would not be relevant, since the hypothesis is applicable at the awareness stage only. An exhibit would then be developed listing the various hypotheses 1, 2, 3, etc., and indented under each hypothesis would be the various tactics, a, b, c, etc.

It is likely that, as the number of hypotheses examined increases, certain tactic stimulators would appear as relevant more often than other stimulators. These would constitute the core tactic stimulators for the context under study, for example, elementary school districts.

It is important to note that there is a reiterative process in Step 6. Tactics generated on the basis of intuitive yet informed intellectual judgment were said to contain basic notions such as number of points of entry into a particular school social system. The points of entry idea then serves as a stimulator for other hypotheses, which in turn may generate additional stimulators and so forth.

CHAPTER

THIRTEEN

Ethics in Social Change

The issue of ethics and values in professional life is significant philosophically. Few professional associations that have put forth ethical guidelines for professional practice have escaped that action without pro-

A preliminary version of this paper was presented by the authors at the conference, "The Implementation of OR/MS Models: Theory, Research and Application," Graduate School of Business, University of Pittsburgh, November 1973. The papers from this conference were published in an edited volume by Schultz and Slevin (1975).

voking considerable response from substantial numbers of their membership. The American Psychological Association (American Psychological Association, 1973; Cook et al., 1971; Cook et al., 1972) and the Operations Research Society of America (1971) are perhaps the two associations that have precipitated the strongest and most sustained reaction, both positive and negative, among their membership with the issuance of professional practice guidelines. For example, some psychologists challenge the view that the subject should be informed of the nature and purpose of the experiment in which he is participating. Resnick and Schwartz (1973) found that subjects who were informed of the purpose of an experiment behaved much differently from those subjects who were unaware of the true purpose of the experiment. Mitroff (1974) has challenged the operations researcher's objectivity and emphasizes that scientists are often biased and partisan to their own theories.

Thus although ethical standards exist for practitioners, it becomes important to ascertain to what extent the practitioner abides by these standards. Menges (1973) examined all articles published in six psychological journals to determine what effect new standards regarding informing subjects of the purpose of experiments had on actual practice. Menges concluded

> The author finds no evidence to indicate that . . . the use of deception will decline, even though they seem inherently incompatible with APA's (1973) call for "openness and honesty" as essential characteristics of the relationship between the investigator and research participant (Menges 1973 p. 1034).

In this chapter a number of ethical issues are discussed concerning the implementation of applied social change. Special emphasis is placed on issues involving the relationship between the interventionist and the client.

This chapter has two main sections. The first section identifies some of the ethical and value dilemmas a change agent might encounter. The second section presents the results of a survey of over 1500 professional change agents. These change agents were asked to indicate their agreement with the issues identified in the first section, how important they believed the issues to be, and how often they encountered the issues as change agents. These issues were cast in normative terms in a short questionnaire.

The reader should be aware that we are challenging the issue of scholarly neutrality in this chapter. All change agents are advocates of some position or technique. Therefore, it is the contention of the authors that the notion of a value-free applied social change is a myth. The OR-MS change agent is certainly not value free, nor is the cultural anthropologist or psychotherapist. Each person strongly believes in the overall rationality of his or her technology and its potential for improving some aspects of the target system.

In studying the scientist's role, Mitroff (1974) indicates that the objectivity stance of the scientists is unrealistic and ". . . if the scientist QUA SCIENTISTS is not a staunch partisan ADVOCATE for his theories, hypotheses, and positions they may be too readily ignored or not taken seriously by the surrounding scientific community" (p. B-614). In a study of the psychology of over 40 Apollo moon scientists, Mitroff (1973) concludes

> that scientists did everything in their power (excluding cheating) to muster every bit of evidence favorable to the theory that they could find. They were not out to falsify their own theories but to confirm them. If they were out to falsify anything, it was the theories of their opponents (1973, p. 15).

The critical question then becomes not one of value versus value-free orientation of the change agent, but rather, "what are the values of the interventionist?" The values of the change agent are critical because they have an important impact on how a problem is defined and diagnosed, and the solutions and strategies that are selected, applied, and evaluated (Guskin and Chesler, 1973). The key issue may thus move from the value versus value-free issue to the one of whose side is the change agent on, and does he know it? For example, the OR-MS change agent may identify or take sides consciously or unconsciously with the advocates of his position in the organization. This partisanship could lead to certain second-order or unanticipated or unintended consequences such as manipulation or coercion of those individuals not sympathetic to the advocated change. Argyris' (1971) study of an OR-MS team in a large organization pointed out how these specialists became aggressive and competitive with clients as they tried to get their techniques adopted (p. 287).

The more appropriate strategy might be for change agents to be more aware of their values and deal with them openly with the client system. In the following discussion we present various value-ethical issues affecting the interaction of a change agent with his client and target system. These issues evolve around the general functions of the change agents role.

VALUE-ETHICAL ISSUES

Compatibility of Change Agent and Target System Goals

The goals of the change agent and the target system may or may not coincide. This depends in part on the level of goals under discussion. For example, a management consultant may be concerned with maximizing his financial gain derived from serving as a consultant. This goal or concern may not be compatible with the target system's desires to limit the outflow of financial resources in all areas of activity. At the same time, both a consultant and target system place high value on improving the financial performance of the organization in one or more areas. For the consultant, this increases the likelihood of being rehired in the future.

The issue involved here is quite complex, and several questions can be raised. Must the client or target system disclose to the consultant the nature of the goals that the consultant's advice is intended to further? Must a consultant be told that his services will be used to secure a contract or the passage of legislation relevant to a military system to which the consultant is opposed? Should a management specialist be informed that a family planning clinic or health delivery system he is helping to design is going to emphasize a birth control technique to which he is opposed by virtue of his religion or other value consideration? Similarly, should the client and target system be fully informed of the values of the consultant? This is particularly important if there is any danger that the means for furthering goals can alter those goals and that the consultant or change agent, with values at variance with those of the target system, may unwittingly make recommendations that would tend to subvert those goals.

Imposition of the Change Agent's Values

One of the first facts a change agent must confront is that he is not engaging in a value-free activity. We have challenged this value-free position as being a myth. The change agent does in fact have values, as pointed out earlier in this discussion.

The issue becomes what values and therefore whose values are to be served by a change. There is a natural tendency for the change agent to promulgate his values, and in such circumstances we must ask whether these values are representative of those possessed by the target system. Often the answer is "yes" because target system or at least the client representing the larger system tends to seek out a change agent with like value schemes. However, sometimes the change agent may discover only belatedly a discrepancy in values between himself and the client or target group. In a specific situation, the client system representative who selects an interventionist may not always share the values of the larger system or of key colleagues and may call in a change agent known by the client to be sympathetic to his view. The change agent is used selectively and possibly unknowingly by the client, while the change agent may be assuming that the client is representative of the larger social system.

For example, a manager who is an advocate of the implementation of a computer-based management information system might select a change agent favorably disposed to the development of an MIS to do a systems analysis to determine the organization's need for this technology. The manager might use this change agent as an advocate of this new technology to sell it to those resisting it in the organization. It would be important for the change agent to be aware of the fact that the manager who contracted him was more favorably disposed to an MIS than others in the system. Being aware of the fact that there are some who resist the innovation would provide the change agent with the knowledge that his first task would be to reduce resistance or potential for resistance and open up the system to considering this change.

It is also both appropriate and necessary for the client to seek clarification from the change agent about the latter's values as they may affect the client or the target system. Seeking such clarification forces the change agent subscribing to the value-free myth to identify his values in an explicit way. Psychotherapists, for example, must undergo therapy as a part of their training. The rationale for this is that is sensitizes the trainee to his own biases which may act as moderators in his dealings with patients. For example, a therapist who has had rather traumatic experiences with his father might tend to perceive psychological problems as an indicator that the patient has experienced a similar traumatic experience with a parent.

Whether psychotherapy is the mechanism or not, it might be constructive for change agents in general to become more aware of their values and motives. Is the change agent really concerned about the welfare of the target system, or does change activity satisfy his or her needs for power and control? If the latter motives are operating, this might cause the change agent to be more manipulative in dealing with the target system (Argyris, 1970). If the change agent is too eager to impress the target system, he or she may not allow the target system to develop its own capabilities for dealing with the problem situation. As a result, the client system becomes overly dependent on the change agent. For example, on OR/MS change agent might be so overly concerned about proving the usefulness of a technique that the target system is not allowed to develop skills through its own trial and error learning so that it can effectively implement this technique when the change agent is no longer on site.

We have been speaking of change agents as if they were single individuals. Clearly, organizations as social entities also maintain value sets, both formally and informally. This can be seen most vividly in family planning organizations. Sometimes formal positions are subverted by informal positions. For example, a vasectomy clinic in Chicago, Illinois has a formal policy not to perform a vasectomy on any unmarried male under the age of

26. At the same time, they refer all such persons to a physician who will perform the vasectomy. Informally, they are still playing an instrumental role in obtaining a vasectomy for that individual.

Table 13.1 provides a summary of key questions concerning the ethical role of government and nongovernment organizations on population. Many of these key questions are basically generic in nature, applying to nearly all organizations. In discussing these questions with others, the authors of this book have found very different but very strongly held responses to the implicit and explicit issues.

In general, questions should also be raised about the values of the change agent that are most relevant to the activities he is to perform. These activities may include problem definition, program planning, implementation, evaluation, and control. It is likely that different values are differentially relevant to different activities. For example, values concerning participatory decision making may be most relevant for program planning problems, whereas values concerning the rights of human subjects may be most relevant when the interventionist is conducting research as part of the problem diagnosis process or program evaluation.

Selection of Target Systems

Is a professional change agent obligated to provide assistance to anyone who requests it? Should a professional change agent refuse his services to clients or target systems whose values are not compatible with his own? In the legal and medical sectors of society, an answer to this is fairly easy and clear: mechanisms do exist for providing these services to persons or groups who can demonstrate a need for legal and medical services and the inability to satisfy their need in conventional ways. It is also clear that vendors of restaurant services, real estate, and other commodities and services cannot refuse clients on the basis of certain values such as those associated with the desire or preference not to have blacks as neighbors or customers. Recently, an interesting and as yet unsettled legal case in Great Britain was presented to the authors in which a manufacturer of industrial goods brought suit against a management consulting group that refused to provide consulting services because the manufacturer refused to hire immigrant Pakistanis for any position but menial, unskilled tasks. Although many details of the case are unknown, it is known that one argument put forth by the complainant is that the consulting agency withheld needed services that it alone was best able to provide and that this was done on the basis of a company (complainant) policy of no relevance to the consulting agency. The consulting agency contended that the hiring policy at issue was symptomatic of the inability of the manufacturer to act reasonably and to make effective use of

Table 13.1   Ethical Issues Faced by Organizations Active in Population Programs*

Do organizations have the obligation to provide information on the consequences (medical, social, etc.) of the programs and methods they support and promote?

Is it right for organizations to assist dictatorial governments in the development of population control programs?

Is it right for organizations to advocate that methods or programs be adopted abroad that are not acceptable at home?

Is it right for organizations to use funds contributed by donors with ulterior motives?

Do organizations working in the population field have any obligation for relateed social and political reforms?

Do population organizations have the right to advocate and promote population control if they know that such control could, in some societies, lead to vast long-run cultural changes?

GOVERNMENT-RELATED ISSUES

Would it be ethically right for a nation experiencing excessive population growth to do nothing at all?

Is it right for a government, in response to excessive population growth, to institute, develop, and implement a voluntary family planning program?

Is it right to include sterilization and abortion in voluntary government family planning programs?

Would it be right for governments to go "beyond family planning" if the gravity of excessive population growth could be shown?

Would it be right for governments to establish involuntary fertility controls?

Would it be right for governments to develop "positive" incentive programs, designed to provide people with money or goods in return for a regulation of their fertility behavior?

Would it be right for governments to institute "negative" incentive programs?

Would it be right for governments to introduce antinatalist shifts in social and economic institutions?

Would it be right for the government of a developed nation to make the establishment of a population control program in a developing nation a condition of extending food aid?

Would it be right for a government to institute a population control program that goes "beyond family planning" for the sake of racist or self-interested political ends?

Would it be right for a government to institute programs that go "beyond family planning"—particularly in a coercive direction—for the sake of future more than of present generations?

Do governments have the right unilaterally to introduce programs that go "beyond family planning"?

* Source: Finnigan, 1972.

329

any advice it might receive. One case in Great Britain is not necessarily an indicator of a trend. But perhaps it is not too soon to begin considering criteria or guidelines for use by a change agent in deciding whether to refuse its publicly offered services to a potential customer. Argyris (1970), for example, suggests that a change agent should work only with client systems that permit free choice, the generation of valid information, and an internal commitment to solve the problem at hand. Argyris' assumption here is that a system cannot effectively create enduring change unless these three criteria exist.

Nature of Change Agent Contract

The contract established between the client system and change agent can be complex, and there may be several key issues that may not be addressed in the contract. Who does the change agent serve? Is it the manager who brought the change agent into the particular organizational unit of the client system organization or is it the organization as a whole? Does the unit to be most directly affected by the change agent's efforts act as a group in seeking his help, or are there certain key members of that group who seek his help? How are conflicts in values and approaches to the problem to be solved? Is the change agent free to opt out when he feels that an unsolvable conflict exists? How responsible is the change agent for dysfunctional effects of his efforts? Should he make his services available to deal with unforeseen problems of a solution after its implementation or does his responsibility simply end after a solution is implemented?

Type of Change

The choice of means to implement change represents another very important area in which value and ethical dilemmas are likely to arise for change agents. Closely related to the kind of manipulation selected is the issue of the degree of manipulation.

There is a wide array of strategies or types of change. These have been classified as coercive, persuasive, reeducative, and facilitative. Manipulation is defined as ". . . (the) deliberate act of changing either the structure of the alternatives in the environment . . . or personal qualities affecting choice without the knowledge of the person involved. . . ." (Warwick and Kelman, 1973, p. 403). If manipulation is defined in this fashion, all strategies involve manipulation to some extent. A change agent is always related to manipulation whether directly in that he may, say as a manager, reorganize a company division, or indirectly, as a consultant, provide information for others to use in deciding where, when, and how to reorganize a firm. Selectively providing information on the positive aspects of a given change reduces the alternatives available to a potential adopter when making a decision. This then is more likely to manipulate a positive attitude on his part toward the given change.

Walton (1965, pp. 167–179) has identified several ethical issues involved in selecting change strategies, although he does not label these as ethical issues but refers to them simply as dilemmas. There are a number of dilemmas a change agent faces if he desires to gain concessions or agreement from another party. One dilemma is whether to exaggerate the desirability of the change or to deemphasize the importance of barriers or problems. Another issue particularly important for change internal to an organization or other social system is whether to build internal cohesion through various socio-emotional functions or to create a strawman out of an external group and stress important differences between the two groups. The change agent can also use power if it is available, or he may choose to develop trust through noncoercive activities. Similarly, a change agent may use threats as opposed to conciliation as an approach to gain compliance. Finally, the change agent may form a coalition with a third group and together force the change target to alter its behavior. An alternative would be to include the change target in all phases of the change process. Wittingly or unwittingly, change agents are making trade-off choices when they are determining the type of intervention to be used. The selection of one type of change generally precludes several other alternative change modes.

Manipulation is not a phenomenon that can be side-stepped either by the change agent, who may claim that he or she only provided information or advice, or by the client, who may claim that it was the change agent who, by word or deed, produced some change. The former case assumes a value-free activity in presenting particular information or recommendations to the client, when in fact the change agent is an advocate. The latter case, at the very least, assumes no value bias in the selection of a change agent and in the cooperation provided him. Neither assumption seems to avoid manipulation.

The basic question here would seem to be "under what conditions are particular forms of manipulation warranted?" The alternative question suggested by others is "under what conditions is any manipulation warranted?" It is interesting to note that colleagues with whom we have discussed this typically posed the alternative question as the most important issue. This implies something negative or perjurative about the concept of manipulation. Why is the alternative question never phrased in the following manner: "under what conditions is manipulation *not* warranted?"

It is difficult to arrive at any universalistic criteria about when coercion or power is to be preferred over facilitation, and reeducation over persuasion, and so forth. The answers seem to be context bound.

We certainly agree with the Population Task Force of the Institute of Society, Ethics and the Life Sciences (1972) that core values such as freedom (the capacity, opportunity, and incentive to make and act on reflective choices), justice (equitable distribution rewards and punishments), and welfare (maintenance and improvement of vital interests of society) should be preserved. Inevitably, however, matters of interpretation arise. Consider the sales manager, whose own position would be enhanced if a particular redesign of sales territories were put into effect that produced more revenue for the firm and certain salesmen, yet less income for other salesmen. The redesign and reallocation of sales territories—say giving more effective salesmen more or better territory—is consistent with the welfare of the manager, the firm, and certain salesmen. At the same time, it is not for the welfare of still other salesmen, who lose by the change. It can be argued that it is only justice to give rewards to those who sell more effectively and punishments to those who do not. But this is a value position unique to only a few cultures.

Responsibility for Diagnosing the Problem

Before discussing some of the issues involved in the responsibility for the outcome of change efforts, it is necessary to pause and consider the diagnosis of the problem. In many ways, the diagnosis of the problem determines the general outline of the solution(s) and in this indirect way is related to the change outcome. For example, diagnosing an organizational problem as not having enough information might oversimplify the apparent solution of providing more information to decision makers. It may be that the problem is somewhat more complex in that organizational decision makers may not trust one another. Thus simply providing more information without dealing with the cause of this mistrust is unlikely to improve decision making.

Possibly, a change agent's responsibilities for the change outcome vary as partial functions of his involvement in the problem definition. The exact same reasoning is applicable to the involvement of the change agent in the various activities related to the implementation of change strategies. In the preceding example, the choice of the strategy selected for dealing with the "information problem" will affect the outcome. A purely technological strategy of generating more information for the decision maker will not resolve the mistrust that underlies the failure of the decision makers to interact cooperatively in decision making. A broader strategy might be required that focuses not only on the technological needs of greater information but also on organizational dynamics involved in processing information.

The client or target system is also a value-laden social unit. Values of the client, for example, may bias the selection of information to be given to the change agent to evaluate. Moreover, client values may cause a particular structuring of information that presents a biased view about the nature and source of problems. This raises the general issue of how much freedom the change agent has in discovering the real problem. The withholding of information from a change agent restricts the number of alternative problems he is able to identify and choose from. In this special sense, his freedom of choice has been narrowed. His freedom of choice is further narrowed if the data provided him are highly structured in a certain way so that certain interpretations are unlikely. Thus an important operating strategy for the change agent is to consider the target system's definition, if any, of the problem as an hypothesis. The change agent's first task is then to collect his own data in the system to determine if, in fact, the client's definition of the problem is correct. Of critical importance here is the change agent's access to information. The change agent must be allowed to gather the information he feels is needed.

The other side of the coin involves ethical issues in the way change agents gather data. Ethical dilemmas in research methodology have received considerable treatment in psychology and much less attention in other disciplines. Tybout and Zaltman (1974, 1975) have compared several professional codes of ethics in terms of their treatment of subject's rights. This comparison is shown in Table 13.2. It appears that the American Marketing Association and American Association for Public Opinion Research have very little direct comment on subject's rights. This is in sharp contrast to the positions of the Marketing Research Society (Great Britain) and American Psychological Association. Tybout and Zaltman (1974) have also classified possible violation of subjects rights and corresponding research questions in terms of their relevance to the basic rights to choice, to safety, and to information. These ethical questions are presented in Table 13.3. These questions and issues are generic and apply to all social settings, including situations in which research is being conducted as a foundation for developing an effective program of social change.

Apart from issues of well-being for participants in the research process, there is a highly practical matter: the resolution of particular ethical issues in research can affect the quality of data. Moreover, good data may not be consistent with "good" ethics as defined in the various codes of professional societies. For example, it has been suggested that, if the name of the research sponsor and the intended use of the data would be likely to cause response bias among subjects, the researchers should not disclose this

Table 13.2   Code of Ethics

Subjects' Rights	AMA	MRS	APA	ASA	POR
A. Right to choose					
1. Awareness of right		X	X		
2. Sufficient information to choose		X	X		
3. Opportunity to choose		X	X		
B. Right to safety					
1. Protection of anonymity	X	X	X	X	X
2. Freedom from stress		X	X	X	
3. Freedom from deception		X	X	X	
C. Right to be informed					
1. Debriefing			X		
2. Dissemination of data	X			X	X

AMA, American Marketing Association; MRS, Market Research Society; APA, American Psychological Association; ASA, American Sociological Association; POR, American Association for Public Opinion Research.

information until after the data are collected. At this point, deception would cease, and the subject would be informed of the sponsor and use of the data. If the subject so chooses, he should be able to withdraw his data. However, this could cause loss of valuable information from certain types of respondents.

Implementation of Change Program

The unavoidability of manipulation has been mentioned earlier. Manipulation, say in the form of a strategy of persuasion, is less onerous when the following techniques are utilized: (1) Getting and keeping attention focused on a particular problem or bottleneck—this is the indispensible first step. The quality of the attention must be active and alert; therefore, the focal issue presented must have high interest potential and ego involvement. (2) Getting and keeping rapport with clients—credibility is essential in establishing rapport. (3) Building credibility—besides implementing basic methods such as truthfulness and accuracy, credibility is often enhanced by utilizing the two-sided approach, sometimes referred to as the "yes, but" technique. (4) Appealing to strong motives—including "emotions." It is acceptable to appeal to people's emotions as long as the appeal is presented rationally. There is often a practical necessity to appeal to strong motives if people's actions are to be changed. (5) Action involvement—it is often necessary in implementing attitude change, for example, to actively involve people in the change process.

Table 13.3 Summary of Ethical Questions for Market Researchers

Subjects' Rights	Possible Results of Violation of Rights	Research Questions
A. The right to choose 1. Awareness of rights	1. Feelings of forced compliance, biased data	1. How do responses from subjects who perceive participation in research to be purely voluntary differ from those of subjects who feel coerced?
2. Adequate information for an informed choice	2. May violate the client's desire for anonymity; may enable subjects to enact subject role	2. Does knowledge of the research client distort subjects' responses? What, if any, subject roles are enacted when subjects are given full information prior to participation? What alternatives to complete disclosure of information prior to subjects' participation are available? How effective are these methods in gaining unbiased data and protecting subjects?
3. Opportunity to make a choice	3. Subjects may avoid environments in which this right is violated	3. If subjects suspect that unobtrusive measures of their behavior are being taken, how is their behavior affected? Do subjects avoid environments in which they suspect unobtrusive measures are taken?
B. The right to safety 1. Protection of anonymity	1. Biased data, refusal to participate in future research	1. How does the degree of anonymity affect subjects' responses? How does violation of promised anonymity affect subjects' willingness to participate in future research?

335

Table 13.3 (Continued)

Subjects' Rights	Possible Results of Violation of Rights	Research Questions
2. Subject stress	2. Biased data, refusal to participate in future research	2. How does the degree of subject stress affect subjects' responses? How do stressful studies affect subjects' willingness to participate in future research? Can debriefing effectively relieve subject stress?
B. The right to safety 3. Deception of subjects	3. Biased data, refusal to participate in future research	3. How does deception influence subjects' responses? Under what conditions is deception essential? What effect does deception have on subjects? How does deception influence subjects' willingness to participate in future research? What alternatives to deception are available and under what conditions can they be employed? What are the implications of using deception on the client's and profession's image?

| | Can debriefing relieve the negative effects of deception? |
| | Can complete disclosure produce valid results when the subject believes he will gain by cooperating with the researcher? |

C. The right to be informed

1. Debriefing

| 1. Unrelieved stress, feelings of being used, refusal to participate in future research | 1. How does debriefing affect subjects' responses in future research? Do subjects believe they gain from debriefing? Conversely, does lack of debriefing make subjects feel used and less willing to participate in future research? Does debriefing perpetuate or reduce marketers' image as manipulators of human behavior? |

2. Dissemination of data

| 2. Subjects may feel that they gain nothing from and are exploited by participating in research and consequently may distort their response and decline to participate in future research | 2. Do subjects want access to research findings? Would giving participants access to research findings increase their willingness to cooperate? Under what conditions would dissemination of research finding to subjects reduce a clients' competitive advantage? |

337

It would also seem that manipulation via a persuasive strategy approach is more onerous when the following techniques are employed: (1) Lying. (2) Innuendo—implying an accusation without risking an argument by actually saying it. This technique is often used when an accusation lacks sufficient evidence. The use of innuendos is often more harmful to one's own credibility than to that of its victim. (3) Presenting opinion as fact—this is perhaps the most common of all the questionable techniques of manipulation. Obviously this technique is counterproductive when one talks to those who know that other interpretations exist. (4) Deliberate omission—one particularly dangerous and common form of deliberate omission is the deliberate refusal to discuss the other side's point of view in a group conflict situation. (5) Implied obviousness—this technique implies that a certain point of view is the only reality and that no further argument or evidence in support of it is needed. Utilization of the above techniques may cause the change agent to be seen as biased and partisan with the result that he is likely to be seen as less credible in his attempts at persuasion.

The change agent can avoid the above problems best when he is informed of all parties both within and without the target system who might reasonably be expected to be affected significantly by the change. This might well be information the change agent could insist on having. At the same ti ne, in providing solutions, the change agent might be expected to provide members of the target system with as many realistic alternatives as possible to choose from. By realistic alternatives we mean simply alternatives with similar likelihoods of similar results. The target system can then select that course of action it considers to have the smallest number of undesirable consequences. It also seems important to us that the client or target system be provided with the rationale for both the advocated solution to a problem as well as the suggested means for implementing that solution. It can be expected that such demands, particularly after the change, will be made by persons in the target system who would be affected by any dysfunctional consequences of the intervention. This suggests one issue not mentioned thus far. The issue concerns value conflicts between the change agent and client in determining the means of implementing a solution. Our own position is that the change agent is obligated to discuss such conflict with the client. If these conflicts are shared openly, the client has an opportunity more directly to affect the decision process, and thus the chance for the client being manipulated by the change agent is reduced. We feel it entirely appropriate, however, for the change agent to be as partisan and persuasive as he chooses. It should be noted, however, that the change agent, particularly if he is outside the target system, cannot be expected to know a priori how his values may differ from those of the client or those of the system the client represents. If the client or some third party initiates the contact leading to, say, a consulting relationship, the change agent may well mistakenly assume that there is compatibility in basic values.

Once a problem solution or change is implemented, the issue of responsibility for its impact becomes important. With respect to the change agent and target system, there are several important issues here. Once a change has been implemented, what responsibility does the change agent have for the unintended dysfunctional consequences that might result? Is the change agent required to provide services to the target system to help resolve these dysfunctional consequences, or is it no longer the change agent's responsibility? For example, the introduction of a management information system may pose a threat to some managers in the system, with the result that they are very resistant to its use. Should the change agents who introduced this MIS be required to come back and help the target system reduce this resistance, or is this someone else's responsibility? One might make the case that, if the change agents were more sensitive to resistance to change, they might have introduced the MIS in a way that might reduce the resistance of those affected.

The issue of responsibility also applies to the target system. In the MIS example, the same issue of responsibility can be raised. What responsibility does the target system have in helping people deal with the psychological impact of change? Should the target system be sensitive and supportive of those affected by the change? Beyond helping people deal with the psychological impact of change, what responsibility does the target system have for relocating personnel who might be displaced as a result of the introduction of some new technological change?

We would like to conclude this section of the chapter with Saul D. Alinsky's eleven rules concerning the ethics of means and ends. They reflect his sharp and unique mind.

First, one's concern with the ethics of means and ends varies inversely with the degree of one's personal vested interest in the issue. Accompanying this rule is the parallel rule that one's concern with the ethics of means and ends varies inversely with one's distance from the scene of conflict.

The second rule of the ethics of means and ends is that the judgment of the ethics of means is and has been dependent upon the political position of those sitting in judgment.

The third rule of the ethics of means and ends is that in war the end justifies almost any means.

The fourth rule of the ethics of means and ends is that judgment must be made in the context of the time in which the action occurred and not from any other chronological vantage point.

The fifth rule of the ethics of means and ends is the concern with ethics increases with the number of means available and vice versa.

The sixth rule of the ethics of means and ends is that the less important the end to be desired the more one can afford to engage in ethical evaluations of means.

The seventh rule of the ethics of means and ends is that generally success or failure is a mighty determinant of ethics.

The eighth rule of the ethics of means and ends is that the morality of a means depends upon whether the means is being employed at a time of imminent defeat or imminent victory.

The ninth rule of the ethics of means and ends is that any means which is effective is automatically judged by the opposition as being unethical.

The tenth rule of the ethics of means and ends is that you do what you can with what you have and clothe it with moral garments.

The eleventh rule of the ethics of means and ends is to phrase goals in general terms like "Liberty, Equality, Fraternity", "Of the Common Welfare", "Pursuit of Happiness", or "Bread and Peace".

SURVEY RESULTS

Nineteen value-ethical issues were developed from the review in the first part of this study. These 19 statements (Table 13.4) are a first attempt to empirically identify to what extent practicing change agents (1) agree with the issues as stated, (2) feel the issues are important or not, and (3) encounter the issues.

Nature of the Sample

Mailing lists (some restricted on a geographical basis) from several organizations were obtained and names selected at random from each list. The associations or document lists were: The Institute for Management Science, The American Anthropological Association, *The Journal of Applied Behavioral Science,* The American Association of Social Workers, and *The Administrative Science Quarterly.* Some professional associations refused our request to purchase their mailing list or some portion of it. All groups or associations were sent a copy of the questionnaire at the time the request was made. An overall response rate of 56% was obtained when miscellaneous unusable responses, undelivered questionnaires, etc. are disallowed. No follow-up letter was used for nonrespondents, and 791 respondents indicated that their activities as change agents were directed outside their employing organizations, 587 indicated they were primarily change agents within their employing organization, and 193 gave no indication.

Table 13.4   Value-Ethical Issues

1. If the interventionist perceives a conflict between his values and the client's values in defining the problem, he is obligated to discuss this with the client.
2. Interventionists should be completely free in their selection of strategies for problem diagnosis.
3. The interventionist should communicate his value orientation to the client system.
4. The interventionist is obligated to perform services even for those client systems with whom he does not share the same values.
5. The client system has an obligation to inform the interventionist of all parties to be affected by the interventionist's activity.
6. In the absence of a written contract covering termination of an interventionist/ client relationship, the interventionist is obligated to work with the client system as long as the client system requests his services.
7. In implementing a solution to a problem, the interventionist should provide members of the client system several options to choose from.
8. Those affected by a given solution have a right to be informed of the rationale for that solution.
9. Those affected by a given solution have a right to have access to the evidence supporting a solution.
10. The interventionist is responsible for making services available for dealing with dysfunctional consequences of his activities.
11. The interventionist is obligated to inform members of the client system as to the goals and objectives of the interventionist's activities.
12. A client hiring an interventionist should make known to others in the client system what the interventionist is doing in the system.
13. The client system should provide any information the interventionist feels he needs to perform his task.
14. If the interventionist perceives a conflict between his values and the client's values in defining the goals and objectives of his activities, he is obligated to discuss this with the client.
15. If the interventionist perceives a conflict between his values and the client's values in determining the means of implementing a solution, he is obligated to discuss this with the client.
16. If the interventionist perceives a conflict between his values and a client's values in determining the means of evaluating a solution, he is obligated to discuss this with the client.
17. Interventionists should be completely free in their selection of strategies for implementing a solution.
18. Interventionists should be completely free in their choice of criteria for selecting a solution.
19. Interventionists should be completely free in their selection of strategies for evaluating a solution.

One of the interesting results of this survey was some of the written comments we received from respondents. Some of the comments were quite vehement in stating the individual felt quite insulted that we should infer that they had values that might affect how they operated as a change agent. The view most often stated was "I am a professional and thus I am value free. . . ." Some respondents felt that if you clearly identified and focused on your values, you were somehow not being professional or scientific. This is a troublesome perspective. Many change agents may consciously ignore their own values and thus unclear about how these values may affect their work.

Limitations of the Study

The reader should be aware of several limitations to the study in interpreting the data: (1) the list of issues presented here is not meant to be exhaustive; rather, this is a preliminary attempt to ascertain how practitioners view value and ethical dilemmas; (2) the sample is biased in that respondents are likely to have a more professional orientation given their membership or reading behavior; (3) the normative wording of the items is likely to create a more favorable response; (4) a survey may oversimplify the issues, since there is no opportunity for the respondent to discuss and perhaps clarify his or her answer.

Importance of Issues

In general, all 19 issues were considered important. The rank order of these issues by importance is presented in Table 13.5. The item (8 in Table 13.4) ranking highest on ascribed importance was considered to be mildly important to very important by approximately 91.3% of the respondents. This item was the statement that "those affected by a given solution have a right to be informed of the rationale for that solution." The item (3 in Table 13.4) having the least ascribed importance was considered important by 78.5% of the respondents. This item read "The interventionist should communicate his value orientation to the client system." The median-ranked item was considered important by approximately 86% of the respondents. In nearly all cases the percentage differences between adjacent ranks were extremely small.

Agreement with Issues

There was considerably more variation in agreement with the issues compared to ascribed importance. The three highest ranking items in importance (Table 13.6) were also the three highest in expressed agreement.

Table 13.5   Rank Order of Items on "Importance" Response

Rank	Percent Impt	Item #	Median	Mean	N
1	91.3	8	5.61	5.186	1545
2	89.7	14	5.44	5.09	1543
3	89.3	15	5.33	5.03	1545
4	88.6	13	5.11	4.90	1536
5	89.8	11	5.27	4.96	1544
6	87.3	9	5.16	4.88	1541
7	87.2	17	4.85	4.68	1534
8	87.0	7	4.98	4.78	1541
9	86.7	19	4.82	4.65	1530
10	86.1	18	4.83	4.65	1529
11	86.0	12	5.11	4.85	1536
12	85.7	16	5.07	4.81	1540
13	84.6	10	4.91	4.72	1525
14	84.1	1	5.22	4.81	1557
15	83.5	2	4.74	4.52	1545
16	82.4	4	4.87	4.62	1532
17	79.4	3	4.78	4.51	1549
18	79.1	5	4.87	4.58	1533
19	78.5	6	4.56	4.36	1541

Range 78.5 – 91.3

This is discussed later. Although 94.2% of the respondents agreed with item 8, only 35.1% agreed with item 6, the least agreed with statement. This latter item was stated in the following way: "In the absence of a written contract covering termination of an interventionist/client relationship, the interventionist is obligated to work with the client system as long as the client system requests his services." It can be seen in Table 13.6 that the median item was supported by 86.7% of the respondents. Between 90 and 94.2% of the respondents were in general agreement with four items. Another eight items were supported by 82.3–89.8% of the respondents, and an additional four items were supported by between 52.3 and 79.3% of the respondents. In general, a substantial number of the items were supported by a large percentage of the respondents.

Commonalities in Issue Agreement and Ascribed Importance

It was noted earlier that the three items receiving the strongest agreement or support were also the three items receiving the highest ratings on

Table 13.6   Rank Order of Items on "Agree" Response

Rank	Percent Agree	Item #	Median	Mean	N
1	94.2	8	1.28	1.59	1553
2	92.5	15	1.46	1.78	1556
3	91.5	14	1.40	1.77	1555
4	90.0	16	1.74	2.01	1550
5	89.8	11	1.50	1.90	1559
6	88.7	12	1.71	2.02	1548
7	84.9	13	1.89	2.20	1552
8	88.4	7	1.89	2.13	1552
9	87.9	9	1.611	2.01	1553
10	86.6	1	1.58	2.01	1562
11	82.8	5	1.93	2.30	1545
12	82.3	10	2.05	2.39	1538
13	79.3	3	2.11	2.47	1559
14	63.3	2	2.83	3.11	1555
15	52.9	4	3.22	3.48	1548
16	52.3	19	3.37	3.46	1546
17	41.1	17	4.06	3.90	1547
18	40.2	18	4.09	3.90	1545
19	35.1	6	4.61	4.13	1553
Range 35.1 − 94.2					

importance. It might be useful to comment further about these items. The three items are:

1. If the change agent perceives a conflict between his values and the client's values in defining the goals and objectives of his activities, he is obligated to discuss this with the client. (14)
2. Those affected by a given solution have a right to be informed of the rationale for that solution. (8)
3. If the change agent perceives a conflict between his values and the client's values in determining the means of implementing a solution, he is obligated to discuss this with the client. (15)

We suggest that these three items are three of the most central norms involving the interventionist–client relationship. There is also a linearity of sorts to the separate norms. One concerns the desirability of openness at the start of the relationship and avoidance of conflict. Another concerns openness in an activity further along in the relationship, namely, the presentation of and justification for an advocated solution to a problem. The third concerns openness with regard to the next step, which is the implementation of a solution.

One of the key questions that should be raised in any study of norms is, "Why are particular norms supported?" Alternatively, "Why are they maintained?" What function do they perform for the individual who participates in them? The intent and design of the questionnaire did not permit exploring such a question. Still, some observations might be made. First, change agent-initiated discussion of value-related information is a means of identifying the value positions the client or client system maintains on a particular issue and how committed he is to that position. This is very useful information for all aspects of the change agents' activity, especially in diagnosing problems. Second, open discussion, particularly of problem solutions and their implementation, is a way of coopting a client in the problem solution and implementation process, thus enhancing the acceptance of particular solutions and implementation techniques. Third, change agent-initiated discussion is a way of clarifying for the client what his or her values are and how those values relate to the problem at hand. Yet another set of reasons are somewhat the reciprocal of those mentioned. Open discussion keeps the change agent more constantly aware of his or her own value positions, and their relevance and effect on his or her involvement with the client problem.

Of course, value and other differences are not necessarily dysfunctional. Some differences may be very desirable. As Rogers and Bhowmik (1971, p. 529) have indicated, homophily-heterophily is concerned with the relationship between source and receiver in a communication exchange. They have indicated that when communicators are homophilous, that is, they share common meanings, attitudes, beliefs, and so forth, communication is likely to be more effective. In heterophilous relationships, the interaction is likely to cause some message distortion, some restriction of communication channels, and so forth. These findings provide the manager with a dilemma in that the interventionist is often seen as somewhat different from the client system, which can lead to a heterophilous type of relationship.

With respect to acknowledgment and being seen as legitimate in this interventionist role, the heterophilous interventionist may have the advantage. This is likely if the agent is perceived as differing, expertise in the area in which the organization is considering change. The most influential change agents are perceived as heterophilous in expertise, while they are seen as sharing many of the same attitudes, values, and beliefs of the client system.

The heterophilous interventionist might be more successful in communicating his or her views to the client if he or she is seen as an expert. The agent is likely to be seen as more legitimate in this role that just "another member" of the client system.

Encountering the Issues

Another dimension of concern focuses on the frequency with which the value-ethical issues are encountered. Respondents were asked to indicate how often they encountered the 19 value-ethical statements using the following response categories "never," "infrequently," "sometimes," "fair amount," "a lot," "all the time." In the analysis of our data, we collapsed responses into four categories. These four categories were constructed from the initial categories as follows:

Very frequently: (a lot and all the time)
Occasionally: (sometimes and a fair amount)
Infrequently: (infrequently)
Never: (never)

Table 13.7 presents the rank order from high to low for the 19 items for each of the four response categories. From Table 13.7 we see that the percentage range for persons encountering issues very frequently ranged from 11.9 to 53.1%, for issues encountered occasionally 33.9 to 57.0%, for issues encountered infrequently 8.2 to 39.6%, and for issues never encountered 2.3 to 17.7%.

The three issues encountered most frequently were items 8, 11, and 7. These items are:

8. Those affected by a given solution have a right to be informed of the rationale for that solution.
11. The change agent is obligated to inform members of the client system about the goals and objectives of the interventionist's activities.
7. In implementing a solution to a problem, the change agent should provide members of the client system several options to choose from.

These items have the common theme of being open with the client system.

The three issues respondents indicated that they encountered infrequently or never were items 10, 6, and 4:

10. The change agent is responsible for making services available for dealing with dysfunctional consequences of his activities.

Table 13.7   Rank Order for Four Response Categories

Very Frequently			Occasionally			Infrequently			Never Encounter		
Rank	Item #	Percent	Rank	Item #	Percent	Rank	Item #	Percent	Rank	Item #	Percent
1	8	53.1	1	1	57.0	1	10	39.6	1	6	17.7
2	11	51.3	2	14	52.7	2	6	35.6	2	10	11.0
3	7	47.5	3	15	52.3	3	4	33.6	3	4	7.5
4	13	43.2	4	16	49.3	4	16	26.8	4	18	5.8
5	12	39.9	5	18	48.5	5	17	25.6	5	19	5.4
6	9	35.3	6	17	48.3	6	19	25.2	6	16&12	5.3
7	3	34.1	7	19	46.9	7	18	24.9	7		
8	5	26.4	8	2	45.7	8	2	23.7	8	5	5.2
9	2	26.1	9	5	45.6	9	15	22.2	9	9	5.1
10	1	22.8	10	3	44.2	10	5	22.1	10	17	5.0
11	19	22.1	11	9	41.3	11	14	22.0	11	11	4.7
12	14	21.5	12	12&4	40.5	12	9	17.9	12	2	4.0
13	15	21.4	13			13	1	17.8	13	15	3.9
14	17	20.4	14	7	38.9	14	3	17.6	14	13	3.8
15	18	20.3	15	13	38.6	15	13	13.9	15	14	3.6
16	16	18.2	16	10	36.3	16	12	13.7	16	3	3.8
17	4	18.0	17	8	35.2	17	7	10.3	17	7	3.2
18	6	12.1	18	11	34.2	18	11	9.4	18	8	3.0
19	10	11.9	19	6	33.9	19	8	8.2	19	1	2.3
Range 11.9–53.1			33.9–57.0			8.2–39.6			8.2–39.6	2.3–17.7	

6. In the absence of a written contract covering termination of an interventionist/client relationship, the interventionist is obligated to work with the client system as long as the client system requests his services.

4. The change agent is obligated to perform services even for those client systems with whom he does not share the same values.

Each of these issues concerns obligations the interventionist may have to work with a client or potential client who demands the interventionist's assistance.

Differences Between Internal and External Change Agents

It was expected that there might be some differences on the value issues between those change agents who are members of the client organization and those who are not members of the client organization. This internal change agent, by virtue of being a member of the system, is likely to experience the following: (1) more pressure to tackle certain problems he or she might not normally choose; (2) less freedom of operation, since more informal pressure can be placed on him or her not to disrupt the system and to meet the more personal needs of the client system; (3) a greater need for awareness of the dysfunctional consequences of his or her activities. If the interventionist is not successful, it will be easier for the system to request that he or she deal with dysfunctional consequences.

Table 13.6 presents a list of the issues on which there was a significant difference between the external and internal change agents on the "Agree," "Important," and "Encounter" dimensions. Table 13.8 indicates relatively few differences between internal and external interventionists. It should also be pointed at that the mean differences reported, although statistically significant, are not very large. On a scale that ranges from 1 to 6, the mean differences reported are less than 1. Their statistical significance may then be an artifact of the large sample size (External, N = 791, Internal, N = 587).

The data in Table 13.8 indicate that most of the significant differences between the internal and external interventionists occurred in encountering the ethical-value dilemmas. In encountering the ethical-value issues, the internal interventionists indicated that they encounter issues 1, 4, 6, 9, 16, and 19 more than did external interventionists.

In looking at issues 1, 4, 6, 9, and 16 as presented in Table 13.8, we can see that they seem to have a common thread running through them. They concern the change agent being sensitive to the needs of the client by providing information and discussing conflicts when they occur. This makes sense from our earlier discussion in which we indicated that the internal interventionist, by being part of the client system, may be operating under more constraints. The internal agent is more accountable for results. Also, by being part of the client system, the internal change agent must be concerned more with longer-term relationships; thus he or she is more likely to deal with conflicts and disputes when they occur.

Internal change agents also indicated that they encounter more often the issue of the change agent being free to select the strategies for evaluating a change program (19). Once again, it is not surprising to find internal change agents encountering this issue more than external change agents. The internal change agent is more likely to be under greater pressure to be successful. The client system may also require some form of evaluation to assess the impact of the program. Thus the internal interventionist will be involved in more evaluation activities than the external change agent.

On agreement with the ethical-value issues, external change agents agree more with issues 4 and 6, whereas internal change agents agree more with issue 7.

Table 13.8 Differences Between External and Internal Interventionists on Ethical-Value Issues

Agree

Item 4: Interventionist should perform services for those with whom he does not share the same values

	X̄	T Value	Significance Level
External	3.60	7.78	$p < 0.001$
Internal	3.03		

Item 6: In absence of contract, interventionist obligated to work as long as client requests

	X̄	T Value	Significance Level
External	4.22	2.30	$p < 0.022$
Internal	4.02		

Item 7: In implementing problem solution, interventionist should provide client options to choose

Importance

Item 3: Interventionist should communicate value orientation to client system

	X̄	T Value	Significance Level
External	4.46	−2.51	$p < 0.012$
Internal	4.64		

Encounter

Item 1: Interventionist discusses value conflicts in problem definition with client

	X̄	T Value	Significance Level
External	3.47	−2.03	$p < 0.042$
Internal	3.60		

Item 4: Interventionist should perform services for those with whom he does not share the same values

	X̄	T Value	Significance Level
External	3.02	−2.53	$p < 0.011$
Internal	3.21		

Item 6: In absence of contract, interventionist obligated to work as long as client requests

Table 13.8 (Continued)

	Agree			Importance			Encounter		
	$\bar{X}$	T Value	Significance Level				$\bar{X}$	T Value	Significance Level
External	2.04	−2.50	$p < 0.012$				2.62	−3.11	$p < 0.002$
Internal	2.20						2.87		

Item 9: Those affected by solution have right to evidence supporting solution

	$\bar{X}$	T Value	Significance Level
External	3.72	−2.26	$p < 0.024$
Internal	3.91		

Item 10: Interventionist's services should be available for dealing with dysfunctional consequences

350

	$\bar{X}$	T Value	Significance Level
External	2.74	−2.97	$p < 0.003$
Internal	2.98		

Item 16: Interventionist discusses value conflicts with client over means of evaluation

	$\bar{X}$	T Value	Significance Level
External	3.20	−2.05	$p < 0.040$
Internal	3.35		

Item 19: Interventionist should be free in selecting strategies for evaluation

	$\bar{X}$	T Value	Significance Level
External	3.27	−2.19	p 0.029
Internal	3.44		

One might initially question why external change agents would agree more with issues 4 and 6, whereas internal change agents indicated that they encountered these issues more. One explanation for why internal interventionists may not have indicated more agreement with issues 4 and 6 follows. Because internal change agents must work with people who do not necessarily share their values (4) and must work as long as the client requests (6), they experience the problems and constraints that this creates for them as change agents, and so they are less likely to agree with these normative statements. The external change agent does not encounter these issues as frequently; it is perhaps easier for him to agree to them in principal.

Internal interventionists did agree more than external change agents on the issue that, in implementing problem solutions, the change agent should provide the client with options from which to choose (7). This fits with our earlier discussion vis-à-vis the internal change agent being under more pressure to produce and maintain a good relationship with clients, given that they are all part of the system. Therefore, it is not surprising that the change agent might want to present the client with more options for implementing a solution with the hope that this choice might increase the client's commitment to and success in implementing a solution. On the other hand, external change agents often are selling a package solution that they market to a variety of client customers. They are thus less likely to want to give the client much opportunity to choose and perhaps require the change agent to modify his packaged solution.

On the importance of the issues, there was only one difference. Internal change agents attached more importance than external change agents to the issue of the change agent communicating his value orientation to the client system. This again fits with the notion of the internal change agent being more sensitive to the needs of the client and working to develop a relationship with the system. It may be more important for the agent to share his values with the client as a way of developing the relationship.

Different Professional Contexts

The next issue of interest concerns differences among change agents in different professional contexts. The following groups were identified by means of respondent self-designation:

1. Organizational Design
2. Manpower Management/Industrial Relations
3. Operations Research/Management Science

4. Organization Change and Development
5. Social Organization
6. Community Development and Organization
7. Counseling
8. General Management and Administration
9. Applied Anthropology
10. Psychotherapy
11. Psychological and Cultural Anthropology

Each group was compared with every other group on the dimensions of each issue. Statistical significance was set at the 0.05 level in the comparison of mean scores on each issue/dimension. The results are reported below.

Table 13.9 presents in summary form the particular groups that differed from each other in a substantive and statistically significant way. Three things stand out from these results.

First, the groups were more similar than dissimilar. This is surprising given the fact that one might expect that, at least in some cases, people engaged in these different activities would receive different training. For example, we would expect operations research people, counseling people, and cultural anthropologists to have received very different training. It may be that the kinds of things professionals are exposed to during their training (books, methodologies) still produce a common philosophical foundation for dealing with client systems. Different training experiences do not produce different value orientations. Therefore, given that professionals coming from these different backgrounds have rather similar responses to these 19 value items means that we should consider these to be generic items. They are not situation specific. Most professionals dealing with applied social change encounter them.

Second, when differences did occur, they primarily concerned the extent of agreement on particular issues and secondarily the importance of the issues. Those working on the most intimate basis with clients—counselors and psychotherapists—feel more strongly about the need to communicate their value orientation to the client. This may be for the following reasons. First, by virtue of working in a one-to-one relationship, it is simply easier to discuss these issues than if the professional were working with a larger group. Second, it may be that counselors and psychotherapists are more sensitive to the importance of values in their relationships with clients. The formal training of these two professional groups is likely to have concentrated on the professional being more sensitive to his values and

Table 13.9

	1[a]	2	3	4	5	6	7	8	9	10	11
2. Manpower	B6										A10
3. Operations research/ management science	B3			A10			A10	A10	A10		
4. Organizational change/development			B3								
7. Counseling	A3, A4, B4, B6, B10, C4, C6	C4	A3, B10, B11, C4	A3, A4, C4	A3, A4, C4, C6, C7	A3		A3	A3, C4, C6		C6, C7
8. General management and administration	B15		B10, B11								
10. Psychotherapy	A1, A3, A4, A12, B4, B6	A1, A12, B12	A1, A3, A12, B6, B10, B10	A1, A3, A4, A12	A1, A3, A4, A12	A1, A3, A12	A12	A1, A3, A12	A1, A3, A12		A1, A3, A12, B6
11. Applied											

[a] 1, organizational design; 2, manpower management/industrial relations; 3, operations research/management science; 4, organization change and development; 5, social organization; 6, community development and organization; 7, counseling; 8, general management and administration; 9, applied anthropology; 10, psychotherapy; 11, psychological and cultural anthropology.

A1 . . . 19 = Agree of issue 1 . . . 19.
B1 . . . 19 = Importance with issue 1 . . . 19.
C1 . . . 19 = Frequency of encountering issue 1 . . . 19.

This Table should be read from left to right. For example, with respect to item B6, manpower had a higher mean score than change agents working in organizational design. Similarly, with respect to item A10, OR/MS had a higher mean score than did organizational change and development change agents.

needs as they affect how he deals with the client. Therefore, these professionals see values as being more salient in the conduct of professional affairs, so they should generally communicate their values. For psychotherapists there is a stronger feeling that, in the event of conflict in values between therapist and client, the conflict should be discussed.

A third observation is that two groups, counselors and psychotherapists, consistently differ from the other groups. This is also true to a lesser extent for the operations research/management science group.

Three items seem to be particularly salient as discrimination between psychotherapists and other groups. The three items, which center on agreement of issues, are:

A1. If the change agent perceives a conflict between his values in defining the problem, he is obligated to discuss this with the client.

A3. The change agent should communicate his value orientation to the client system.

A12. A client hiring an interventionist should make known to others in the client system what the interventionist is doing in the system.

Psychotherapists were in more agreement with the first statement (A1) than any other group. Scales ranged from 6.0 for strongly agree (with statement) to 1.0 for strongly disagree. Psychotherapists, on the average, were in only mild disagreement with the statement, whereas five groups fell between full and mild disagreement, and five groups fell between full and strong disagreement. Psychotherapists were also in significantly more agreement (3.49) than eight of the other ten groups with regard to a statement that change agents should make their value orientation known to clients. Counselors (3.27) differed significantly from six of the other ten groups on this issue. Like psychotherapists, they expressed mild agreement. The other groups had means ranging from 1.87 to 2.72, with an overall mean of 2.45. Psychotherapists also differed significantly from every other group in expressing less disagreement (a mean of 2.81 versus a group mean of 1.83) with the statement that a client hiring an interventionist should make known to others in the client system what the interventionist is doing in the system.

Conclusion

A relatively large number of value-ethical issues have been identified as important among varied change agents. These issues, as articulated in a survey, received varying degrees of agreement and varying degrees of encounter. Overall, several issues stand out as important and widely held norms. These are put forward here as a subject for further analysis. For example, why do individual interventionists participate in these norms? Some answers were put forward very tentatively. Much closer scrutiny is needed based on additional empirical research to address this question fully. A more intriguing question that should also be addressed at the same time is how positions taken with regard to the various value-ethical dilemmas discussed in this paper affect the ways in which an interventionist behaves. We feel that issues such as those identified in this paper are confronted at one time or another by an interventionist in his career (particularly early in his career and training). The resolution of this confrontation provides the personality to his professional behavior. This personality is anything but neutral and value free. It is highly partisan, which we feel is a very healthy state. What is not healthy for the interventionist, his profession, or client is the denial of this state and the refusal to assess its practical impact on the helping relationship.

This chapter can be summarized by the following principles:

1. The change agent must recognize the potential for conflict between his values and the client's in defining the problem. The change agent should also make an explicit decision about whether to discuss conflicts with the client when they occur.

2. The change agent should have control over the selection of strategies for problem diagnosis.

3. The change agent should be aware of his value orientation and should make an explicit decision about whether to discuss conflicts with the client when they occur.

4. The change agent should be sensitive to the potential obligation to perform services for client systems whose values he does not share.

5. The client system should consider informing the change agent of all parties to be affected by the change agent's activity.

6. In the absence of a written contract covering termination of a change agent–client relationship, the change agent should make an explicit decision about whether he feels he is obligated to work with the client as long as the client requests the change agent's services.

7. In implementing a solution to a problem, the change agent should make an explicit decision about whether to provide members of the change target system several options from which to choose.

8. The change agent should be prepared to receive and answer questions regarding the rationale and supporting evidence for a given solution from those in the change target system.

9. The change agent should consider if it is in the best interests of members in the change target system to have the rationale and supporting evidence for the solution.

10. The change agent should consider his responsibility-obligation in making his services available for dealing with the possible dysfunctional consequences of his activities.

11. The change agent should consider his obligation to inform members of the target system about his goals and objectives.

12. A client hiring a change agent should consider his obligation to inform others in the target system about what the change agent is doing in the system.

13. The client should consider his obligation to provide the change agent with any information the change agent feels he needs to perform his task.

14. The change agent must recognize the potential for conflict between his values and the client's in defining the goals and objectives of the change agent's activities. The change agent should also make an explicit decision about whether to discuss these conflicts with the client when they occur.

15. The change agent must recognize the potential for conflict between his values and the client's in determining the means of implementing a solution. The change agent should make an explicit decision about whether to discuss these conflicts with the client.

16. The change agent must recognize the potential for conflict between his values and the client's in determining the means of evaluating a solution. The change agent should make an explicit decision about whether to discuss these conflicts with the client when they occur.

17. The change agent should, at the beginning of his relationship with a client, clearly define the scope of the change agent's freedom to select the strategies for problem diagnosis.

18. The change agent should, at the beginning of his relationship with a client, clearly define the scope of the change agent's freedom to select the criteria for selecting a solution.

19. The change agent should, at the beginning of his relationship with a client, clearly define the scope of the change agent's freedom to select the strategies for implementing a solution.

20. The change agent should, at the beginning of his relationship with a client, clearly define the scope of the change agent's freedom to select the strategies for evaluating a solution.

Bibliography

Abcarian, Gilbert, "Political Deviance and Social Stress: The Ideology of the American Radical Right," in J. P. Scott and S. F. Scott (Eds.), *Social Control and Social Change,* Chicago: The University of Chicago Press, 1971, pp. 137–161.

Abueva, Jose V. and others, *A Comparative Study of Administrative Perceptions: Rice Administration in Indonesia,* Honolulu: East-West Center, University of Hawaii, September, 1973.

Alinsky, Saul D., "Citizen Participation and Community Organization in Planning and Urban Renewal," Chicago: The Industrial Areas Foundation, 1962.

Alinsky, Saul D., "Of Means and Ends," *Union Seminar Quarterly Review,* Vol. XXII, No. 2, 1967.

American Psychological Association, "Ethical Standards of Psychologists," *American Psychologist,* Vol. 18, 1963, pp. 56–60.

Anderson, K., "An Experimental Study of the Interaction of Artistic and Non-Artistic Ethos in Persuasion," unpublished Ph.D. dissertation, 1961, Madison: University of Wisconsin, 1967.

Anderson, K. and T. Clevenger, Jr., "A Summary of Experimental Research in Ethos," *Speech Monographs,* Vol. 30, 1963, pp. 59–78.

Angell, J., "Some Effects on the Truth-in-Lending Legislation," *Journal of Business,* Vol. 44, January, 1971, pp. 79–84.

Apodaca, Anachto, "Corn and Custom: the Introduction of Hybrid Corn to Spanish American Farmers in New Mexico," in E. H. Spicer (Ed.), *Human Problems in Technological Change,* New York: Russell A. Sage Foundation, 1952, pp. 35–39.

Arensberg, Conrad M. and Arthur H. Niehoff, *Introducing Social Change: A Manual for Community Development,* Chicago: Aldine-Atherton, 1964.

359

Arensberg, Conrad M. and Arthur H. Niehoff, *Introducing Social Change: A Manual for Community Development*, 2nd edition, Chicago: Aldine-Atherton, 1971.

Argyris, Chris, *Interpersonal Competence and Organization Effectiveness*, Homewood, Ill.: Irwin-Dorsey, 1962.

Argyris, Chris and Donald A. Schon, *Theory in Practice: Increasing Professional Effectiveness*, San Francisco: Jossey-Bass, 1974.

Argyris, Chris, *Management and Organizational Development*, New York: McGraw-Hill, 1971.

Argyris, Chris, *Organizational and Innovation*, Homewood, Ill.: Dorsey Press, 1965.

Argyris, Chris, *Intervention Theory and Method*, Reading, Mass.: Addison-Wesley, 1970.

Arndt, Johan, "Profiling Consumer Innovators," in J. Arndt (Ed.), *Insights into Consumer Behavior*, Boston: Allyn and Bacon, 1968.

Atkyns, R. L. and G. J. Hannerman, "Illicit Drug Distribution and Dealer Communication Behavior," *Journal of Health and Social Behavior*, Vol. 15, No. 1, March, 1974, pp. 36–43.

Bandura, Albert, *Principles of Behavior Modification*, New York: Holt, Rinehart and Winston, 1969.

Barneck, J. and D. Payne, "The Lackland Accident Counter Measure Experiment," in W. Haddon, E. Suchman, and D. Klein (Eds.), *Accident Research: Methods and Approaches*, New York: Harper & Row, 1969, pp. 504–522.

Barnett, Homer G., *Innovation: The Basis of Culture Change*, New York: McGraw-Hill, 1953.

Beal, George M. and Joe M. Bohlen, "The Diffusion Process," *Special Report* No. 18, Ames, Iowa: Iowa State University Press, 1962.

Bean, Alden S. and Enrique Romagosa, "Organizational Structure and Technology Transfer in a Large Decentralized Firm," working paper, National Science Foundation, Office of National R & D Assessment, July 1975.

Bean, Alden, Rodney D. Neal, Michael Radnor, and David A. Tansik, "Structural and Behavioral Correlates of Implementation in U.S. Business Organizations," in Randall L. Schultz and Dennis Slevin (Eds.), *Implementing Operations Research/Management Science*, New York: American Elsevier, 1975.

Beckhard, Richard, *Organization Development: Strategies and Models*, Reading, Mass.: Addison-Wesley, 1969.

Belk, Russel and Ivan Rose, "An Investigation of the Nature of Word-of-Mouth Communication Across Adoption Categories for a Food Innovation," *Proceedings of 2nd Annual Conference Association for Consumer Research*, 1971.

Bell, William E., "Consumer Innovators: A Unique Market for Newness," In S. A. Greyser (Ed.), *Toward Scientific Marketing: Proceedings of the 1963 Winter Conference of the American Marketing Association*, Chicago: American Marketing Association, 1964.

Benne, Kenneth D., Leland P. Bradford, Jack R. Gibb, and Ronald O. Lippitt (Eds.), *The Laboratory Method of Changing and Learning*, Palo Alto: Science and Behavior Books, 1975.

Bennett, Claude F., "Diffusion Within Dynamic Populations," *Human Organization*, No. 28, Fall 1969, pp. 85–95.

Bennis, Warren G. and Edgar H. Schein, "Principles and Strategies in the Use of Laboratory Training for Improving Social Systems," in *Personal and Organizational Change Through Group Methods*, New York: John Wiley & Sons, 1965, pp. 201–233.

Bennis, Warren, *Changing Organizations*, New York: McGraw-Hill, 1966.

Berelson, Bernard (Ed.), *Family Planning and Population Programs*, Chicago: Aldine-Atherton, 1966.

Berlyne, D. E., "Curiosity and Exploration," *Science*, Vol. 153, No. 1, July 1966, pp. 25–33.

Berscheid, Ellen, "Opinion Change and Communicator-Communicatee Similarity and Dissimilarity," *Journal of Personality and Social Psychology*, Vol. 4, 1966, pp. 670–680.

Bhola, Harbans Singh, *Categories of Social Change*, unpublished paper, Columbus: Ohio State University, School of Education, 1965.

Black, T. R. L., "Ten Institutional Obstacles to Advances in Family Planning," in M. Potts and C. Woods (Eds.), *New Concepts in Contraception*, Honolulu: University of Hawaii Press, 1972.

Blake, Brian, Robert Perloff, and Richard Heslin, "Dogmatism and Acceptance of New Products," *Journal of Marketing Research*, Vol. VII, November 1970.

Blau, Peter, *Exchange and Power in Social Life*, New York: John Wiley & Sons, 1964.

Blau, Peter, "Parameters of Social Structure," *American Sociological Review*, Vol. 39, No. 5, October 1974, pp. 615–635.

Blum, Henrik L., *Planning for Health: Development and Applications of Social Change Theory*, New York: Human Sciences Press, 1974.

Blumberg, Arthur and William Wiener, "One from Two: Facilitating an Organizational Merger," *Journal of Applied Behavioral Science*, Vol. 7, 1971.

Bonilla de Ramos, Elssy, *Fatalism and Modernization in Columbia*, M.S. thesis, Michigan State University, 1966.

Boone, L. E., "Personality and Innovative Buying Behavior," *The Journal of Psychology*, No. 86, 1974, pp. 197–202.

Boyan, Norman J., "Problems and Issues of Knowledge Production and Utilization," in Eidell and Kitchel (Eds.), *Knowledge Production and Utilization in Educational Administration*, Eugene, Ore.: CASEA and UCEA, 1968, pp. 21–36.

Bradshaw, Barbara and C. Bernell Mapp, "Consumer Participation in a Family Planning Program," *American Journal of Public Health*, Vol. 62, No. 7, July 1972, pp. 972–989.

Brager, G. A., "Institutional Change: Perimeters of the Possible," *Social Work*, January 1967, pp. 78–90.

Bridges, Edwin M., "A Model for Shared Decision Making in the School Principalship," *Educational Administration Quarterly*, Vol. 3, 1967, pp. 51–61.

Bryk, J., *Research Report: Learning Performance in the Defensive Driving Course (DDC) and the DDC Self Instruction Program*, National Safety Council Research Department, November 1973.

Bryk, J. and S. Schupack, *Research Report: Supervisory Safety Training in the Coal Mining Industry*, National Safety Council Research Department, November 1972.

Burnkrant, R. E. and A. Cousineau, "Informational and Normative Social Influence in Buyer Behavior," working paper, Berkeley: University of California, 1974.

Burns, Tom and G. M. Stalker, *The Management of Innovation*, London: Tavistock Publications, 1961.

Campbell, Angus and Philip E. Converse, *The Human Meaning of Social Change*, New York: Russell Sage Foundation, 1972.

Campbell, Donald T., "Reforms as Experiments," *American Psychologist*, Vol. 24, 1969, pp. 409–429.

Campbell, John and Marvin Dunnett, "Effectiveness of T-Group Experiences in Managerial Training and Development," *Psychological Bulletin*, Vol. 70, No. 2, 1968, pp. 73–108.

Campbell, Rex R., "A Suggested Paradigm of the Individual Adoption Process," *Rural Sociology*, Vol. 31, December 1966, pp. 458–466.

Cancian, "Stratification and Risk-Taking: A Theory Tested on Agricultural Innovation," *American Sociological Review*, Vol. 32, 1967.

Carlson, Richard O., *Adoption of Educational Innovations*, Eugene, Ore.: Center for the Advanced Study of Educational Administration, 1965.

Carter, Charles and Bruce R. Williams, *Industry and Technical Progress: Factors Governing Speed of Application of Science*, London: Oxford University Press, 1957.

Carter, Launor and Harry Silberman, *The Systems Approach, Technology and the School*, Professional Paper SP-2025, Santa Monica: System Development Corporation, April 1965.

Carter Launor and Harry Silberman, "Management Information Systems: The Challenge to Rationality and Emotionality," *Management Science*, Vol. 17, No. 6, pp. B275–B292.

Cernada, E. C., J. L. Lee, and M. Y. Lin, "Family Planning Telephone Services in Two Asian Cities," *Studies of Family Planning*, Vol. 5, No. 4, pp. 111–114.

Chen, Lincoln D. (Ed.), *Disaster in Bangladesh: Health Crises in a Developing Nation*, London: Oxford University Press, 1973.

Churchman, C. W. and A. H. Schainblatt, "The Researcher and the Manager: A Dialectic of Implementation," *Management Science*, Vol. II, No. 4, February 1965, pp. B69–B87.

Clark, B. R., "Interorganizational Patterns in Education," *Administrative Science Quarterly*, Vol. 10, 1965, pp. 224–237.

Clark, Peter, A., *Organizational Design Theory and Practice*, London: Tavistock, 1972.

Coch, L. and J. R. P. French, Jr., "Overcoming Resistance to Change," *Human Relations*, Vol. 1, 1948, pp. 512–532.

Cohen, J. B. and E. Golden, "Informational Social Influence and Product Evaluation," *Journal of Applied Psychology*, Vol. 56, 1972, pp. 54–59.

Coleman, James S., *Equality of Educational Opportunity*, Washington, D.C., U.S. Office of Education, 1966.

Coleman, Raymond J. and M. J. Riley, *Mis: Management Dimensions*, San Francisco: Holden-Day, 1973.

Cook, S., L. A. Kimble, L. M. Hicks, W. J. McGuire, P. H. Schoggen, and M. B. Smith, "Ethical Standards for Psychological Research," *APA Monitor*, Vol. 2, No. 7, 1971, pp. 9–28.

Cook, S., L. A. Kimble, L. M. Hicks, W. J. McGuire, P. H. Schoggen, and M. B. Smith, "Ethical Standards for Research with Human Subjects," *APA Monitor*, Vol. 3, No. 5, pp. i–xix, 1972.

Cooke, Robert and Giorgio Inzerilli, "Trust Relationships in an Academic Environment," working paper, Evanston, Ill.: Northwestern University, 1971.

Coughlan, Robert J., Robert Cooke, and L. Arthur Safer, Jr., "Educational Organization Development: The Survey Feedback and Problem Saving Strategy," Evanston, Ill.: Northwestern University, 1973.

Coughlan, Robert J., Robert Cooke, and L. Arthur Safer, Jr., *An Assessment of a Survey Feedback—Problem Solving—Collective Decision Intervention in Schools*, Final Report to the Office of Education, U.S. Dept. of Health, Education and Welfare, 1972.

Cunningham, James V., *The Resurgent Neighborhood*, Notre Dame, Ind.: Fides Publishers, 1965.

Czepiel, "The Diffusion of Major Technological Innovation in a Complex Industrial Community: An Analysis of the Social Processes in the American Steel Industry," Ph.D. dissertation, Graduate School of Management, Northwestern University, June 1972.

Dahl, Robert, "Further Reflections on the Elitist Theory of Democracy," *American Political Science Review*, Vol. 60, No. 2, June 1966, pp. 296–305.

Dalton, Gene W., Louis B. Barnes, and Abraham Zalesnak, *The Distribution of Authority in Formal Organizations*, Harvard University, Division of Research, Graduate School of Business Administration, Boston, 1968.

Darden, William and Fred Reynolds, "Backward Profiling of Male Innovators," *Journal of Marketing Research*, XI, February 1974.

Dasgupta, R., "Village (or Community) Factors Related to the Level of Agricultural Practice," paper presented at Southern Sociology Society, New Orleans, 1966, p. 23.

Day, George and W. K. Brandt, "Consumer Research and the Evaluation of Information Disclosure Requirements: The Case of Truth-in-Lending," *Journal of Consumer Research*, Vol. 1, June 1974, pp. 21–32.

Day, George and W. K. Brandt, *A Study of Consumer Credit Decisions: Implications for Present and Prospective Legislation*, Palo Alto: Stanford University, 1972.

Doktor, R. H. and W. F. Hamilton, "Cognitive Style and the Acceptance of Management Science Recommendations," *Management Science*, Vol. 19, No. 8, April 1973, pp. 884–894.

Donnelly, James and Michael Etzel, "Degrees of Product Newness and Early Trial," *Journal of Marketing Research*, Vol. V, August 1973.

Downs, Anthony, *Inside Bureaucracy*, Boston: Little, Brown, 1967.

Duncan, Robert, "Characteristics of Organizational Environments and Perceived Environmental Uncertainty," *Administrative Science Quarterly*, Vol. 17, No. 3, 1972a, pp. 313–327.

Duncan, Robert, "Organizational Climate and Climate for Change in Three Police Departments: Some Preliminary Findings," *Urban Affairs Quarterly*, Vol. 8, No. 2, 1972b, pp. 205–245.

Duncan, Robert, David Gilfillan, William Money, and Harold Welsch, *Multihospital Systems: An Evaluation—Part Two—Organizational Studies*, Evanston, Ill.: Health Services Research Center, Northwestern University, 1975.

Eaton, J. W., "Symbolic and Substantive Evaluative Research," *Administrative Science Quarterly*, IV, 1962, pp. 421–442.

Ehrlich, Howard and James Rinehart, "A Brief Report of the Methodology of Stereotype Research," *Social Forces*, Vol. 43, 1965, pp. 564–575.

Eicholz, Gerhard C. and Everett M. Rogers, "Resistance to the Adoption of Audiovisual Aids by Elementary School Teachers: Contrast and Similarities to Agricultural Innovation," in Matthew Miles (Ed.), *Innovation in Education*, New York: Teachers College Press, Columbia University, 1964, pp. 299–316.

Eicholz, Gerhard C., "Why Do Teachers Reject Change?" *Theory into Practice*, Vol. 2, December 1963, pp. 264–268.

Engel, James, Roger Blackwell, and Robert J. Kegerreis, "How Information is Used to Adopt an Innovation," *Journal of Advertising Research*, Vol. 9, No. 4, 1973.

Erikson, Erik, "Identity and the Life Cycle," *Psychological Issues*, Vol. 1, Monograph 1, 1959.

Etzioni, Amitai, "Toward a Theory of Guided Societal Change," *Social Science Quarterly*, Vol. 50, No. 3, December 1969, pp. 749–754.

Etzioni, Amitai and Eva Etzioni-Halery, (Eds.) *Social Change: Sources, Patterns and Consequences*, 2nd edition, New York: Basic Books, 1973.

Eugster, C., "Field Education in West Heights: Equipping a Deprived Community to Help Itself," *Human Organization*, Vol. 23, No. 3.

Evans, Franklin B., "Selling as a Dyadic Relationship—A New Approach," *American Behavioral Scientist*, May 1963.

Felling, W. E., "How Six Regions Organized to Deal with Their Environmental Management Problems," IREM/USIU National Conference, San Diego, February 26–27, 1973.

Filstead, William J., *Qualitative Methodology: Firsthand Involvement with the Social World*, Chicago: Markham Publishing Co., 1970.

Finnigan, Oliver D., *Incentive Approaches in Population Planning Programs*, Office of Health and Public Services, USAID Philippines, c/o American Embassy, Manila, Philippines, June 1972.

Fliegel, Frederick C., "Differences in Prestige Standards and Orientation to Change in a Traditional Agricultural Setting," *Rural Social Society*, Vol. 30, 1965, p. 288.

Flinn, William L., "Influence of Community Values on Innovations," *AJS*, Vol. 76, No. 6, May 1970, pp. 983–991.

Foster, G. M., *Traditional Cultures: And the Impact of Technological Change*, 1962, pp. 75–76.

French, John P. and Bertrand Raven, "The Bases of Social Power," in D. Cartwright (Ed.), *Studies in Social Power*, Ann Arbor: Institute for Social Research, 1959.

Fromkin, Howard, "A Social Psychological Analysis of the Adoption and Diffusion of New Products and Practices from a Uniqueness Motivation Perspective," *Proceedings of 2nd Annual Conference Association for Consumer Research*, 1971.

Frye, N., "The Ethics of Change: The Role of the University," *The School Guidance Worker*, October 1969, pp. 1–12.

Gallaher, Art, Jr., "Directed Change in Formal Organizations: The School System," in Carlson et al. (Eds.), *Change Processes in the Public Schools*, Eugene, Ore.: Center for the Advanced Study of Educational Administration, 1965, pp. 37–54.

Gallaher, Art, Jr., "The Role of the Advocate and Directed Change," in Wesley (Ed.), *Media and Educational Innovation*, Lincoln, Neb.: University of Nebraska Press and Extension Division, 1964.

Gerlach, Luther P. and Virginia H. Hine, *Lifeway Leap: The Dynamics of Change in America*, Minneapolis: University of Minnesota Press, 1973.

Gibb, Jack R., "Defensive Communication," *Journal of Communication*, Vol. XI, No. 3, September 1961, pp. 141–148.

Giffen, Kim, "The Contribution of Studies of Source Credibility to a Theory of Interpersonal Trust in the Communication Process," *Psychological Bulletin*, Vol. 68, 1967, pp. 104–120.

Goldberg, Marvin, "A Cognitive Model of Innovative Behavior: The Interaction of Product and Self-Attitudes," *Proceedings of 2nd Annual Conference Association for Consumer Research*, 1971.

Golembiewski, Robert T. and Arthur Blumberg (Eds.), *Sensitivity Training and Laboratory Approach*, 2nd edition, Itasca, Ill.: F. E. Peacock Publishers, 1973.

Griener, Larry, "Antecedents of Planned Organizational Change," *Journal of Applied Behavioral Science*, Vol. 3, No. 1, 1967, pp. 51–87.

Grinstaff, L., "New Ideas: Conflict and Evolution," *International Journal of Psycho-Analysis*, Vol. L, 1969, pp. 517–528.

Gross, Neal, Joseph B. Giacquinta, and Marilyn Bernstein, *Implementing Organizational Innovation: A Sociological Analysis of Planned Change*, New York: Basic Books, 1971.

Guba, Egon G. and David L. Clark, *The Configurational Perspective: A View of Educational Knowledge Production and Utilization*, Washington, D.C.: Council for Educational Development and Research, Inc., November 1974.

Guba, Egon G., "Development, Diffusion and Evaluation," in Eidell and Kitchel (Eds.), *Knowledge Production and Utilization in Educational Administration*, Eugene, Ore.: CASEA and UCEA, 1968, pp. 37–63.

Guskin, A. and M. Chesler, "Partisan Diagnosis of Social Problems," in Gerald Zaltman (Ed.), *Processes and Phenomena of Social Change*, New York: Wiley-Interscience, 1973, pp. 353–376.

Hage, Jerald and Michael Aiken, *Social Change in Complex Organizations*, New York: Random House, 1970.

Hagen, E. E., *On the Theory of Social Change*, Homewood, Ill.: Dorsey Press, 1962.

Hall, Douglas and Benjamin Schneider, *Organizational Climates and Careers: The Work Lives of Priests*, New York: Seminar Press, 1973.

Hamblin, Robert L., R. Brooke Jacobsen, and Jerry L. L. Miller, *A Mathematical Theory of Social Change*, New York: John Wiley & Sons, 1973.

Hardin, John H., H. Proshansky, B. Kutner, and I. Chein, "Prejudice and Ethnic Relations," in Lindzey and Aronson (Eds.), *The Handbook of Social Psychology*, Vol. V, Reading, Mass.: Addison-Wesley, 1969, pp. 1–76.

Havelock, Ronald G., "Dissemination and Translation Roles," in Eidell and Kitchel (Eds.), *Knowledge Production and Utilization in Educational Administration*, Eugene, Ore.: CASEA and UCEA, 1968, pp. 64–119.

Havelock, Ronald G., *Planning for Innovation*, Ann Arbor: Center for Research on Utilization of Scientific Knowledge, University of Michigan, 1971.

Havelock, Ronald G., "What Do We Know From Research About the Process of Research Utilization?", presented at the International Conference on Making Population/Family Planning Research Useful: The Communicator's Contribution, Honolulu, HI: East-West Communication Institute, December 3–7, 1973.

Havelock, Ronald G. et al., *Planning for Innovation Through Dissemination and Utilization of Knowledge*, CRUSK, Institute for Survey Research, University of Michigan, 1971.

Havelock, Ronald G. and Mary C. Havelock, *Training for Change Agents*, Ann Arbor: Institute for Social Research, University of Michigan, 1973.

Havens, A. Eugene, "Increasing the Effectiveness of Predicting Innovativeness," *Rural Sociology*, Vol. 30, 1965, p. 158.

Heider, Fritz, *The Psychology of Interpersonal Relations*, New York: John Wiley & Sons, 1958.

Hendershot, Gerry E. and James W. Grimm, "Abortion Attitudes Among Nurses and Social Workers," *American Journal of Public Health*, Vol. 64, No. 5, May 1974, pp. 438–441.

Hillman, A. and F. Seever, "Elements of Neighborhood Organization," *Making Democracy Work*, New York: National Federation of Settlements and Neighborhood Centers, 1968.

Homans, George, *Social Behavior: Its Elementary Forms*, 2nd edition, New York: Harcourt, Brace, Jovanovich, 1974.

Hopkins, David S. and Early J. Bailey, "New-Product Pressures, *The Conference Board Record*, June 1971, pp. 16–24.

Hornstein, Harvey, Barbara Bunter, Warner Burke, Marion Gindes, and Roy Forwicke (Eds.), *Social Intervention: A Behavioral Science Approach*, New York: The Free Press, 1971.

Horowitz, E. L., "'Race' Attitudes," in Klineberg (Ed.), *Characteristics of the American Negro*, New York: Harper & Row, 1944, pp. 139–247.

Hovland, C. I., I. L. Hanis, and H. H. Kelley, *Communication and Persuasion*, New Haven: Yale University Press, 1953.

Hovland, C. I., A. Lumsdaine, and F. Sheffield, "Experiments on Mass Communication," Vol. 3, *Studies in Social Psychology* in World War II, Princeton: Princeton University Press, 1949.

Howard, John and Jagdish Sheth, *The Theory of Buyer Behavior*, New York: John Wiley & Sons, 1969.

Huntington, Samual, "Reform and Political Changes," *Political Order in Changing Societies*, New Haven: Yale University Press, 1968, pp. 344–362, quoted on page 279 of G. Zaltman, P. Kotler, I. Kaufman, *Creating Social Change*, New York: Holt, Rinehart & Winston, 1972.

Inzerilli, Giorgio, "An Analysis of Erickson's Concept of Trust," working paper, Evanston, Ill.: Northwestern University, 1970.

Jacoby, Jacob, "Personality and Innovation Proneness," *Journal of Marketing* Research, Vol. 8, May 1971, pp. 244–247.

Jacoby, Jacob, "Multiple-Incident Approach for Studying New Product Adopters," *Journal of Applied Psychology*, Vol. 55, August 1971, pp. 384–388.

Jacobson, Paul B., "The Use of Inter-Institutional Agencies in the Dissemination and Implementation of Education Research," in Goldhammer and Elam (Eds.), *Dissemination and Implementation*, Bloomington, Ind.: Phi Delta Kappa, 1962.

Johnson, D. W., "Influence on Teachers' Acceptance of Change," *Elementary School Journal*, Vol. LXX, December 1969, pp. 143–153.

Jones, Garth N., *Planned Organizational Change: A Study in Change Dynamics*, New York: Frederick A. Praeger, 1969.

Jung, Charles, "Two Kinds of Linkage for Research Utilization in Education," paper presented at the American Educational Research Association Meeting, New York, February 1966.

Jung, Charles and Ronald Lippett, *Utilization of Scientific Knowledge for Change in Education*, Ann Arbor: University of Michigan, CRUSK, 1966.

Kadushin, C., "Individual Decisions to Undertake Psychotherapy," *Administrative Science Quarterly*, Vol. 3, December 1958–59, pp. 379–411.

Kahn, Robert, Donald Wolfe, J. Q. Snoek, and Richard Rosenthal, *Organizational Stress: Studies in Role Conflict and Ambiguity*, New York: John Wiley & Sons, 1964.

Kahneman, D. and E. O. Schild, "Training Agents of Social Change in Israel: Definition of Objectives and a Training Approach," *Human Organization*, Vol. XXV, No. 1, Spring 1966, pp. 71–77.

Kanter, Rosabeth Moss, *Communes: Creating and Managing the Collective Life*, New York: Harper & Row, 1973.

Kasulis, Jack J., *Cognitive Structure, Segmented Analysis and Communication Effectiveness: A Field Study of Cognitive Complexity Theory,* unpublished Ph.D. thesis, Northwestern University, 1975.

Katz, Daniel and Kenneth Braley, "Racial Stereotypes of One Hundred College Students," *Journal of Abnormal and Social Psychology,* Vol. 28, 1933, pp. 280–290.

Katz, Daniel and Robert Kahn, *The Social Psychology of Organizations,* New York: John Wiley & Sons, 1966.

Kelman, H. C., "The Relevance of Social Research to Social Issues: Promises and Pitfalls," in P. Halmos (Ed.), *The Sociology of Sociology* in the Sociological Review Monograph, Monograph No. 16, The University of Keele, Great Britain, 1970, pp. 77–100.

Kelman, Herbert C., "Compliance, Identification, and Internalization: Three Processes of Attitude Change," *Journal of Conflict Resolution,* Vol. 2, 1958, pp. 51–60.

Kelman, Herbert and Donald P. Warwick, "Bridging Micro and Macro Approaches to Social Change: A Social-Psychological Perspective," in G. Zaltman (Ed.), *Processes and Phenomena of Social Change,* New York: Wiley Interscience, 1973, pp. 13–60.

Kennedy, Dennis, "The Semantic Differential Technique and the Measurement of Adoption Stages," unpublished working paper, Northwestern University, 1974.

Kiesler, Charles, *The Psychology of Commitment,* New York: Academic Press, 1972.

Kiesler, Charles A. and Sara B. Kiesler, *Conformity,* Reading, Mass.: Addison-Wesley, 1970.

King, Bert T. and Elliott McGinnies, *Attitudes, Conflict, and Social Change,* New York: Academic Press, 1972.

King, T. R., "An Experimental Study of the Effect of Ethos upon the Immediate and Delayed Recall of Information," *Central States Speech Journal,* Vol. 17, 1966, pp. 22–28.

Kivlin, Joseph E., *Correlates of Family Planning in Eight Indian Villages,* East Lansing, Mich.: Michigan State University, Diffusion of Innovations Research Department, 1968, p. 38.

Klein, Donald, "Some Notes on the Dynamics of Resistance to Change: The Defender Role," in Watson (Ed.), *Concepts for Social Change,* Washington, D.C.: National Training Laboratories NEA for COPED, 1967.

Klineberg, O., *Tensions Affecting International Understanding,* Bulletin 62, New York: Social Science Research Council, 1950.

Kloghan, Gerald and Walter Coward, Jr., "The Concept of Symbolic Adoption: A Suggested Interpretation," *Rural Sociology,* Vol. 30, March 1970, pp. 77–83.

Kohn, Carol A. and Jacob Jacoby, "Patterns of Information Acquisition in New Product Purchases," *Journal of Consumer Research,* Vol. 1, No. 2, 1974.

Kolb, David A., Irwin M. Rubin, and James M. McIntyre (Eds.), *Organizational Psychology—A Book of Readings,* 2nd edition, Englewood Cliffs, N.J.: Prentice-Hall, 1974.

Kothandapani, V., "Validation of Feeling, Belief and Intention to Act as Three Components of Attitude and Their Contribution to Prediction of Contraceptive Behavior," *Journal of Personality and Social Psychology,* Vol. 19, September 1971, pp. 321–333.

Kotler, Philip and Gerald Zaltman, "The Theory of the Best Project," *Journal of Advertising Research,* February, 1976.

Krebs, Dennis, "Empathy and Altruism," *Journal of Personality and Social Psychology,* Vol. 32, No. 6, 1975, pp. 1134–1146.

Kurtz, R. A., H. P. Chalfont, and K. Kaplan, "Inner City Residents and Health Decision-

Makers: Perceptions of Health Problems and Solutions," *American Journal of Public Health*, Vol. 64, No. 6, June 1974, pp. 612–613.

Lake, Dale and Daniel Callahan, "Entering and Intervening in Schools," in R. Schmuck and M. Miles (Eds.), *Organization Development in Schools*, Palo Alto, Calif.: National Press Books, 1971, pp. 139–154.

Lambert, Zarrel, "Perception Patterns, Information Handling, and Innovativeness," *Journal of Marketing Research*, Vol. IX, November 1972.

Landy, D., "Problems of Persons Seeking Help in Our Culture," in M. N. Zald (Ed.), *Social Welfare Institutions*, New York: John Wiley & Sons, 1965, pp. 559–574.

Lauer, Robert H., *Perspectives on Social Change*, Boston: Allyn & Bacon, 1973.

Lavidge, R. J. and G. A. Steiner, "A Model for Predictive Measurements of Advertising Effectiveness," *Journal of Marketing*, Vol. 25, October 1961, pp. 59–62.

Lawler, Edward E., III, *Motivation in Work Organizations*, Monterray, Cal.-Brooks/Cole Publishing Co., 1973.

Lawrence, Paul R. and Jay W. Lorsch, *Organization and Environment*, Homewood, Ill.: Richard D. Irwin, 1969.

Lazarsfeld, Paul F. and Jeffrey G. Reitz, *An Introduction to Applied Sociology*, New York: American Elsevier, 1975.

Lee, Kam Han, "Social Marketing Strategies and Nutrition Education," unpublished Ph.D. dissertation, Northwestern University, 1975.

Lemert, J. B., "Dimensions of Source Credibility," paper presented at the meeting of the Association for Education in Journalism, 1963.

Lerner, Daniel, "is International Persuasion Sociologically Feasible?" *The Annuals*, Vol. 398, November 1971, pp. 44–49.

Lerner, D., *The Passing of Traditional Society*, New York: The Free Press, 1958.

Levenson, Harry, Janice Molinare, and Andrew G. Spahn, *Organizational Diagnosis*, Cambridge, Mass.: Harvard University Press, 1972.

Likert, Rensis, *The Human Organization: Its Management and Value*, New York: McGraw-Hill, 1967.

Lin, Nan and Gerald Zaltman, "On the Nature of Innovations," in G. Zaltman (Ed.), *Processes and Phenomena of Social Change*, New York: Wiley Interscience, 1973.

Lin, Nan and Ronald Burt, "Differential Effect of Information Channels in the Process of Innovation Diffusion," *Social Forces*, Vol. 54, No. 1, September 1975, pp. 256–274.

Lippitt, Gordon, *Visualizing Change: Model Building and the Change Process*, Fairfax, Va.: NTL Learning Resources Corp., 1973.

Lippitt, Ronald, Jean Watson, and Bruce Wesley, *The Dynamics of Planned Change*, New York: Harcourt, Brace & Jovanovich, 1958, p. 77.

Lippitt, Ronald and Ronald Havelock, "Needed Research on Research Utilization," in *Research Implications for Educational Diffusion*, Lansing, Mich.: Michigan Department of Education, 1968.

Lippitt, Ronald, et al., "Design for a Research Utilization System for the Social and Rehabilitation Service," in R. Havelock (Ed.), *A Report to the Social and Rehabilitation Service*, Center for Research on Utilization of Scientific Knowledge, University of Michigan, September 1973.

Liska, A. E., "Emergent Issues in the Attitude-Behavior Consistency Controversy," *American Sociological Review*, Vol. 39, April 1974, pp. 261–272.

Liska, A. E., *The Impact of Attitude on Behavior: The Consistency Controversy*, Cambridge, Mass.: Schenkman, 1975.

Littaner, D., A. F. Wessen, and J. Goldman, "Evaluating Change in Systems of Child Feeding," in R. M. Coe (Ed.), *Planned Change in the Hospital*, New York: Praeger, 1970.

Litwak, "An Approach to Linkage in 'Grass Roots' Community Organization," paper delivered at the Leadership Institute of the National Conference on Social Welfare.

Loewenberg, Frank M., "Toward a Systems Analysis of Social Welfare Manpower Utilization Patterns," in F. M. Loewenberg and R. Dolgoff (Eds.), *The Practice of Social Intervention: Roles, Goals, and Strategies*, Itasca, Ill.: F. E. Peacock, 1972.

Loewenberg, Frank M. and Ralph Dolgoff (Eds.), *The Practice of Social Intervention: Roles, Goals, and Strategies*, Itasca, Ill.: F. E. Peacock, 1972.

Logan, M. H., "Humoral Medicine in Guatemala and Peasant Acceptance of Modern Medicine," *Human Organization*, Vol. 32, No. 4, Winter 1973, pp. 385–396.

Loomis, Charles P., "In Praise of Conflict and Its Resolution," *American Sociological Review*, Vol. 32, 1967, p. 25.

Lowenthal, J., "The Technical Assistance Process Reconsidered," Graduate School of Management, Vanderbilt University, March 1973.

Lyons, Gene M. (Ed.), *Social Research and Public Policy*, Hanover, N.H.: The Public Affairs Center, Dartmouth College, 1975.

Mackie, R. R. and P. R. Christensen, *Translation and Application of Psychological Research*, Technical Report 716-1, Personnel and Training Branch, Psychological Services Division, Office of Naval Research, 1967.

Mahler, Walter, *Diagnostic Studies*, Reading, Mass.: Addison-Wesley, 1974.

Maloney, John C. and Eugene P. Schonfeld, "Social Change and Attitude Change," in G. Zaltman (Ed.), *Processes and Phenomena of Social Change*, New York: Wiley Interscience, 1973.

Management of New Products, 4th edition, New York: Booz, Allen & Hamiltin, 1965, p. 9.

Mancuso, Joseph, "Why Not Create Opinion Leaders for New Product Introductions?" *Journal of Marketing*, Vol. 33, July 1969.

March, James G. and Herbert A. Simon, *Organizations*, New York: John Wiley & Sons, 1958.

Marrow, A., D. Bowers, and S. Seashore, *Management by Participation*, New York: Harper & Rows, 1967.

Maslow, Abraham, *Motivation and Personality*, 2nd edition, New York: Harper & Row, 1970.

Mason, Deborah, "Egypt Pushes Drive for Family Planning," *The Christian Science Monitor*, Tuesday, April 23, 1974, p. F10.

McClelland, David C., *The Achieving Society*, Princeton, N.J.: Van Nostrand, 1961.

McCrosky, J. C., "Scales for the Measurement of Ethos," *Speech Monographs*, Vol. 33, 1966, pp. 65–72.

McElvaney, Charles and Matthew Miles, "Using Survey Feedback and Consultation," in R. Schmuck and M. Miles (Eds.), *Organizational Development in Schools*, Palo Alto, Calif., National Press Books, 1971, pp. 113–138.

McGuire, William J., "Attitudes and Opinions," *Annual Review of Psychology*, Vol. 17, 1966, pp. 475–514.

McKinlay, J. B., "Some Approaches and Problems in the Study of the Use of Services—An

Overview," *Journal of Health and Social Behavior*, Vol. 13, No. 2, June 1972, pp. 115–121.

Mehrabian, Albert, *Tactics of Social Influence*, Englewood Cliffs, N.J.: Prentice-Hall, 1970.

Menges, R. J., "Openness and Honesty Versus Coercion and Deception in Psychological Research," *American Psychologist*, Vol. 28, No. 12, 1973, pp. 1030–1034.

Merton, Robert K., *Social Theory and Social Structure*, New York: The Free Press, 1968, enlarged edition.

Merton, Robert K., Patricia S. West, and M. Jahoda, *Social Fictions and Social Fact: The Dynamics of Race Relations in Hilltown*, New York: Columbia University Bureau of Applied Social Research, 1949.

Miles, Matthew B., "On Temporary Systems," in M. Miles (Ed.), *Innovation in Education*, New York: Bureau of Publications, Teachers College, Columbia University, 1964.

Miles, Matthew B., H. A. Hornstein, D. M. Callahan, P. H. Calder, and R. S. Schiavo, "The Consequence of Survey Feedback: Theory and Evaluation," in W. Bennis, K. Benne, and D. Chin (Eds.), *The Planning of Change*, New York: Holt, Rinehart & Winston, 1969, pp. 457–467.

Miro, Carmen A. and Donald J. Bogue, "A Comparative Study of Fertility in Major Cities of Latin America," working paper, Community and Family Study Center, University of Chicago.

Mogulof, M. B., "Federal Support for Citizen Participation in Social Action," *The Social Welfare Forum 1969*, New York: Columbia University Press, 1969, pp. 86–107.

Monteiro, L., "Expense is No Object. . . : Income and Physician Visits Reconsidered," *Jour al of Public Health*, Vol. 14, No. 2, June 1973, pp. 99–114.

Morrison, Denton, "Some Notes Toward Theory on Relative Deprivation, Social Movements, and Social Change," in G. Zaltman (Ed.), *Processes and Phenomena of Social Change*, New York: Wiley Interscience, 1973, pp. 153–164.

Myers, James and Thomas S. Robertson, "Stability of Self-Designated Opinion Leadership," in S. Ward and P. Wright (Eds.), *Advances in Consumer Research*, 1974.

Myers, Sumner and D. G. Marquis, *Successful Commercial Innovations*, Washington, D.C.: National Science Foundation, 1969.

Mitnoff, Ian, "On Mutual Understanding and the Implementation Problem: A Philosophical Case Study of the Psychology of the Apollo Moon Scientists," paper presented at the Implementation of OR/MS Models: Theory, Research and Application, University of Pittsburgh, Graduate School of Business, November 15–17, 1973.

Newcomb, T. M., "Autistic Hostility and Social Reality," *Human Relations*, Vol. 1, 1947, pp. 69–86.

Niehoff, Arthur and J. Charnel Anderson, "The Process of Cross-Cultural Innovation," *International Developments Review*, Vol. 6, 1964, pp. 5–11.

Niehoff, Arthur (Ed.), *A Casebook of Social Change*, Chicago: Aldine-Atherton, 1966.

Niehoff, Arthur and J. Charnel Anderson, "Peasant Fatalism and Socio-Economic Innovation, *Human Organization*, Vol. 25, 1965, pp. 273–283.

Niehoff, Arthur and J. Charnel Anderson, *A Selected Bibliography of Cross-Cultural Change Projects*, Alexandria, Va.: George Washington University Human Resources Office, 1964.

O'Brian, Terrance, "Stages of Consumer Decision Making," *Journal of Marketing Research*, Vol. VIII, August 1971, pp. 283–290.

O'Donnell, E. J. and C. S. Chilman, "Poor People on Public Welfare Boards and Committees: Participation in Policy-Making?", Welfare in Review, May–June 1969, pp. 1–10.

Okediji, Francis O., "Overcoming Social and Cultural Resistances," International Journal of Health Education, Vol. 15, No. 3, July–September 1972, pp. 3–10.

O'Neill, B., et al., "Automobile Head Restraints—Frequency of Neck Injury Claims in Relation to the Presence of Head Restraints," American Journal of Public Health, Vol. 62, March 1972, pp. 399–406.

Operations Research Society of America, "Guidelines for the Practice of Operations Research," Operations Research, Vol. 19, No. 5.

Ostlund, Lyman, "Identifying Early Buyers," Journal of Advertising Research, Vol. 12, No. 2, April 1972.

Pareek, Udai and Y. P. Singh, "Communication Nets in the Sequential Adoption Process," Indian Journal of Psychology, Vol. 44, 1969, pp. 33–55.

Pessemier, Edgar A., New-Product Decisions: An Analytical Approach, New York: McGraw-Hill, 1966.

Pessemier, Edgar A. and H. Paul Root, "The Dimensions of New Product Planning," Journal of Marketing, January 1973, pp. 10–18.

Pizam, Abraham, "Psychological Characteristics of Innovators," European Journal of Marketing, Vol. 6, No. 3, 1972.

Pope, C. R., S. S. Yoshioka, and M. D. Greenlick, "Determinants of Medical Care Utilization: The Use of the Telephone for Reporting Symptoms," Journal of Health and Social Behavior, Vol. 12, June 1971, pp. 155–162.

Population Task Force of the Institute of Society, Ethics and the Life Sciences, Ethics, Population and the American Tradition, a study prepared for the Commission on Population Growth and the American Future by the Institute of Society, Ethics and the Life Sciences, Hastings-on-Hudson, New York, 1970.

Pressman, Jeffrey L., "Decision Makers and Evaluations: Some Differences in Perspective and Possible Directions for the Future," in G. J. Lyons (Ed.), Social Research and Public Policy, Hanover, N.H.: The Public Affairs Center, Dartmouth College, 1975.

Presthus, R., Men at the Top: A Study in Community Power, New York: Oxford University Press, 1964.

Pugh, D. S., D. Hickson, C. Hinings, K. MacDonald, C. Turner, and T. Lupton, "A Conceptual Scheme for Organizational Analysis," Administrative Science Quarterly, Vol. 8, 1963, pp. 289–315.

Radnor, Michael and Robert Coughlan, "A Training and Development Program for Administrative Change in School Systems," paper presented at the American Educational Research Association Meeting, Chicago, April 1972.

Ram, N. V. Raghu, Perspectives in Family Planning, Bella Vista: Administrative Staff College of India, Hyderabad-4, March 1972, pp. 89–90.

Report of the Third Advisory Group Meeting, Inter-Governmental Coordinating Committee (IGCC), Kuala Lumpur, Malaysia, July 1973.

Resnick, J. H. and T. Schwartz, "Ethical Standards as an Independent Variable in Psychological Research," American Psychologist, Vol. 28, No. 2, 1973, pp. 134–139.

Reynolds, Fred and William Darden, "Predicting Opinion Leadership for Women's Clothing Fashions," Combined Proceedings: Marketing Education and the Real World and Dynamic Marketing in a Changing World, American Marketing Association, 1972.

RITE Number 1, Philadelphia: Research for Better Schools, Inc., 1967.

Robertson, Leon, "Factors Associated with Safety Belt Use in 1974 Starter-Interlock Equipped Cars," *Journal of Health and Social Behavior*, Vol. 16, No. 2, June 1975, pp. 173–177.

Robertson, Leon S. et al., "A Controlled Study of the Effect of Television Messages on Safety Belt Use," *American Journal of Public Health*, Vol. 64, No. 11, November 1974, pp. 1071–1080.

Robertson, Leon S. and R. Hadden, Jr., "The Buzzer-Light Reminder System and Safety Belt Use," *American Journal of Public Health*, Vol. 64, No. 8, August 1974, pp. 814–815.

Robertson, Thomas S., "Determinants of Innovative Behavior," paper presented at the American Marketing Association, Washington, D.C., 1967.

Robertson, Thomas S., *Innovative Behavior and Communication*, New York: Holt, Rinehart & Winston, 1971, pp. 73–77.

Robertson, Thomas S. and James N. Kennedy, "Prediction of Consumer Innovators: Application of Multiple Discriminant Analysis," *Journal of Marketing Research*, Vol. 5, February 1968, pp. 64–69.

Robertson, Thomas S. and John R. Rossiter, "Fashion Diffusion: The Interplay of Innovator and Opinion Leader Roles in College Social Systems," unpublished paper, Graduate School of Business Administration, University of California, 1968.

Robinson, John P., Robert Athanasion, and Kendra Head, *Measures of Occupational Attitudes and Occupational Characteristics*, Ann Arbor: Institute for Social Research, University of Michigan, 1969.

Rodman, Hyman and Ralph Kolodny, "Organizational Strains in the Researcher–Practitioner Relationship," *Human Organization*, Vol. 23, 1964, pp. 171–182.

Roethlisberger, F. J. and W. J. Dickson, *Management and the Worker*, Cambridge, Mass.: Harvard University Press, 1939.

Rogers, Everett M., Nan Lin, and Gerald Zaltman, "Design for a Research Utilization System for the Social and Rehabilitation Service," in Ronald G. Havelock, *Ideal Systems for Research Utilization: Four Alternatives*, Social and Rehabilitation Service, U.S. Department of Health, Education and Welfare, March 1974.

Rogers, Everett M., *Communication Strategies for Family Planning*, New York: The Free Press, 1973.

Rogers, Everett M. and Rekha Argawala-Rogers, *Communication in Organizations*, New York: The Free Press, 1976.

Rogers, Everett M., J. D. Eveland, and Alden Bean, "The Agricultural Extension Model and Its Extension to Other Innovation-Diffusion Problems," working paper prepared for the National Science Foundation, May 1974.

Rogers, Everett M., *Diffusion of Innovations*, New York: The Free Press, 1962.

Rogers, Everett M., *Family Planning Communication Strategies*, New York: The Free Press, 1973.

Rogers, Everett M., *Modernization Among Peasants: The Impact of Communication*, New York: Holt, Rinehart & Winston, 1969.

Rogers, Everett M., "Strategies for Linking Research and Use," *What Do We Know From Research About the Process of Research Utilization?*, presented at the International Conference on Making Population/Family Planning Research Useful: The Communicator's Contribution, East-West Communication Institute, 1973.

Rogers, Everett M. and D. K. Bhowmik, "Homophily-Heterophily: Relational Concepts for Communication Research," *Public Opinion Quarterly,* Vol. 34, 1971, pp. 523–538.

Rogers, Everett M. and E. B. de Ramos, "Prediction of the Adoption of Innovations: A Progress Report," paper presented at the Rural Sociology Society, Chicago, 1965, p. 7.

Rogers, Everett M. and Nemi Jain, "Needed Research on Diffusion Within Educational Organizations," in *Research Implications for Educational Diffusion,* Lansing, Mich.: Michigan Department of Education, 1968.

Rogers, Everett M. and F. Floyd Shoemaker, *Communication of Innovations: A Cross-Cultural Approach,* New York: The Free Press, 1971.

Rokeach, Milton, Beliefs, *Attitudes, and Values,* San Francisco: Jossey Bass, 1968.

Rostow, W. W., *The Stages of Economic Growth,* New York: Cambridge University Press, 1960.

Rothman, Jack, *Planning and Organizing for Social Change: Action Principles from Social Science Research,* New York: Columbia University Press, 1974.

Rothman, Jack, John L. Erlich, and Joseph G. Teresa, *Promoting Innovation and Change in Organizations and Communities,* New York: John Wiley & Sons, 1976.

Rubin, Irwin et al., "Initiating Planned Change in Health Care Systems," *JABS,* Vol. 10, No. 1, 1974, p. 108.

Sandhu, H. S. and D. E. Allen, "The Village Influence on Punjabi Farm Modernization," *AJS,* Vol. 79, No. 4, January 1974, pp. 967–980.

Sarbin, T. R., R. T. Taft, and D. E. Bailey, *Clinical Inference and Cognitive Theory,* New York: Holt, Rinehart & Winston, 1960.

Schaller, L. S., *Community Organization: Conflict and Reconciliation,* Nashville: Abdington Press, 1966.

Schein, Edgar H., *Organizational Psychology,* Englewood Cliffs, N.J.: Prentice-Hall, 1965.

Schein, Edgar H. and Warren G. Bennis, *Personal and Organizational Change Through Group Methods: The Laboratory Approach,* New York: John Wiley & Sons, 1965.

Schein, Edgar H., "The Mechanisms of Change," in W. G. Bennis, K. D. Benne and R. Chin (Eds.), *The Planning of Change,* 2nd edition, New York: Holt, Rinehart & Winston, 1969.

Schein, Edgar H., *Process Consultation: Its Role in Organization Development,* Reading, Mass.: Addison-Wesley, 1969.

Schein, Edgar H., *Organization Psychology,* Englewood Cliffs, N.J.: Prentice-Hall, 1970.

Schewing, Eberhard E., *New Product Management,* Hinsdale, Ill.: Dryden Press, 1974.

Schiffman, Leon, "Perceived Risk in New Product Trial by Elderly Consumers," *Journal of Marketing Research,* Vol. IX, February 1972.

Schiffman, Leon and Vincent Gaccione, "Opinion Leaders in Institutional Markets," *Journal of Marketing,* Vol. 38, No. 2, April 1974.

Schmuck, Richard and Philip Runkel, *Organizational Training for a School Faculty,* Eugene, Ore.: CASEA, 1970.

Schmuck, Richard, "Social Psychological Factors in Knowledge Utilization," in K. Eidell and R. Kitchel (Eds.), *Knowledge Production and Utilization in Educational Administration,* Eugene, Ore.: CASEA and UCEA, 1968, pp. 143–173.

Schulberg, Herbert C. and Frank Baker, "Program Evaluation Models and the Implementation of Research Findings," *American Journal of Public Health,* Vol. 58, 1968, pp. 1248–1255.

Schulberg, Sheldon and F. Baker (Eds.), *Program Evaluation in the Health Fields,* New York: Behavioral Publications, 1969.

Schultz, Randall L. and Dennis Slevin (Eds.), *Implementing Operations Research/Management Science,* New York: American Elsevier, 1975.

Scott, Joh. P. and Sarah F. Scott (Eds.), *Social Control and Social Change,* Chicago: The University of Chicago Press, 1971.

Scurrah, Martin, Moshe Shani, and Carl Zipfel, "Influence of Internal and External Change Agents in a Simulated Educational Organization," *Administrative Science Quarterly,* Vol. 16, No. 1, 1971, pp. 113–121.

1972 Service Analysis, Department of Human Resources, City of Chicago, Chicago, Ill.

Sherif, Carolyn W., Muzafer Sherif, and Roger E. Nebergall, *Attitude and Attitude Change,* Philadelphia, Pa.: W. B. Saunders, 1965.

Sieber, Sam D., K. S. Louis, and L. Metzger, *The Use of Educational Knowledge,* New York: Bureau of Applied Social Research, Columbia University, 1972.

Slocum, John and Herbert Rand, "Human Relations Training for Middle Management: A Field Experiment," *Academy of Management Journal,* Vol. 13, No. 4, 1971, pp. 403–410.

Smith, Anthony D., *The Concept of Social Change,* London: Routledge & Kegan Paul, 1973.

Smith, Donald R., "A Theoretical and Empirical Analysis of the Adoption-Diffusion of Social Change," Ph.D. thesis, Baton Rouge, La.: Louisiana State University, 1968.

Spindler, George (Ed.), *Education and Culture: Anthropological Approaches,* New York: Holt, Rinehart & Winston, 1963.

Stankey, G. H., J. C. Hendee, and R. N. Clark, "Applied Social Research can Improve Participation in Resource Decision Making," *Rural Sociology,* Vol. 40, No. 1, Spring 1975, pp. 67–74.

Steffire, Volney, *New Products and New Enterprises: A Report on an Experiment in Applied Social Science,* Irvine: University of California, March 31, 1971.

Stein, Bruno and S. M. Miller (Eds.), *Incentives and Planning in Social Policy,* Chicago: Aldine-Atherton, 1973.

Stephenson, Robert W., Benjamin S. Gantz, and Clara E. Erickson, "Development of Organizational Climate Inventories for Use in R&D Organizations," *IEEE Transactions on Engineering Management,* EM-18, No. 2, 1971.

Stiles, Lindley and Beecham Robinson, "Change in Education," in G. Zaltman and R. Schwartz (Eds.), *Perspectives on Social Change,* New York: Wiley Interscience, 1973.

Stogdill, Ralph M. (Ed.), *The Process of Model Building,* Columbus, Ohio: Ohio State University Press, 1970.

Suchman, Edward A., *Evaluative Research: Principles and Practice in Public Service and Social Action Programs,* New York: Russell Sage Foundation, 1967.

Summers, Gene F., "Introduction," in Gene F. Summers (Ed.), *Attitude Measurement,* Chicago: Rand McNally, 1970.

Sweetzer, D. A., "Attitudinal and Social Factors Associated with Use of Seat Belts," *Journal of Health and Social Behavior,* Vol. 8, June 1967, pp. 116–125.

Szybillo, George J., "A Situational Influence on the Relationship of a Consumer Attributed to New-Product Attractiveness," *Journal of Applied Psychology,* Vol. 60, No. 5, 1975, pp. 652–655.

Tauber, E. M., "Reduce New Product Failures: Measure Needs as Well as Purchase Interest," *Journal of Marketing,* Vol. 37, No. 3, July 1973, pp. 61–64.

Taylor, James, "The Role of Risk in Consumer Behavior," *Journal of Marketing*, Vol. 38, No. 2, April 1974.

Telfer, R. G., "Dynamics of Change," *The Clearing House*, Vol. XLI, November 1966, pp. 131–135.

Theodore, C., "The Demand for Health Services," in R. Anderson (Ed.), *A Behaviorial Model of Families' Use of Health Services*, Series No. 25, Chicago: University of Chicago, Center for Health Administration Studies Research, 1968.

Thompson, James D., *Organizations in Action*, New York: McGraw-Hill, 1967.

Triandis, Harry C., "Cultural Influences Upon Cognitive Processes," in Leonard Berkowitz (Ed.), *Advances in Experimental Social Psychology*, Vol. 1, New York: Academic Press, 1964.

Triandis, Harry C., "The Impact of Social Change on Attitude," in B. T. King and E. McGinnies (Eds.), *Attitudes, Conflict & Social Change*, New York: Academic Press, 1972, pp. 127–136.

Tripodi, T. and H. Miller, "The Clinical Judgment Process: A Review of the Literature," *Social Work*, Vol. 11, No. 3, 1966, pp. 63–69.

Turnbull, Brenda J., Lorraine I. Thom, and C. L. Hutchins, *Promoting Change in Schools: A Diffusion Casebook*, San Francisco: Far West Laboratory for Educational Research and Development, 1974.

Turner, Howard M., Jr., *The People Motivators*, New York: McGraw-Hill, 1973.

Twedt, Dik Warren, "How Important to Marketing Strategy is the 'Heavy User'?", *Journal of Marketing*, January 1964, pp. 71–72.

Tybout, A. and G. Zaltman, "Ethics in Marketing Research: Their Practical Relevance," *Journal of Marketing Research*, Vol. XI, November 1974, pp. 357–368.

Uhl, K., R. Andrus and L. Poulsen, "How Are Laggards Different? An Empirical Inquiry," *Journal of Marketing Research*, Vol. VII, February 1970, pp. 51–54.

Unman, David B., *New Product Programs: Their Planning and Control*, New York: American Management Association, 1969.

Urban, Glen L., "SPRINTER MOD III: A Model for the Analysis of New Frequently Purchased Consumer Products," *Operations Research*, September–October 1970, pp. 805–854.

USAID, "Selecting Effective Leaders of Technical Assistance Teams," Bureau for Technical Assistance, Agency for International Development, Washington, D.C., March 1973.

Van Willigren, J., "Concrete Means and Abstract Goals: Papago Experiences in the Application of Development Resources," *Human Organization*, Vol. 32, No. 1, Spring 1973, pp. 1–8.

Walton, Richard E., "Two Strategies of Social Change and Their Dilemmas," *The Journal of Applied Behavioral Science*, Vol. ii, April–June 1965, pp. 167–179.

Warwick, D. and H. Kelman, "Ethical Issues in Social Intervention," in Gerald Zaltman (Ed.), *Processes and Phenomena of Social Change*, New York: Wiley Interscience, pp. 377–418.

Watson, Goodwin, "Resistance to Change," *American Behavioral Scientist*, Vol. 14, May–June 1971, pp. 745–766.

Watson, Goodwin, "Resistance to Change," in Watson (Ed.), *Concepts for Social Change*, Washington, D.C.: National Training Laboratories NEA for COPED, 1967, pp. 10–25.

Watson, Jeanne, "Some Social and Psychological Situations Related to Change in Attitude," *Human Relations*, Vol. 3, 1950, pp. 15–56.

Weber, Max, *The Theory of Social and Economic Organization*, A. M. Henderson and Talcott Parsons (translators), New York: The Free Press, 1947.

Weinstein, A. G., "Predicting Behavior from Attitudes," *POQ*, Vol. 36, Fall 1974, pp. 355–360.

Weiss, Robert S. and Martin Rein, "The Evaluation of Broad-Aim Programs: Experimental Design, Its Difficulties, and an Alternative," *Administrative Science Quarterly*, Vol. 16, 1971, pp. 97–109.

Westinghouse Population Center, Health System Division, *Distribution of Contraceptives in the Commercial Sector of Selected Developing Countries: Summary Report*, April 1974, pp. 38–49.

Whitford, W. C., "The Functions of Disclosure Regulation in Consumer Transactions," *Wisconsin Law Review*, Vol. 31, No. 2, 1973, pp. 400–470.

Wicker, A. W., "Attitudes Versus Actions: The Relationship of Verbal and Overt Behavior Responses to Attitude Objects," *Journal of Social Issues*, Vol. 25, Autumn 1966, pp. 41–78.

Wicker, A. W., "An Examination of the 'Other Variables' Explanation of Attitude-Behavior Inconsistency," *Journal of Personality and Social Psychology*, Vol. 19, July, pp. 18–30.

Williams, A. F., "Personality and Other Characteristics Associated with Cigarette Smoking Among Young Teenagers," *Journal of Health and Social Behavior*, Vol. 14, December 1973, pp. 374–380.

Woodward, Joan, *Industrial Organization: Theory and Practice*, London: Oxford University Press, 1965.

Zaltman, Gerald, "Introduction," in B. Sternthal and G. Zaltman (Eds.), *Broadening the Concept of Consumer Behavior*, monograph of the Association for Consumer Research, 1974.

Zaltman, Gerald, "Research Utilization Techniques Borrowed from Marketing Relevant to Communication Research in Population/Family Planning," presented at the *International Conference on Making Population/Family Planning Research Useful: The Communicator's Contribution*, East-West Communication Institute, Honolulu, Hawaii, December, 1973.

Zaltman, Gerald, "An Assessment of a Proposed Trade Regulation Rule on Food Advertising," unpublished report, University of Pittsburgh, July 1975.

Zaltman, Gerald, David Florio, and Linda Sikorski, *Creating Educational Change*, New York: The Free Press, 1977.

Zaltman, Gerald, Philip Kotler, and Ira Kaufman (Eds.), *Creating Educational Change*, New York: Holt, Rinehart & Winston, 1972.

Zaltman, Gerald, Christian Pinson, and Rinehart Angleman, *Metatheory and Consumer Research*, New York: Holt, Rinehart & Winston, 1973.

Zaltman, Gerald, *Marketing: Contributions from the Behavioral Sciences*, New York: Harcourt, Brace and Jovanovich, 1965.

Zaltman, Gerald, "New Perspectives on Diffusion Research," in David Gardner (Ed.), *Proceedings of the Association for Consumer Research*, 1971.

Zaltman, Gerald and Bernard Dubois, "New Conceptual Approaches in the Study of Innovation," in D. Gardner (Ed,), *Proceedings of the Second Annual Conference Association for Consumer Research*, 1971.

Zaltman, Gerald, Robert Duncan, and Jonny Holbek, *Innovations and Organizations*, New York: Wiley Interscience, 1973.

Zaltman, Gerald and Nan Lin, "On the Nature of Innovations," *American Behavioral Scientist*, Vol. 14, May–June 1971, pp. 651–673.

Zaltman, Gerald and Christian Pinson, "Empathetic Ability and Adoption Research," working paper, Northwestern University, 1974.

Zaltman, Gerald and Christian Pinson, "Perception of New Product Attributes," working paper, Northwestern University, 1974.

Zaltman, Gerald and Ronald Stiff, "Theories of Diffusion," in S. Ward and T. Robertson (Eds.), *Consumer Behavior: Theoretical Sources,* Englewood Cliffs, N.J.: Prentice-Hall, 1973.

Zimbardo, Philip G., "The Tactics and Ethics of Persuasion," in B. T. King and E. McGinnies, Attitudes, Conflict, and Social Change, New York: Academic Press, 1972.

PRINCIPLES OF PLANNED SOCIAL CHANGE

This appendix collects all the principles of planned social change which are found at the end of each chapter.* These are far from exhaustive. Undoubtedly some readers will find principles they know to be important missing from the list. Undoubtedly, too, some readers will have an example or experience that provides an exception to certain principles. These principles are provided as rules of thumb. They represent the experience of many people: each principle has been found important in at least a few different settings. Many of the principles may appear too obvious to mention. Some of them strike us this way. However, it is surprising how often a seemingly obvious principle has been ignored, with very unfortunate consequences in a planned social change setting. If some principles are seemingly obvious, they are apparently also easy to forget.

Chapter 1: Introduction to Social Change

This chapter can be summarized by the following principles:

1. In defining the concept of change and innovation, the process is one of relearning on the part of an individual or group (1) in response to newly perceived requirements of a given situation requiring action and (2) that results in changes in the structure and/or functioning of social systems.
2. In planning for change, the change agent and client system should be aware that, in addition to planned change, there is also unplanned change. Unplanned change occurs as a result of the interaction among forces in the social system. However, it is brought about with no apparent deliberateness and no coordinated goal setting on the part of those involved in it.
3. The change agent should distinguish between those affected by the change and that subpart of those affected who may be seeking assistance. The change target system is that unit in which the change agent is trying to alter the status quo such that the individual, group, or organization must relearn how to perform their activities. The change client system is comprised of individuals requesting assistance in altering the status quo.
4. An individual or group can be considered to be operating in the change agent role when they are working to change the status quo in the

---

* Chapter 12 does not contain summary principles.

change target system such that the individuals involved must relearn how to perform their roles.

5. Change agents and the client system should avoid the rationalistic bias in designing change programs. The tenuous assumption here is that when rational individuals are presented with the logic for a change, they will accept it. The fact that what the change target considers rational may not be the same as what the change agent considers rational is ignored.

6. In planning for change and innovation, the change agent and client system should try to be clear about the change objectives; otherwise, ambiguity and uncertainty may occur that can cause resistance.

7. Change agents and client systems should be wary of poorly defined change problems. Too often symptoms are mistaken for causes for problems, resulting in the misdirection of remedial efforts and ineffective uses of resources.

8. Change agents and client systems should avoid focusing change programs only at the individual level and forgetting that an individual's behavior is quite dramatically affected by those around him and in the larger culture.

9. Change agents and the client system should avoid technocratic bias in designing change programs. The error here is overemphasis on developing a change program without an accompanying plan for implementing the program. This neglects the need for a well-developed strategy for implementing the change.

10. A stimulus to change through innovation is created by performance gaps—a perceived discrepancy between what the change target is performing and how the change target or someone outside the change target believes it could be performing.

11. To avoid discrepancies between attitude and behavior regarding change, it is important that the change target have the means for converting favorable attitudes regarding change into actual behavior.

12. Discrepancies between attitudes and behavior can be reduced by minimizing the time lag between attitude formation and its behavioral expression.

13. Discrepancies between attitudes and behavior can be reduced by minimizing the lack of knowledge or awareness of how to implement the logical behavioral consequences of an attitude.

14. The change agent should identify the most salient need a target system has and establish awareness within the target group that the advocated

change better meets those needs (assuming it does). This stresses the relative advantage of the change.

15. The change agent should identify those social relationships most likely to be affected by the change and minimize conflict by adapting the change to be compatible with those relationships and/or prepare the target group well in advance to receive a potentially disruptive change.

16. The change agent must exercise considerable care in making the change as compatible as possible with existing values, beliefs, capabilities, and situational factors surrounding the change target.

17. It is highly desirable to stimulate word-of-mouth communications to augment other communication channels.

18. Change agents should think in terms of optimum rather than minimum time for introducing change.

19. To the extent possible, a change should be developed to permit limited use, that is, it should be divisible.

20. The introduction of a change should occur in such a way that the discontinuance of it, because of bad experiences, enables the target system to return to an earlier preferred state. This gives the change reversibility.

21. Every effort should be made to simplify the change to make it less complex or difficult to understand.

Chapter 2: Defining Social Problems

This chapter can be summarized by the following general principles:

1. The change agent should realize that, once he accepts a particular definition of a problem, he has taken a major step in determining what the problem solution will be. Thus considerable care should be given to the problem definition task.

2. The change agent should expect to encounter problem definitions at variance with his own, particularly as the symptoms of a problem become more severe and numerous and as the number of different groups involved in the problem area increase.

3. In the great majority of cases—indeed, perhaps in all—total objectivity is impossible; hence the change agent should try to clarify his own partisanship.

4. The use of causal images or models is very helpful in diagnosing prob-

lems. The change agent should try to identify what forces under what conditions produce the factors of immediate concern.

5. The change agent should also seek to understand how one force or factor produces another. This understanding facilitates manipulation of forces to achieve a desired goal.

6. A distinction should be made between forces or factors the change agent can manipulate directly and those to which he must adapt.

7. Among the factors that must be considered in diagnosing problems are the past history of a group or person having a problem, the relevant environment, organizational structure and processes (in the case of a group), and characteristics of individuals. Each of these considerations has a large number of subsidiary factors.

8. Techniques for gathering problem-related data are numerous and diverse. It is often useful to employ more than one technique as a way of allowing for different perspectives on problems.

9. The change agent should attempt to specify the nature of the problem as a policy problem, organizational structure problem, and so forth, or as a combination of different kinds of problems, each possibly having different importance weightings.

10. The change agent must carefully document the manifestations or symptoms of the problem and map its scope and severity.

11. The change agent should also examine past remedial efforts concerning the problem at hand and the reasons for their successes and failures.

Chapter 3: Resistance to Change

This chapter can be summarized by the following general principles:

1. Change should be made as compatible as possible with the cultural values of the change target in order to reduce resistance to change.

2. Change should not be presented in such a way as to threaten the change target's self-esteem or image in the eyes of those around it, or resistance to change will be encountered.

3. Change should be presented in a way that does not threaten the cohesiveness of the change target system, or resistance will be encountered.

4. Change should not threaten or jeopardize an individual's position vis a vis reference groups, or resistance to the change-innovation will be enountered.

5. Change should avoid altering the balance or interdependence between units involved in the change, or resistance will be encountered.

6. Change should be presented in a way to avoid creating conflict among individuals or groups involved, or resistance will be encountered.

7. Change should be implemented in a way that does not increase a persons's uncertainty or ambiguity about how the change or innovation will affect the way he performs his role, or resistance will be encountered.

8. Change should be presented in a way which minimizes the degree to which it is seen as threatening to the power and influence of individuals or groups in the target system, or resistance will be encountered.

9. Change should be implemented in a manner that does not create role conflict for individuals in the target system, or resistance will be encountered.

10. Change should be implemented in a way which avoids increasing intergroup competition in the target system, or resistance will be encountered.

11. There should be top-level support in the system for the proposed change or innovation, or resistance will be encountered.

12. The system should try and provide rewards-incentives to participants for adopting the change or innovation that are attractive to them as a way of reducing their resistance.

13. Change should be implemented in such a way that members of the change target system perceive that there is a good chance of successfully achieving the change, or resistance will be encountered.

14. The proper climate for minimizing resistance to change is one in which members have a high need for change, are open to change, have a commitment to change, feel that there is a potential for change in the system, and feel that they have some control over the change process.

15. Those who are going to use the change or innovation should realize that they have or will be provided with the necessary technical skills to implement the change, or resistance will be encountered.

16. Proponents of change should empathize with opponents of change in order to understand the causes of resistance so that some informed action can be taken to reduce or eliminate those causes.

17. Proponents of change should strive to reduce resistance by reducing the change target's ignorance regarding the change.

18. Proponents of change should be aware of the personality characteristics (such as dogmatism, low tolerance for ambiguity, low risk-taking

propensity) of members of the change target system that can create resistance to change.

Chapter 4: Facilitative Strategies

This chapter can be summarized by the following principles:

1. Facilitative strategies can be used when the client system recognizes a problem, agrees that remedial action is necessary, is open to external assistance, and is willing to engage in self-help.
2. A facilitative approach must be coupled with a program of creating awareness among the target groups of the availability of assistance.
3. Facilitative strategies making it very easy to change may be necessary to compensate for a low motivation to change: great ease compensates for low motivation.
4. Facilitating change through solution diversification is desirable when members of a client system desire different ways of satisfying a common need.
5. Change will be more likely if the resource-providing institution is located within the client system.
6. Long-run, persuasive social change is more apt to be achieved if resources are applied to the community rather than the individual.
7. The more general the goals to which a resource is committed, the more likely it is to be used effectively.
8. Tying a resource to a specific time period inhibits community participation in a change program and thus reduces the effectiveness of resources committed to the program.
9. The creation of new roles within a client system is desirable if existing roles are inadequate to utilize a needed resource.
10. Facilitative strategies, such as the provision of funds or capabilities, are necessary when the client system lacks resources to continue a change.
11. The change agent should assess the client's ability to sustain change itself and its own ability to provide continued assistance if the client does not have the ability to sustain change.
12. Different subgroups within the client system may require different facilitative strategies at any given point in time as well as over time.
13. The larger the magnitude of the intended change, the more important it is to undertake facilitative efforts.

14. The greater the resistance to change, the less effective facilitative strategies will be.

15. Certain attributes of the change object, such as complexity, accessibility, and divisibility, may require offsetting facilitative efforts.

16. When change must occur quickly and an openness to change does not exist, a facilitative approach is unlikely to be effective.

17. When the change objective involves altering a firmly held attitude or firmly entrenched behavior, a facilitative strategy alone is unlikely to be helpful.

Chapter 5: Reeducative Strategies

This chapter can be summarized by the following principles:

1. Reeducative strategies are feasible, other things being equal, when change does not have to be immediate.

2. Reeducative strategies can be effective in providing the foundation for future action by establishing an awareness of a need (general or specific) for change. It may be desirable not to mention a specific change if it is potentially controversial until a clear need has been established.

3. The stronger the degree of commitment a change requires to be effective, the less impactful reeducational strategies alone will be.

4. Reeducational strategies are effective in inoculating people against appeals to resist change or to revert back to the previous situation.

5. Reeducational strategies can be effective in (a) connecting causes with symptoms, (b) creating awareness of a problem, and (c) establishing that a known problem can be resolved.

6. Reeducational strategies are necessary when the use of the advocated change requires skills and knowledge the client system does not possess.

7. When a change agency does not possess the resources to sustain a needed long-term involvement, a reeducative strategy alone is not indicated.

8. Reeducational strategies are particularly useful at the awareness stage of the adoption process.

9. Generally, reeducation strategies alone are insufficient for accomplishing large-scale change in the short run, particularly where motivation to change is low relative to the magnitude of the change.

10. The higher the anticipated level of resistance, the more necessary it is to

initiate educational programs well in advance of the actual introduction of the change.

11. Reeducational strategies are essential when the change involves a radical departure from past practices.

12. Reeducational strategies are feasible when little control over the client is necessary and the rationale for change is clearly presented in terms of the client's perspective.

Chapter 6: Persuasive Strategies

This chapter can be summarized by the following principles:

1. Persuasive strategies are indicated when a problem is not recognized or not considered particularly important, or when a particular solution to a problem is not perceived to be potentially effective.

2. Persuasive strategies are desirable when the client is not committed to change.

3. Persuasive strategies are desirable when it is necessary to induce a client system to reallocate its resources from one program or activity to the activity advocated by the change agent.

4. Persuasive strategies are not feasible when the client system has no access to resources to sustain a change effort.

5. Persuasive strategies are often necessary when the change agent has no direct control over the client system through the manipulation of resources of value to the client system.

6. Persuasive strategies are particularly appropriate when the client is at the evaluation or legitimation stages of the adoption process.

7. When a persuasive effort is appropriate to one subgroup but not to another, care must be exercised to prevent persuasive content from reaching the subgroup that would respond negatively to such messages.

8. Persuasive strategies are appropriate where the magnitude of change is great and the change is perceived to be risky and socially disruptive.

9. The greater the time constraints and the lower the ability to use power, the more desirable or necessary it may be to use persuasive strategies.

10. Persuasive strategies are indicated when the change cannot be implemented on a trial basis, is difficult to understand, and has no very visible relative advantage.

11. Persuasive strategies are especially effective in combating resistance to change, although the strategies used to combat resistance to initiating change may not be those used to combat resistance after the change has been implemented.

Chapter 7: Power Strategies

This chapter can be summarized by the following principles:

1. Although power strategies may be desirable when commitment by the client system is low, they are also unlikely to produce commitment.
2. The lower the perceived or felt need for change within a client system, the greater the need for a power-oriented strategy.
3. A power strategy will be ineffective if the client system does not have the requisite resources to accept change and the change agency cannot provide them. A power strategy may be effective, on the other hand, in getting a client system to reallocate resources to initiate and sustain change.
4. Power strategies are desirable when a protracted adoption decision-making process is likely but change must be immediate or at best initiated well before a protracted decision would finally be made.
5. Power strategies can be effective in overcoming resistance by creating change rapidly before resistance can be mobilized.
6. The less susceptible to modification a change is the greater the need for a power strategy to force changes among the client system. Power strategies may also be useful in securing a trial use of the change.

Chapter 8: The Use of Multiple Strategies

This chapter can be summarized by the following principles:

1. Social change should be presented to a client system as being in direct response to a strongly felt need that can be satisfied without undue social, psychological, or financial cost.
2. A change should not be introduced unless the client system has or can very readily acquire the necessary capacity—social and psychological—to accept it and sustain it.
3. Before undertaking a change effort, the change agent should forecast the resources he needs to create change and match this forecast with his available or expected resources.
4. The lower the degree of commitment to change in a client system, the greater the effort the change agent must expend to heighten commitment.
5. The change agent must carefully assess the nature of the advocated change and determine how it may be adapted to fit more closely to client system needs and behavior. It is also necessary to determine

whether particular attributes of the change as perceived by the client require special promotional or introductory strategies.

6. The larger the magnitude of change required for the adoption of an innovation or change, the greater the need for inducement or incentives beyond so-called rationale justifications.

7. When designing an innovation and developing a program to gain its adoption and diffusion, it is highly desirable to determine whether the client system can be subdivided or segmented into meaningful groups that might benefit from a differentiated innovation and/or implementation plan.

8. Diagnosis of potential cultural, social, and psychological sources of resistance to change should be undertaken early in the planning for social change.

9. Every effort should be made to identify potential undesirable consequences of particular strategies and to develop contingency plans for countering these consequences should they materialize.

10. The change agent should develop in the change plan the optimal timing sequences for various strategies and tactics.

11. The selection of strategies should consider the adoption decision-making stage of relevant groups within the client system.

12. The change agent must determine whether the client system is at all aware of a potential change that could solve a problem or satisfy a need, or are aware of a change possibility but do not relate it to their problem.

13. The goals of a change effort should be stated very explicitly and operationally to facilitate implementation, evaluation, and control activities.

Chapter 9: The Change Agent

This chapter can be summarized by the following general principles:

1. The change agent should be sensitive to the needs and perspectives of the change target system in designing solutions to change situations.

2. The change agent should institutionalize change (build capabilities within the change target system) so that a vacuum is not created when the change agent leaves the system.

3. The change agent should always seek the simplest solution when working with a change problem.

4. The change agent should have administrative capabilities (planning, managing change team personnel) so that the change process can be managed effectively.

5. The change agent should strive to maintain good interpersonal relations with personnel in the change target system during the change process.

6. The change agent should consider the change problem as stated by the target system as a hypothesis and then should seek additional information regarding the change situation before selecting a particular course of action.

7. The change agent should be sensitive to and tolerant of the constraints that govern the change situation.

8. The change agent should be prepared to operate under stressful conditions without getting defensive when confronted by the client system as it deals with its anxieties and frustrations associated with the change.

9. The change agent should have the self-confidence and positive self-image to accept setbacks with poise and not project anger or frustration onto the change target.

10. The change agent should be able to define the change program in a manner that is attractive to the various constituencies involved in the change program.

11. The change agent should strive to maximize credibility in the eyes of the change target system in terms of motives, competence and truthfulness.

12. The change agent should work through opinion leaders in the change target system.

13. The change agent should expect an initial period in which change efforts will not be rewarded in terms of quick change in the target system.

14. In order to maximize the change target system's cooperation with the change agent, the change target should perceive that it had some freedom or free choice in entering into a relationship with the change agent.

15. The change agent should get the change target system involved in problem definition and need specification in order to develop target system commitment to and trust in the change agent.

16. In order to maximize the change target system's cooperation with the change agent, the target system should perceive that the change agent helps to gain more influence over the change process.

17. The more similar (homophilous) in attitudes, values, and beliefs the change agent is to the change target system, the more motivated will the target system be in working cooperatively with the change agent.

18. There should be an exchange of expectations regarding the change

process between the change agent and the change target system so that no misunderstandings occur regarding the change.

19. The change agent should strive to create as broad a power base as possible (i.e., referent, legitimate, expertise) for a position in the change target system.

20. In determining change objectives, the change agent should consider the nature and scope of the change, the target of the change (attitudes, values, beliefs) who is going to be affected, and the initial key people with whom to work in the change target system.

21. The optimally structured change agent would be a team consisting of an internal and external change agent who are homophilous with the change target system.

Chapter 10: Characteristics of Change Targets: Individuals

This chapter can be summarized by the following principles:

1. The change agent must help establish awareness and recognition of a need to change and connect the advocated change to this awareness.

2. The change agent should not confuse or equate his perception with that of the target group.

3. The change agent should monitor the target group's perception about the change throughout the decision and postdecision process, since perception may frequently change.

4. Where appropriate, the change agent should stress the availability of a change or the ease with which a suitable change can be developed to address a need. This can stimulate the target system's motivation to change. The change agent can stimulate this further by demonstrating how the target system can control the change effort (if indeed it can).

5. The change agent should be wary of using attitudes to predict behavior and vice versa.

6. Every effort should be made to reinforce the target system's plans. Stressing the legitimacy of the change is one way of doing this.

7. The change agent should make every effort to have the target system try the advocated change on a restricted basis. Trial may be symbolic or behavioral.

8. The change agent should be particularly concerned with the target system's evaluation processes after the trial stage. The change agent may provide assistance to the target system while it evaluates the trial experience and information from other sources.

9. After change has been adopted, it is very important to provide support

for the decision or action to reinforce enthusiasm or to counteract dissonance.

10. The change agent should provide assistance to help the target system to cope with its environment as it affects goal attainment.

11. The change agent must consider explicitly the strategic implication of each stage in the overall adoption/rejection decision process and adapt his actions according to the requirements and special processes of each stage.

12. The change agent should target initial efforts towards those persons who adopt innovations early after awareness, and try to foster early awareness for such persons who are otherwise late knowers.

13. The change agent should also distinguish between light users of a change or innovation versus heavy users, with relatively more emphasis on the latter.

14. It is also desirable for the change agent to focus on interpersonal relations rather than individuals alone.

15. When analyzing the target group, the change agent should attempt to identify individuals or types of individuals who have high influence propensity, that is, who are active opinion- or fact-giving people or who are role models. This group must be among the very first to be reached.

16. A useful activity for a change agent, perhaps working with a group of other change agents or persons knowledgeable about given target groups, is to construct an index or rating chart to predict the likelihood of change in a target group. This requires joint consideration of individual factors associated with change (Table 10.1) and factors affecting the climate for change (Table 10.3).

Chapter 11: Characteristics of Change Targets: Organizations

This chapter can be summarized by the following principles:

1. It is important for the organization to scan its environment, since this is a critical source of ideas for change and innovation.

2. The organization must scan its environment, since this is an important source of information regarding the technology or means that may be required for change.

3. The organization must be sensitive to the pressures existing in its environment with respect to how these environmental pressures might support or resist some change the organization is contemplating.

4. Organizations can enter into joint programs with other organizations as a way of stimulating change and reducing its financial costs.

5. The organization must be aware that there are two phases to the change process, each with its own set of problems. During the initiation stage, the organization must generate much information regarding the change. During the implementation stage, the problem becomes one of integrating the change into the organization.

6. Organizations with complex environments that are changing in terms of their composition and the kinds of demands being made on them should be prepared to closely monitor their environment and be ready to change.

7. The organization should shift its structure for the initiation and implementation stages of the change-innovation process. During the initiation stage, a higher degree of complexity, lower formalization, and lower centralization should be used, since these increase the information-gathering and processing capabilities of the system. During implementation, the organization should shift its structure to lower complexity, higher formalization, and higher centralization, since this ensures that those using the change will have a clearer understanding as to how to use the change.

8. To facilitate the shifting of its structure from the initiation to implementation stages, three processes must exist in the organization: (1) the ability to deal with conflict, (2) effective interpersonal relations, (3) the institutionalization of the dual structures.

9. The greater the need for change, the more the organization should differentiate its structure for the initiation and implementation of change.

10. The greater the uncertainty associated with the change situation, the more the organization should differentiate its strucutre for the initiation and implementation of change.

11. The more radical the change, the more the organization should differentiate its structure for the initiation and implementation of change.

12. Utilizing the dual structures for the initiation and implementation of change must be institutionalized in the organization. This process must be seen as legitimate and supported by the top-level people in the organization.

Chapter 13: Ethics in Social Change

This chapter can be summarized by the following principles:

1. The change agent must recognize the potential for conflict between his values and the client's in defining the problem. The change agent should also make an explicit decision about whether to discuss these conflicts with the client when they occur.

2. The change agent should have control over the selection of strategies for problem diagnosis.

3. The change agent should be aware of his value orientation and should make an explicit decision about whether to discuss any conflicts with the client when they occur.

4. The change agent should be sensitive to the potential obligation he has to perform services even for those client systems whose values he does not share.

5. The client system should consider informing the change agent of all parties to be affected by the change agent's activity.

6. In the absence of a written contract covering termination of a change agent–client relationship, the change agent should make an explicit decision about whether he feels he is obligated to work with the client as long as the client requests the change agent's services.

7. In implementing a solution to a problem, the change agent should make an explicit decision about whether to provide members of the change target system several options from which to choose.

8. The change agent should be prepared to receive and answer questions regarding the rationale ard supporting evidence for a given solution from those in the change target system.

9. The change agent should consider if it is in the best interests of members in the change target system to have the rationale and supporting evidence for the solution.

10. The change agent should consider his responsibility-obligation in making his services available for dealing with the possible dysfunctional consequences of his activities.

11. The change agent should consider his obligation to inform members of the target system about his goals and objectives.

12. A client hiring a change agent should consider his obligation to inform others in the target system about what the change agent is doing in the system.

13. The client should consider his obligation to provide the change agent with any information the change agent feels he needs to perform his task.

14. The change agent must recognize the potential for conflict between his values and the client's in defining the goals and objectives of the change agent's activities. The change agent should also make an explicit decision about whether to discuss with the client these conflicts when they occur.

15. The change agent must recognize the potential for conflict between his

values and the client's in determining the means of a solution. The change agent should also make an explicit decision about whether to discuss these conflicts with the client when they occur.

16. The change agent must recognize the potential for conflict between his values and the client's in determining the means of evaluating a solution. The change agent should also make an explicit decision about whether to discuss these conflicts with the client when they occur.

17. The change agent should, at the beginning of his relationship with a client, clearly define the scope of the change agent's freedom to select the strategies for problem diagnosis.

18. The change agent should, at the beginning of his relationship with a client, clearly define the scope of the change agent's freedom to select the criteria for selecting a solution.

19. The change agent should, at the beginning of his relationship with a client, clearly define the scope of the change agent's freedom to select the strategies for implementing a solution.

20. The change agent should, at the beginning of his relationship with a client, clearly define the scope of the change agent's freedom to select the strategies for evaluating a solution.

SUBJECT
INDEX

Shoemaker, F. F., 9, 209, 220, 228, 229,
302
on adopter research, 239
on change agents, 16, 197–203, 289
on diffusion, 7, 159
on normative change, 11
on resistance to change, 86
Sieber, S. D., 287
Sikorski, L., 5, 119, 287
on information use, 290–291
Silberman, H., 88
Simon, H. A., 24
Simulation, 311
Sinding, 114
Singh, Y. P., 232, 233
Slocum, J., 27
Smith, A. D., 8
Social barriers to change, 72–75
Social change, characteristics of, 13–16
conceptualization of, 6–16
defining problems for, 32–57
definitions of, 6–10, 29
ethics in, 323–356
facilitative strategies for, 90–108
factors associated with, 241–246
goals of, 26–29
innovation and, 12–13
multiple strategies for, 166–180
in organizations, 248–279
participants in, 16–19
persuasive strategies for, 134–151
pitfalls in, 19–23
planned, 10, 29
power strategies for, 152–165
recent works in, 5
reeducative strategies for, 111–131
research in, 283–295
resistance to, 59, 61–89
stimuli for, 23–26, 30
strategies for, 59–60
translation from theory into practice in,
295–311
types of, 10–12, 330–332
unplanned, 10, 29
see also Innovation
Social indicators, 37
Social problems, categories of, 55–56
defining process of, 32–37
manifestations of, 54–55
metatheory approach to, 34–40

open systems approach to, 40–44
optaining data on, 44–50
perceptions of, 50–52
social change relieving, 5, 30
types of, 52–54
Social relations, impact of change on, 13–14,
30
Social workers, 84–86
Solution diversification, 96, 108
Solzhenitzyn, A. I., 68
South East Alabama Self-Help Association,
106–108
Soviet Union, 111
Stalker, G. M., 77, 261
Stankey, G. H., 283
Stanton, 190
Status differentials, 77–78
Status quo, change differentiated from, 6–7
in resistance to change, 63
in social research, 284
Stein, B., 136
Steiner, G. A., 226
Stephanson, R. W., 193–194
Sterilization, 70–71. See also Birth Control;
Family planning; Vasectomies
Stiff, R., 9, 226, 230
Stiles, L., 88
Stimulation techniques, 304–310
Stimuli for change, 23–26, 30
Strategies for change, 59–60
dilemmas in selection of, 331, 358
facilitative, 90–108
multiple, 166–180
persuasive, 134–151
power, 152–165
premature commitment to, 204–205
reeducative, 111–131
Stress, social change relieving, 5
Subjects (of research), rights of, 333–337
Sullivan, L., 105
Suppliers, in organizational change, 276–279
Surveillance, 153, 160
Susceptibility to modification, 16
Sweetzer, D. A., 155
Switching rules, 270–272
Symbolic adoption, 233–236
Systemic coherence of social groups, 73

Taft, R. T., 298
Tanker, 228